Grid Connected Converters

Grid Connected Converters

Modeling, Stability and Control

Hassan Bevrani
University of Kurdistan, Kurdistan, Iran

Toshiji Kato
Doshisha University, Kyoto, Japan

Toshifumi Ise
Osaka University, Osaka, Japan

Kaoru Inoue
Doshisha University, Kyoto, Japan

ELSEVIER

Dedicated to our families.

Elsevier
Radarweg 29, PO Box 211, 1000 AE Amsterdam, Netherlands
The Boulevard, Langford Lane, Kidlington, Oxford OX5 1GB, United Kingdom
50 Hampshire Street, 5th Floor, Cambridge, MA 02139, United States

ISBN: 978-0-323-99902-1

For information on all Elsevier publications visit our website at
https://www.elsevier.com/books-and-journals

Publisher: Charlotte Cockle
Acquisitions Editor: Graham Nisbet
Editorial Project Manager: Aleksandra Packowska
Production Project Manager: Prasanna Kalyanaraman
Cover Designer: Christian J. Bilbow

Typeset by TNQ Technologies

Contents

Foreword

The global energy situation in 2022 is under critical pressure and future predictions are saying that a large shift of energy mix is needed—better today than tomorrow. A main reason is the challenging predictions of a future climate crisis, which includes a global average temperature increase of more than 2°, or even higher due to the carbon emission from fossil fuels—if nothing is done on the planet. A major need is to change the fossil fuel–based electricity production to become based on renewables (especially wind and solar), which directly gives a significant reduction in carbon emission, but there is also a need to transform the transportation sector to become electrified—which also will give a very direct and clear reduction of the emission. The transportation sector is fully based on fossil fuel today—but electrifying it—a large carbon emission reduction can be achieved. A third area is to save energy either by using less energy in our daily living or make the conversion processes much more efficient. One example seen in the past has been the classical bulbs, which today all are based on LEDs with an amazing energy reduction. The world will in the future need to be based on electricity instead of fossil fuel and my prediction is that electricity will cover above 60 % of all the global energy consumption achieved within the next 40 years to come.

To realize such energy system, we need to convert electrical power from one stage to another—e.g., from ac to dc or vice versa. This is done by the modern power electronics technology, which has been present for more than 60 years. In the last decades, the most dominant power converter topologies have been the voltage source converter–based systems and they are on a very rapid expansion in many different applications (i.e., wind, solar, storage, EV-chargers, EVs, adjustable speed drives, HVDC power transmission). The voltage-sourced systems can achieve very fast dynamics thanks to the used capacitors and fast switching power devices, high power density, low cost, and they are very scalable in power. Further on—due to the high dynamic ability in the system—the systems where they are applied into will also become fast—and here often large energy savings can be achieved or a high performance. We are seeing millions of such units around in the world and that will soon be billions in the global power system.

This book with the title "*Grid Connected Converters—Modeling, Stability, and Control*" by Profs. Hassan Bevrani, Toshiji Kato, Toshifumi Ise, and Kaoru Inoue is addressing this important technology in order to give researchers and industry a unique understanding and propose methods to improve the performance of grid connected converters. This book comprises 9 chapters with Part I discussing "Concepts, fundamentals, modeling, and dynamics analysis" and a Part II having the title "Control synthesis for stabilizing and performance enhancement". This book gives the fundamentals about the grid-connected converters and comes up with many interesting methods on how to analyze the converters connected to the grid. It also improves the performance on them if certain problems are expected to appear.

I enjoyed reading this book where the topics are presented in a new, easy, and understandable way—I hope you enjoy the work too.

Frede Blaabjerg, IEEE fellow
Professor in Power Electronics and Drives
Aalborg University, Denmark

Preface

Increased needs for electrical energy and environmental concerns besides growing attempts to reduce dependency of power generation to fossil fuel resources have caused power grid industries to set an ambitious target of renewable generation. Therefore, the capacity of installed converter-based distributed generators (DGs) and renewable energy sources (RESs) over the world is rapidly growing; and this increases the significance of *grid connected converters* (GCCs) from *stability, modeling, and control* points of view as challenging issues.

The most of DGs/RESs require power converters to interface their primary power sources to the grid. It is well known that the quality of control and functions of these power converters in different operation modes significantly affect the grid dynamic performance. The relevant technical challenges are more important when the penetration of converter interfaced generating units increases. It is investigated that poor operation of GCCs may adversely affect the grid frequency, voltage and power fluctuation, and leads to degrading the grid control performance and even stability.

In response to mentioned challenges, the present book provides comprehensive coverage of modeling, dynamic analysis, and controller synthesis of the GCCs for a wide range of applications. Moreover, the increasing penetration of converter-interfaced DGs/RESs and the control flexibility of GCCs motivate the use of these converters to develop additional ancillary services to control undesired system dynamics in the power grids. Advanced control of GCCs, however, has the potential to offset the intermittent nature of distributed energy resources and provide control support to the host utility during abnormal conditions. To this end, new trends in GCCs modeling and dynamic equivalencing should be discussed.

The modern GCCs structure, inertia challenge requirements, and control levels are discussed. Recent advances in the visualization of virtual synchronous generators and the associated effects on the grid performance are explained. The physical constraints and engineering aspects of the advanced control schemes are considered. Optimal and robust control strategies are explained using real-time simulations and experimental studies.

This book summarizes a long-term academic/research outcomes and contributions, influenced by the authors' practical experiences on the GCCs, power grids/microgrids dynamics and stability, and power electronic systems in several countries, universities, and power electric companies. The book also provides a thorough understanding of the basic principles of the GCCs structure, modeling, dynamics analysis, and control synthesis. The book could be useful for engineers and operators in the electric industry, as well as academic researchers. Since the book describes GCCs issues from introductory to the advanced steps; it could be also useful as a supplementary text for university students in electrical engineering at both undergraduate and postgraduate levels.

The book is fundamentally split into two parts. The first part in five chapters deals mostly with concepts, fundamentals, modeling, and dynamics analysis; while the second part in four chapters is concentrated on control synthesis for stabilizing and performance enhancement. In addition to dynamic modeling, stability analysis, and controller synthesis, inertia challenge requirements, and virtual dynamics are extensively addressed.

Chapter 1 discusses the main characteristics and challenges of modern power grids in the presence of high penetration of converter-based RESs and DGs. The overall structure and significant control issues of renewable integrated power grids are presented. The concept and application of GCCs in integration of RESs/DGs and their important role in the grid stability and performance are discussed. Then, the current trends and future directions in the relevant issues are emphasized.

Chapter 2 deals with the structure and essentials of GCC as the significant block of future renewable integrated power grids. Basic control loops of two main topologies, grid-following and grid-forming GCCs, are introduced. Then dynamic characteristics emulation as an important capability of grid-forming GCCs is emphasized. The necessity of developing required standards and grid codes are explained.

Chapter 3 addresses some simple and effective modeling methods for different dynamics and applications of the GCCs. The obtained models can be used to evaluate the dynamic performance and behavior of the GCCs in connection with the grid, as well as in the analysis and synthesis of the required control systems. The proposed modeling methods can provide the state-space models for different applications of active power-frequency and reactive power-voltage response analysis.

Chapter 4 explains the GCC stability assessment and sensitivity analysis. After presenting an overview on the most popular stability analysis issues, several stability analysis methods in both frequency and time domains are introduced. The discussion is started from the basic stability analysis method using the eigenvalues/poles graph and then extended to address more effective approaches based on the GCC frequency characteristics and the Poincare' map. Finally the significance of sensitivity analysis in the stability study is emphasized.

Chapter 5 examines the impacts of GCCs penetration on the overall power grid. For evaluating the GCCs integration impacts, an effective modeling method is needed to handle the main dynamics of GCCs, conventional synchronous generators and the power grid. Two simple modeling methods are presented to analyze the grid integration effect of the GCCs in active-power and frequency points of view.

Chapter 6 describes the control structure and main control loops of the GCCs. The role of control system and the GCC dynamic characteristics in both grid-following and grid-forming types are introduced. The control objectives to ensure the system stability and desirable performance are discussed, and an example for application of feedback and feedforward control loops is explained. The importance of dynamic characteristics emulation capability in GCCs is emphasized.

Chapter 7 investigates the impacts of time delay in GCC dynamic performances and frequency characteristics. A simple delay compensation method is addressed, and virtual resistors are used to improve the oscillation damping characteristic. Finally, a combination of feedback and feedforward control schemes is designed to ensure GCC stability and a desirable tracking characteristic over a wide range of operating frequencies.

Chapter 8 describes five controller synthesis methodologies for the GCCs based on continuous and discrete optimal control, Lyapunov energy function, MPC, and robust $H\infty$ control theorems. The addressed optimal control methods are focused on the output voltage regulation, while Lyapunov-based digital control emphasizes the overall system stability. The $H\infty$ controller provides a damping control by emulating the virtual inertia and damping characteristic; and the MPC-based control system covers both GCC output voltage and current controls.

Chapter 9 presents the general concept of flexible GCC-based grid dynamic shaping with some applications to improve power grid performance and dynamics. Using this concept, some new perspectives to enhance the power grid stability and performance are discussed. The role of GCC-based dynamic shaping in the grid ancillary service support is addressed, and the application of GCC-based dynamic shaping for HVDC and low-frequency transmission systems is discussed.

Hassan Bevrani
Professor, University of Kurdistan

Toshiji Kato
Professor, Doshisha University

Toshifumi Ise
Professor Emeritus, Osaka University

Kaoru Inoue
Professor, Doshisha University

January 2022

Acknowledgments

Most of the contributions, outcomes, and insight presented in this book were achieved through a long-term teaching and research cooperation on the grid connected converters over the years. The materials given in the present book are mainly the research outcomes and original results of authors in Smart/Micro Grids Research Center-*SMGRC*, University of Kurdistan (Kurdistan, Iran); Doshisha University (Kyoto, Japan), and Osaka University (Osaka, Japan). It is a pleasure to acknowledge the received supports from these sources, and the awards from the Alexander von Humboldt (AvH) Foundation.

The authors would like to thank Dr. Jia Liu (Xi'an Jiaotong University, China), Dr. Jonggrist Jongudomkarn (Khon Kaen University, Thailand), Dr. Mustafa Al-Tameemi (Kawamura Electric Inc., Japan), and Prof. J. Raisch (Technische Universität Berlin, Germany) for their kind supports. Special thanks to S. Rehimi (University of Kurdistan), for reviewing the whole book. Finally, the authors offer their deepest personal gratitude to their families for their patience during the preparation of this book.

Part I

Concepts, fundamentals, modeling, and dynamics analysis

An introduction to renewable integrated power grids

1

1.1 Modern power grids

Operation and control in today's power grids face new challenges arising from the growing integration of power electronic-based distributed generator (DGs) and loads. Conventional power grid is challenged by the losses, environmental concerns, and increasing need to electric energy. Thus, this grid is in transition from a conventional hierarchical centralized structure with one-way communication for power and data flows between generation and demand sides to a modern and smart power grid.

In a conventional power grid, the system is not fully observable and controllable due to a limited number of sensors, flexible actuators, and appropriate control systems. Most of operation and control process even system restoration procedure are accomplished manually. Unlike conventional power grid, a modern power grid makes a compromise between environmental needs, economic and efficiency issues, as well as system reliability. There is a bidirectional network for both power flow and data. It fully sensors throughout system with selfmonitoring and selfhealing characteristics.

Modern power grid performs a network of DGs and renewable energy options, with decentralized and distributed pervasive control and operation systems. There are many customer choices in marketing and deregulation policy point of views. Remote and automatic monitoring/control is dominant with a high diversity in both generation and demand. In addition to fully generation/load control and bidirectional flows of power and data; emerging numerous DGs/renewable energy source (RESs), wide network of monitoring units such as phasor measurement units (PMUs) and intelligent electronic devices, as well as high complexity can be considered as new characteristics of modern power grids. A modern power grid uses advanced metering infrastructure which facilitates remote reading/monitoring, remote connection/disconnection, remote configuration; increases the visibility to the customer; and opens a gateway to home/network automation. The structures of a conventional and modern power grid are conceptually shown in Fig. 1.1.

Need to fast numerical calculating and data processing algorithms, facing with a highly decentralized control structure with significant uncertainty, intermittent nature of RESs, reducing of system inertia, as well as the necessity of updating grid codes and conventional dynamic analysis and control synthesis methods are the main challenges associated with the modern power grid systems. On the other hand, using smart wide-area monitoring and adaptive control systems, flexible demand response capabilities, regulation and ancillary services supports from the DGs/RESs, and constructive virtual dynamics emulation are some promising solutions in response to the mentioned challenges.

Grid Connected Converters. https://doi.org/10.1016/B978-0-323-99902-1.00001-X

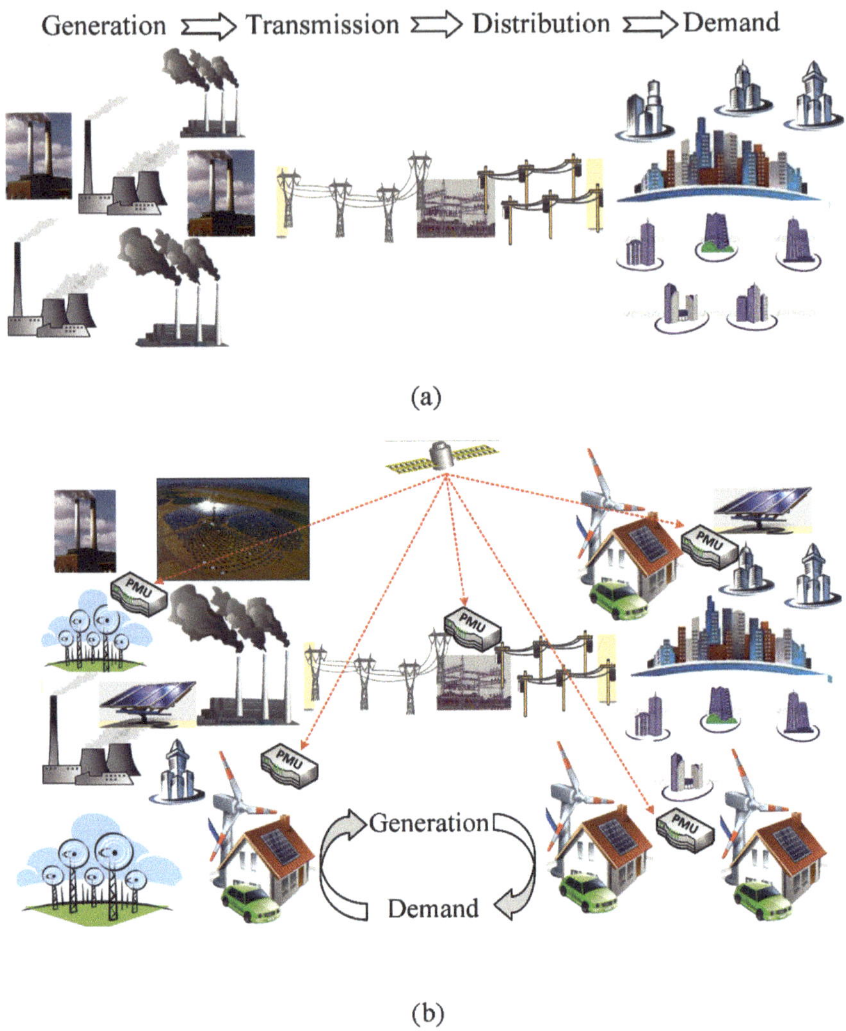

Figure 1.1 (A) Conventional power grid, (B) Modern power grid.

The main dynamic performance and stability problems in a modern power grid are caused by the reduction of the system rotational inertia, as DGs and RESs with grid connected converter (GCCs) gradually replace synchronous generators (SGs). Reducing rotational inertia in a power grid can negatively affect the system response and may degrade the control capability and performance. This may lead to significant fluctuation in grid frequency, angle and voltage, and even system instability [1].

Decreasing power grid inertia due to high penetration of GCCs makes load-power balancing and frequency regulation extremely challenging. The intermittency of renewable power generation significantly magnifies this issue.

Emulation of desirable dynamics such as inertia and proper shaping of injected power from the controlled power sources via GCCs is a promising solution. The regulation power provided by DGs and microgrids (MGs) through the GCCs may support the grid robustness against various disturbances and reduces power fluctuations and parameters perturbations. Due to the fast response of GCCs, this supplementary regulation power makes an effective impact in a short period [1,2].

Controlling loads and flexible demand side units, which is known as demand response, also provides a suitable scheme for a smooth control and regulation issues in the power grids. The switching-based control ability of load blocks enables the demand to respond faster to system disturbances, in comparison with conventional power plants. This ability together with recent advances in monitoring, computing, and communication technologies makes load-side units ideal candidates for grid stability improvement and control.

The reliance of the modern power grid on information technologies and its transformation to a complex cyber-physical system has made it as a main target from the cyber-attacks. Although rapid progress in the cyber part offers many advantages in operation and control issues, it also increases the attack risks of cyber intrusion causing unreliability and performance degradation problems [1].

The high penetration of RESs using GCCs in power grids introduces technical challenges due to their high uncertainty, intermittency, and nonsynchronous grid connection. Although the GCCs increase flexibility in operation and regulation power requirements, the replacement of SGs by converter-based DGs/RESs reduces system rotational inertia. This condition is more critical in islanded power grids with a few SGs and low kinetic energy [1].

The increased penetration of renewable power may lead to other challenges for the grid stable operation. For example, utilizing photovoltaic (PV) power to maintain the demand-supply balancing in daytime with a high PV power supply is a challenging issue. The flexibility of GCCs for maintaining grid generation-load balance should be increased. The installed PV systems usually use the maximum power point tracking control because of its unstable power output characteristic. This characteristic depends on irradiance variation which then increases the fluctuation of the residual electricity load in the overall power system. Despite the increased fluctuation, high penetration of PV systems leads to less contribution of the conventional thermal power plants in the overall grid power generation, and hence their ability to regulate power will also be reduced [3].

Low level and time-varying nature of inertia in modern power grids with significant penetration of converter-based DGs/RESs is causing faster and larger power fluctuations making the dynamic analysis and control of grid challenging. Wide-area control schemes incorporating synchrophasor data have the potential to improve grid performance and stability, and lead to more efficient and coordinated control actions. The model identification using PMUs data for the purpose of renewable integrated grid and the aggregated dynamic impact of GCCs and other parameters estimation issues are addressed in several works [1]. The data recorded by PMUs provide valuable information on the dynamics of the power grid with numerous GCCs that can be used for data-driven modeling and control. The data from PMUs can be used for estimation of some important parameters such as electromechanical modes of a power grid and their confidence intervals [4].

However, wide-area monitoring and control would not be applicable without a robust, real-time, reliable, and secure communication architecture happening in the whole network. As mentioned before, with the advent of a modern grid, the flow of power and data in the power system alternated to a bidirectional style. Deploying internet of things technology facilitates communication based on the internet infrastructure in a wide area network. However, transmission protocols play a crucial role in fulfilling a precondition of the grid penetration increment to provide robust, real-time, reliable, and secure communication. In this direction, the IEC 61850 standard series, which are initially communication protocol of the grid substations automation, have been developed to make whole horizons of grid intelligent. IEC 61850−7−420 and IEC 61850−90−7 introduce used cases about how to exchange information among the DGs and supervisory of the power grid [5,6].

1.2 Renewable energy sources and distributed generators

The RESs convert natural energy sources such as sunlight, wind, ocean, hydropower, biomass, geothermal resources, bio-fuels, and hydrogen into consumable electric energy forms. The energy generation technologies based on these energy sources can be considered to be clean and renewable alternatives in comparison with conventional technologies based on fossil fuels and nuclear fission. Solar power comes from the radiant light and heat of the sun. Airflow can run wind turbines for generating electricity. Installing RESs like wind and solar power units to power systems is becoming bulky, such that the total installed capacity of wind energy is expected to be more than 800 GW until 2023. The same growth for solar power through PV technologies is expected. These show a transition and an evolution for the energy paradigms, which needs a comprehensive remodeling and dynamic analysis of the GCCs as the most important actuator block in the modern renewable integrated power grids in a multi-time scale manner.

Water power can be exploited in a form of kinetic energy. The biomass can be used directly as fuel or to produce biofuels which can be burned in internal combustion engines or boilers. Geothermal power is from the heat of the earth itself, which can be mostly used for thermal power production. Among the clean technologies, solar and wind turbines have experienced a great growth in the last years [2].

Although the initial investment in renewable energy-based power generating systems is considerable, it will be reduced with the fast developing of the relevant technologies. It is noteworthy that the outputs of RESs are intermittent, and the generated power depends on the amount of primary energy at the place, for instant solar power and wind power depends on the amount of solar radiation and wind speed, respectively. Most RESs may not individually participate to the grid management effectively, because they may not work in their nominal capacity most of the time. This degrades the system reliability and performance. That is why, to ensure a high reliability and efficiency for a power grid, energy storage systems (ESS) and more power generation options are required [2].

Planning the required power reserve due to the fast growth of intermittent renewable generation and its effects on power grid control and performance is a significant issue in modern power grids operation and control. However, the most beneficial approach is the contribution of RESs themselves in the ancillary/regulation services to provide the regulation power reserve. For example, solar and wind farms can respond to a received dispatching function from the dispatching center or the supervisory control and data acquisition center for seconds, while it may take minutes for a conventional SG with a slower output power ramp rate. Therefore, like conventional generating units, these variable generation resources can support regulation issues in modern electric power grids [1].

The GCCs can receive the required operation/control set-points and references from the corresponding electric utility or market operator to produce the required regulation power support. These references are distributed between the GCCs of the existing RESs/DGs to determine the amount of contribution for each participant power source in the grid power regulation. The required amount of regulation power is mainly determined by considering the amount of reserve from those SGs that can be relocated.

According to the initial objectives of MG advent in version 2003 of the IEEE 1547 standard, the MG with RESs works in grid connected mode through the GCCs and supplies consumers by the grid. However, in the case of any failure, which could result in deviation of frequency and the voltage of the main grid, MG should disconnect from the grid and supply whole or part of loads autonomously by the ESSs, which have been charged by RESs during the grid connected mode [7]. In 2018 version of IEEE 1547 standard, this strategy improved by adding voltage and frequency ride-through capabilities to the RES interfaces in abnormal situations during the grid connected mode [8]. These characteristics will be served by smart GCCs. In addition to their typical responsibility of converting resources output power to the grid in a desired level, smart GCCs control voltage and frequency of sources by adjusting the relevant set points in the grid. Applying this enhancement in the grid control methodologies accelerates the pace of RES/MG integration to the utility grid as a service provider [6].

The RESs power curtailment method is another challenging issue which currently is applied to maintain the balance between supply and demand when a high level of renewable power is predicted. The quality of this method is directly depends on the control quality and performance of GCCs in off normal conditions. Curtailment of excessive power, similar to the load shedding, can be considered as an effective method to support the grid control and to improve the system flexibility. It can be also used to allow RESs to generate at reduced levels which can ramp up quickly to balance the system. The amount of curtailed power, duration, and method of curtailment is discussed through examples of several countries that apply this process in different manners. Power curtailment is crucial for the installation of different scales of RESs. For example, in Ref. [9], large-scale integration of PV power in a distribution grid signifies the application of power curtailment in controlling voltage, feeder currents, distribution substation overloading, and the grid frequency. Moreover, based on the real power curtailment of GCCs [10], proposed a strategy to enhance the performance of significantly unbalanced low voltage distribution networks with high residential PV penetrations [3].

The GCCs for wind turbines have also showed the applicability of the fast frequency impact of curtailment response control. For example, when a fast primary frequency regulation mechanism is activated in a wind farm system, the frequency response is highly enhanced. The participation of wind power in primary frequency regulation by its reserved capacity is discussed in Refs. [11,12].

1.3 Grid connected converters

This section is mainly focused on the application and the role of GCCs in integration of RESs such as PVs and wind turbines to the grid. The GCC structure and its control loops are discussed in Chapter 2.

In general, most of DGs including RESs use GCCs to connect the primary sources to the grid. For example, Fig. 1.2 depicts two structures for connecting a wind power

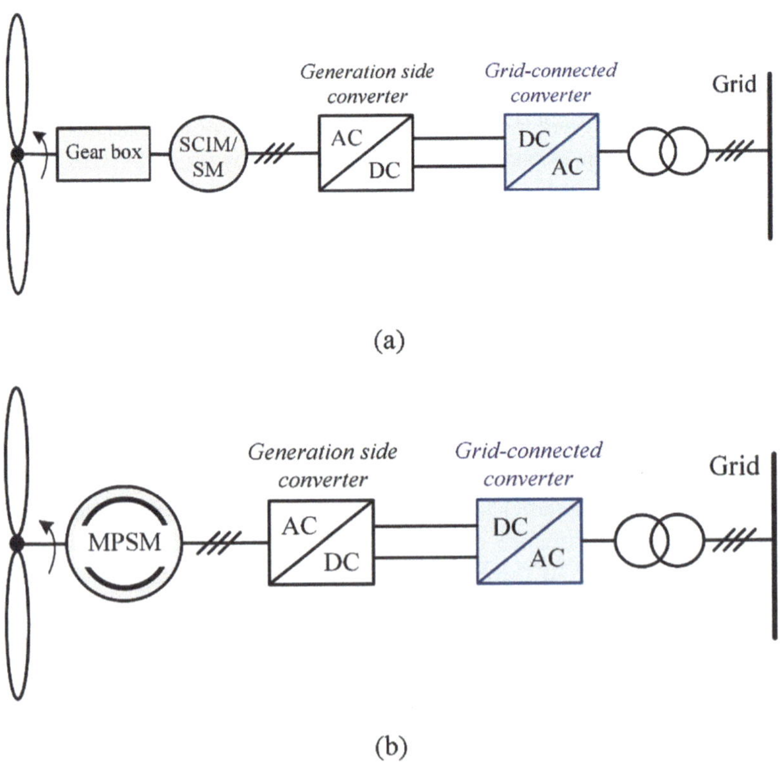

Figure 1.2 GCC-based wind generator; (A) with gearbox, and (B) without gearbox.

generator to the grid by implementing a GCC between the electrical machine and the utility grid. Normally, as shown in Fig. 1.2A, a squirrel-cage induction machine (SCIM) or a synchronous machine (SM) is used in this configuration. By using a multi-pole wound rotor (or permanent-magnet) synchronous machine (MPSM) in the generation-side block, the gearbox can be eliminated as shown in Fig. 1.2B.

The wind energy conversion system transfers power from the wind power generator to the grid. Two power converters are used to regulate the power exchange. A hierarchical structure can be considered to implement the control system. As shown in Fig. 1.3, two control systems are used separately for the generation side and grid-connected converters. A more detailed block diagram for the control system of the GCC is given in Chapter 2.

Similar to the wind power generator systems, the solar power generators are becoming economically viable. The PV system capacity is increasing continuously all over the world reaching values of hundreds of megawatts, thus making these plants a crucial part of future renewable integrated power grids. The large and medium PV units operate mostly in rural areas where require enough surface for installing. They can be connected to radial distribution grids as shown in the example of Fig. 1.4. A general structure for a PV with a GCC is shown in Fig. 1.5.

Because of the intermittency of PV power generation, using it as the only available power source in the grid may not be considered as a stable, reliable, and controllable power source and may not provide ancillary services like conventional generators. One solution is to upgrade the PV system with an embedded ESS and a local energy management system for the coordination of inner power flows. The ESSs can be used to store or to release electrical power like an energy buffer. Therefore, they can help to solve problems due to renewable energies' intermittent availabilities and fast transients. The performed hybrid system can be considered as an active PV generator for the grid since it can supply ancillary services as conventional generators. Moreover, it can be dispatched and then provide a power reference that is demanded by the grid operator. A general structure that has been widely used for integrating the ESSs in the grid is based on the direct connection of a battery bank to the DC-bus of the GCC. A PV controller is used to extract the maximum power from PV panels and send it to the battery bank [2].

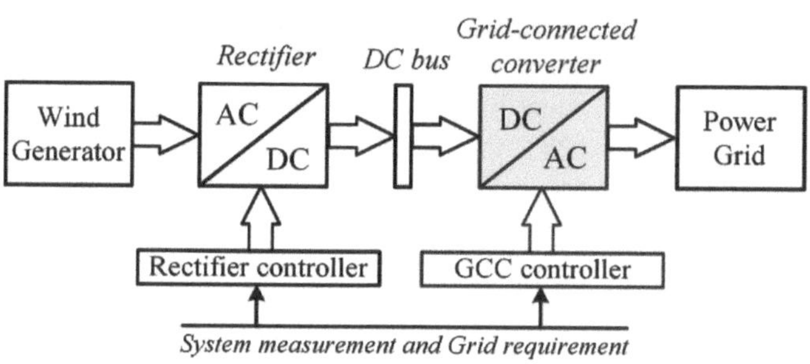

Figure 1.3 Wind energy conversion system.

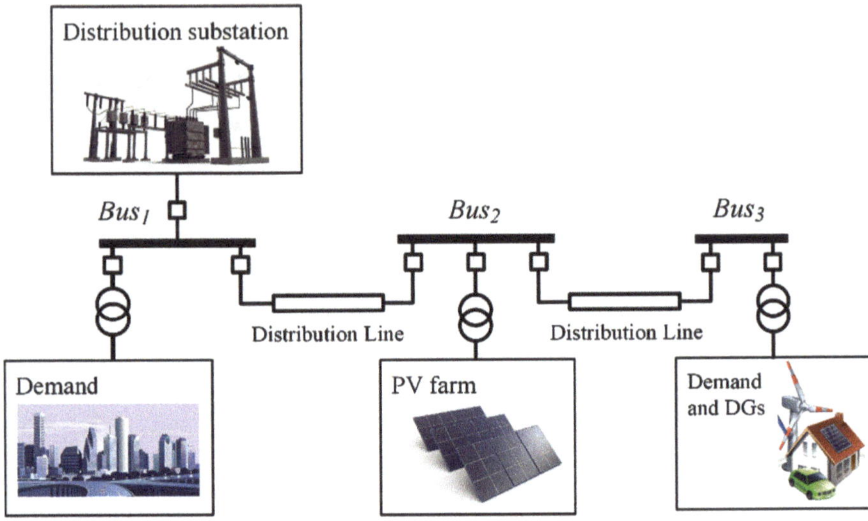

Figure 1.4 A distribution grid with a PV farm.

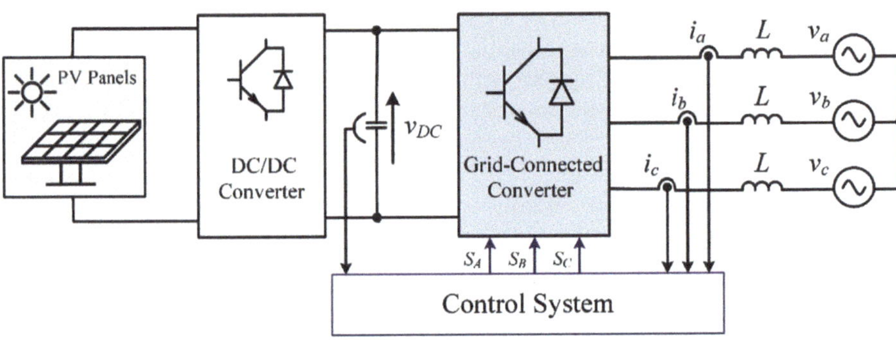

Figure 1.5 A general structure for a grid connected PV.

The power flows between the different sources of the mentioned hybrid system must be controlled in order to supply the real and reactive power required by the grid operator commands. This is performed by the control of different power electronic converters, which must be then coordinated as shown in Fig. 1.6. In this example, three sources of PV, the ESS, and the grid are considered. Three power electronic converters are used to regulate the power exchanges among them. Therefore, in the control system block, three main control loops are used for the control of the three sources. The control system uses appropriate feedback signals such as currents, voltages, frequency, state of charge (SoC) of ESS, as well as grid commands.

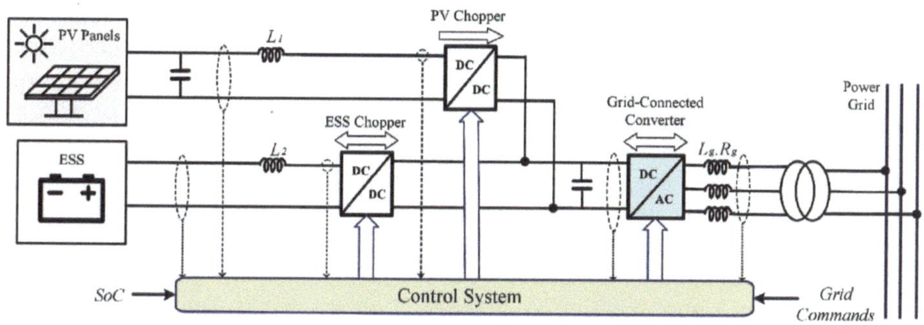

Figure 1.6 Grid connected PV and ESS example with the control system.

1.4 Renewable integrated power grids: characteristics and challenges

The capacity of installed converter-based DGs and RESs in power grids is rapidly growing, and a high penetration level is targeted for the next few decades. In most countries, significant targets are considered for using the distributed DGs/RESs in their power grids for near future. A schematic diagram of a renewable integrated power grid with distributed PMUs that are synchronized by global positioning system (GPS) is shown in Fig. 1.7.

The increase in converter-based power sources in the power grids has a significant impact on CO_2 reduction; however, recent studies have shown that relatively high

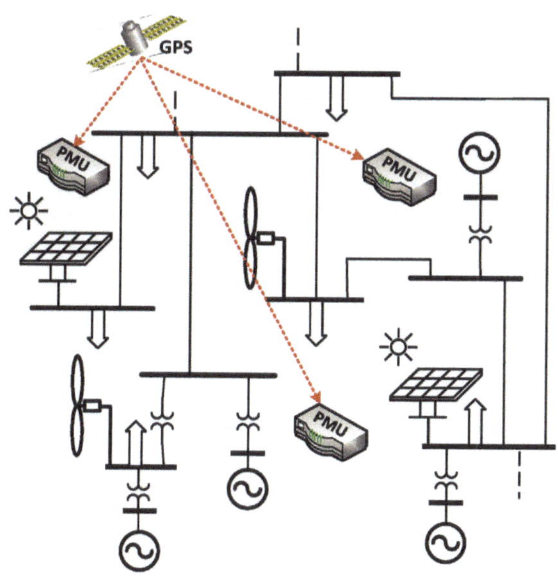

Figure 1.7 PMU-based renewable integrated power grid.

DGs/RESs integration will have some undesirable impacts on power grid dynamics, frequency and voltage regulation, as well as other control and operational issues. Decreasing system inertia and highly variable dynamic nature of RESs are known as the main reasons. These impacts may increase for the dynamically weak power grids at the penetration rates that are expected over the next decade [13].

Dynamic impacts of converter-based DGs and high penetration of RESs on power grids dynamic response, performance, and stability are discussed in numerous reports over the recent two decades. It is investigated that a low inertia due to replacing of conventional SGs with the GCC-based DGs can negatively affect the grid frequency dynamic performance and stability. A number of recent works have suggested the emulation of virtual inertia by the existing GCCs to overcome this challenge. Furthermore, several research works have been recently focused on the use of electric vehicles, storage devices, and controllable loads to provide frequency control supports in the power grids. In addition to frequency control, voltage and damping controls using the GCCs in renewable integrated power grids have attracted many attentions [13].

As mentioned, using significant number of distributed GCC-based DGs into power grids adds new technical challenges. As the electric industry seeks to reliably integrate large amounts of DGs/RESs into the power grid in a secure environment, considerable effort is needed to accommodate and effectively manage the installed distributed GCC-based resources. A key aspect is how to handle changes in topology and dynamics caused by penetration of numerous DGs/RESs in the grid and how to make the power grid robust and able to take advantage of the potential flexibility of GCCs. In a modern control framework, a part of power produced by available microsources in the grid are used as a primary energy source of inertia emulator by the GCCs to provide virtual inertia as a supporting control for abovementioned controllers (like a fine tuner) to improve power grid stability and performance [13].

In future, a significant share of converter-based DGs in the electric power grids is expected. This increases the total system generation power, while does not contribute to the system rotational inertia. System dynamics are faster in power grids with low rotational inertia, making control and power grid operation more challenging [14]. The performance of traditional control schemes can be degraded, and in turn, may result in large deviations and, potentially, grid instability.

For the control design purpose, usually a reduced-order linearized model around an operating point is used and it is assumed that all system parameters are known and time invariant. These assumptions, however, are not valid in a real power grid with dominated GCC-based DGs. The main dynamic modes of the system are varying stochastically during a day because of the variation of load and aggregated inertia. The dynamic modes will change more significantly by integration of new RESs into the power grid, and a fixed linearized time-invariant model will not correctly represent the behavior of the power grid [13].

Reduction in the cost of electricity transmission infrastructure from utilizing distributed energy resources has increased the penetration of DGs and MGs. However, the nature of RES is intermittent. Therefore, the primary role of GCCs, that is, power conversion and energy feeding, has been developed to provide dispatchable sources, able to overcome RES randomness as well as meet tied power system frequency and voltage deviations. Smart GCCs offer services such as harmonics, voltage, and frequency control to achieve cooperation between RESs and the utility grid. Along with smart GCCs, the presence of ESSs and responsive loads allows the DGs to participate in the electricity market. These characteristic has brought the opportunity of employing converter-based DGs as ancillary service providers of utility grid, including voltage control, black-start aid, and especially frequency regulation support [6].

With proper design of GCCs control system, they can contribute to power grid control and provide the vast majority of the services which traditionally provided by conventional SGs such as frequency/active power control, voltage/reactive power control, and ride-through for both frequency and voltage disturbances.

1.5 Current trends and future directions

Most of the current research activities in the field of GCCs stability and control rely on an assumption that they are working in nominal operating points. However, due to uncertainties, and changes in system parameters and operation state, the GCC dynamic response is becoming more changeable. Consequently, the GCC dynamics characteristics may affect the operation of power grid relaying, power quality management and monitoring, and other renewable integrated power grid operation and control issues.

To this end, the proper GCC dynamics estimation and modeling play an important role in modern power grids. It is also important to consider the concept of resilience in GCC operation and control issues. This attempt aims to increase the resilience of GCC control systems against cyber-attacks and the resulting manipulations. To facilitate a higher penetration level of RESs, it is also necessary to take benefit of the additional virtual inertia created by GCCs for improving the grid dynamic performance and stability.

The unpredictability of load is also an important challenge in a GCC integrated power grid control. Loads are becoming more and more erratic especially in the distribution grids with electric transportation systems. These grids, without a global control system, can only rely on appropriate control of GCCs set-points for desirable dynamic shaping of output active and reactive power as will be discussed in Chapter 9.

To have a secure and reliable power grid with high penetration of RESs, it is vital to use the significant potential of GCCs in providing regulation power support and ancillary services. Recent works show the high ability of the fundamental GCC blocks of future smart grids to contribute to the power grid performance enhancement and control. In practice, several utilities have already revised their grid codes for this purpose, and some recent research activities have been conducted to develop more effective controllers for control support of existing GCCs.

The capability of GCCs contribution to regulation power reserve is now needed by the grid code of some utilities. Furthermore, if due to congestion management, sufficient inertia, and reserve provision, some GCCs are disconnected; their energy can be used to produce upward energy reserves. The allocation of the regulation reserve by the GCCs in future smart communities will be different from the same method for the SGs, despite flexibility in control, since their outputs experience high intermittency. Here, two important directions in application of GCCs in demand response and virtual inertia are briefly emphasized.

1.5.1 Dynamic impact of GCCs

Due to the different dynamic behavior of GCCs compared to that of the conventional SGs, the increasing share of GCCs in modern power system leads to new classification in power system stability phenomena. The high integration of the GCCs provides a wider range of dynamic time scales in a modern power system. The time scale of GCC control loops ranges from a few microseconds to several milliseconds, which belong to the wave and electromagnetic phenomena. Therefore, for analyzing the dynamic impacts of GCC-based DGs on a modern power grid stability and performance, there is a need to extend the bandwidth of the power system stability phenomena to cover this time scale (some impacts are discussed in Chapter 5).

In addition to the GCC output filter, the associated inner-current control loop and phase-locked loop (PLL) may affect the power grid dynamics, mostly due to incapability of PLL to quickly synchronize with the grid voltage and also due to high gains in the inner-current control loop and PLL [15]. The high-level control loops are also important to form the overall dynamic performance of GCCs in a power grid.

Therefore, the operation and control of modern power grids must be based on a combination of physical body and control system responses of SGs and numerous GCC-based DGs. Stabilizing, protection, ride-through capability, and the ability to model these hybrid systems are basic important technical issues in a renewable integrated power system.

1.5.2 GCC-based demand response

Although the contribution of demand side to the power grid control support using the GCCs is an important issue, it is still difficult to consider demand-side control support as a fixed participator in the overall power grid regulation requirement.

The demand response is defined as the ability of the system operator to perform a flexible control of the grid loads and easy possibility of turning the existing load blocks off or on in response to coming up contingencies, economic/technical limitations, as well as regulation and protection requirements. This ability effectively supports the power system in all fields of voltage/frequency control, active/reactive control, reliability, security, and power quality. The demand response can reduce the generation participation rate in the grid regulation, and thus it helps to reduce the amount of required generation, power reserve, as well as operation cost, and CO_2 emission.

1.5.3 GCC-based virtual inertia

As mentioned before, high integration of GCCs reduces the overall inertia of a power grid. Low inertia can negatively affect the grid frequency dynamic performance and stability. A solution to this issue is to fortify the system with virtual inertia.

The contribution of GCCs in regulation task refers to the ability of these systems to regulate their power output, by an appropriate control action. This can be regarded as adding virtual inertia to the grid and considered as a solution.

Virtual inertia emulation requires the GCC to be able to store or release an amount of energy depending on the grid frequency deviation from its nominal value, analogous to the inertia of a conventional generator. This setup, which is known as virtual synchronous generator (VSG), will then operate to emulate desirable dynamics, such as inertia and damping properties, by flexible shaping of its output active and reactive powers as conceptually shown in Fig. 1.8. This VSG provides a promising solution to improve power grid stability and performance in the presence of a high penetration of RESs.

Virtual inertia can be produced using an ESS with a GCC under an appropriate control scheme to emulate the required inertia. Reinterpretation of this concept in the frequency domain is represented in Fig. 1.9. More details on internal virtual control are given in Chapters 3, 5, and 9. The GCC control system includes the dynamics of virtual inertia control loop, $K(s)$, and inertia constant M^{GCC}. It is noteworthy that the $K(s)$ uses the SoC of the ESS and frequency deviation as input feedback signals and

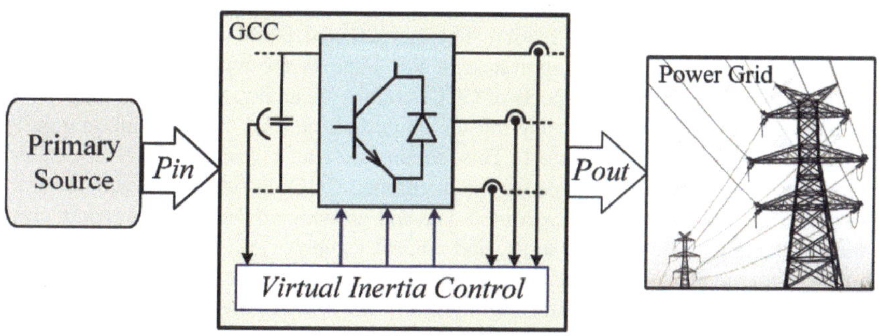

Figure 1.8 Conceptual structure of a GCC-based virtual inertia emulator.

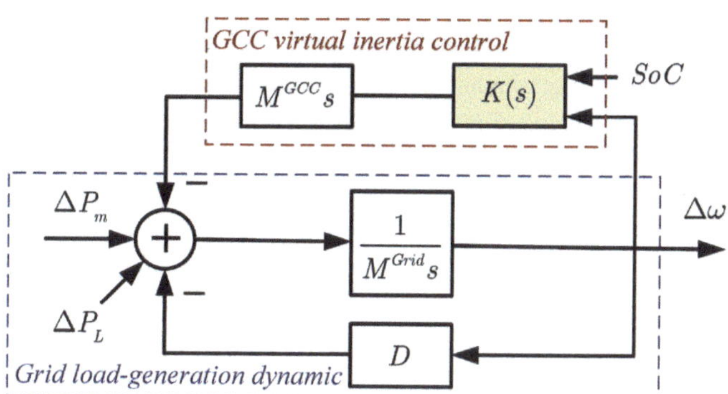

Figure 1.9 Frequency response representation of a power grid with GCC-based virtual inertia emulator.

provides the dynamic of the virtual inertia control loop. While the inertia of SGs in power grid in Fig. 1.9 refers to M^{Grid}. In the block diagram, D, ΔP_L, and ΔP_m demonstrate the damping factor, load change, and mechanical power change in the power grid, respectively.

The VSG is not only applicable for improving of frequency regulation and oscillations damping, particularly during the transient state following a disturbance, but also it is useful to support the voltage stability. The VSG system can use the available GCCs and their primary sources to participate in power oscillation damping by adjusting their active and reactive power generations.

1.5.4 Future research needs

Based on the current issues concerning power grid control and the impact of distributed GCCs, relevant research priorities and future works can be considered as follows:

(i) Effective control solutions for the GCCs in a power grid with high integration of RESs, particularly in islanded grids due to their relatively low inertia, significant power fluctuation, and various uncertainties are needed. In this direction, providing appropriate coordination between the GCCs with the SGs and ESS is important.

(ii) Measurement-based dynamics identification and modeling for adaptive control and online parameter tuning of the GCCs is vital. The increasing size and diversification of demand/power sources magnify the importance of this issue in the renewable integrated power grids. Online computational aspects of GCCs control are an important issue, and the online tuning of GCCs set-points considering the unpredictable load changes can be quite challenging in operation and control. This emphasizes the significant role of data-driven modeling and control techniques for the distributed GCCs in future relevant studies.

(iii) Considering a huge amount of recorded data and growing advances in computer, communication, intelligent electronic tools, and control technologies, artificial intelligence, and machine learning can play a significant role in the control of distributed GCCs of the future grids. This direction also requires highlighting the cyber-security in the control and operation of GCC systems.

(iv) New control mechanisms for GCCs in renewable integrated power grids, considering new demand response and virtual inertia-based control approaches are required. Additional flexibility may be required from various GCC control levels so that the power grid can continue to balance supply and demand in the presence of numerous RESs.

(v) A complete understanding of reliability considerations via effective modeling/aggregation techniques is vital to identify a variety of ways that GCCs in power grids can accommodate the large-scale integration of the renewable power in future. An accurate dynamic model is needed for the GCC stability analysis and control synthesis in a renewable integrated power grid. A proper dynamic modeling and aggregation of the GCCs, for performance and stability studies, is a key issue to understand the dynamic impact of distributed microsources and simulate their functions in new environment. Furthermore, the overlap between GCC voltage and frequency/active power dynamics in a renewable integrated power grid has not been highlighted so far, and it needs to be clearly addressed to design more effective and generalized control systems.

1.6 Summary

This chapter addresses an introduction to use GCCs in modern power grids. The integration of renewable energy sources and the fundamentals of renewable integrated power grids are described. The relevant characteristics and general control structures are discussed. The impacts of GCCs in grid operation and control are emphasized, and finally, a brief review with current trends and future directions is presented.

References

[1] H. Bevrani, et al., Power system frequency control: an updated review of current solutions and new challenges, Elec. Power Syst. Res. 194 (2021).

[2] H. Bevrani, B. Francois, T. Ise, Microgrid Dynamics and Control, Wiley, NJ, USA, July 2017.

[3] N. Harag, et al., Autonomous dual active power frequency control in power system with small-scale photovoltaic power generation, J. Mod. Power Syst. Clean Energy (2021) (in press).

[4] H. Ghasemi, C.A. Canizares, Confidence intervals estimation in the identification of electromechanical modes from ambient noise, IEEE Trans. Power Syst. 23 (2) (2008) 641–648.

[5] Communication Networks and Systems for Power Utility Automation—Part 7-420: Basic Communication Structure—Distributed Energy Resources Logical Nodes. Document IEC 61850-7-420, first ed., 2009. Available online: https://webstore.iec.ch/preview/info_iec61850-7-420%7Bed1.0%7Den.pdf (Accessed on 2 July 2021).

[6] L. Tightiz, H. Yang, H. Bevrani, An interoperable communication framework for grid frequency regulation support from microgrids, Sensors 21 (2021).

[7] IEEE standard for interconnecting distributed resources with electric power systems, IEEE (2003) 1–28, https://doi.org/10.1109/IEEESTD.2008.4816078.

[8] IEEE standard for interconnection and interoperability of distributed energy resources with associated electric power systems interfaces, IEEE (2018) 1–138, https://doi.org/10.1109/IEEESTD.2018.8332112.

[9] R. Luthander, D. Lingfors, J. Widén, Large-scale integration of photovoltaic power in a distribution grid using power curtailment and energy storage, Sol. Energy 155 (2017) 1319–1325. ISSN 0038-092X.

[10] X. Su, M.A.S. Masoum, P.J. Wolfs, Optimal PV inverter reactive power control and real power curtailment to improve performance of unbalanced four-wire LV distribution networks, IEEE Trans. Sustain. Energy vol. 5 (3) (July 2014) 967–977.

[11] J. Morren, S.W.H. de Haan, W.L. Kling, J.A. Ferreira, Wind turbines emulating inertia and supporting primary frequency control, IEEE Trans. Power Syst. vol. 21 (1) (February 2006) 433–434.

[12] Y. Wang, et al., Methods for assessing available wind primary power reserve, IEEE Trans. Sustain. Energy vol. 6 (1) (January 2015) 272–280.

[13] H. Golpira, A.R. Messina, H. Bevrani, Renewable Integrated Power System: Stability and Control, IEEE-Wiley, 2021.
[14] H. Bevrani, Robust Power System Frequency Control, 2nded., Springer, Gewerbestrasse, Switzerland, 2014.
[15] IEEE Task Force, Stability Definitions and Characterization of Dynamic Behavior in Systems with High Penetration of Power Electronic Interfaced Technologies, PES-TR77, April 2020.

Grid connected converters: fundamentals and configurations

2

2.1 General structure and essentials

Power electronic converters have been developed for many years with different types of applications. The application in the drive of electrical machines for industrial tasks was the main one at the beginning, then they started to use in many household appliances. The renewable energy opened a new chapter for wide application of power electronic grid connected converters (GCCs) in power and energy industry. Although still some distributed generators (DGs), such as the old small wind turbines, are directly connected to the grid to convert mechanical power to electrical power using the induction machines characteristics; because of many benefits, most of DGs are connected to the grid using the GCCs.

2.1.1 General structure and classification

The GCCs are effectively used to convert the primary energy source to the electric energy and transfer it to the power grid as shown in Fig. 2.1. Beside renewable energy, the development of many high-voltage direct current (HVDC) systems also requires the application of power converters in the grid.

As shown in Fig. 2.1, for many DGs in addition to a typical source-side DC/DC or AC/DC power electronic converter to connect the primary source to a DC bus, there is a GCC consisting of a DC link; a set of switching semiconductor devices; and a grid-side passive filter that suppresses switching harmonics from propagating into the grid. Since the GCC power stage is built solely with switching devices and passive filters, an effective control system is required for a desirable system operation. In modern GCCs, most of control systems are digital type, which are fully programmable and exhibit a high degree of flexibility for retuning and updating of the used control algorithms.

The impact of GCCs on the power grid performance cannot be neglected, and many studies confirm that the desirable dynamics of the new renewable integrated power grids are directly depend on the design and operation quality of the employed GCCs. It is usually assumed that the grid is strong enough to behave as a voltage source at the point of common coupling (PCC) of the GCC. Therefore, the GCCs are connected to the grid as current injectors, while they follow the voltage and frequency of the grid. These GCCs are operating in the *grid-following* mode.

For the GCC systems operate on the grid-following mode, usually a phase-locked loop (PLL) is used to align with the grid voltage at the PCC. Thus, they follow the measured voltage (as a reference) by aligning. These converters are not expected to response the grid frequency changes.

Grid Connected Converters. https://doi.org/10.1016/B978-0-323-99902-1.00006-9

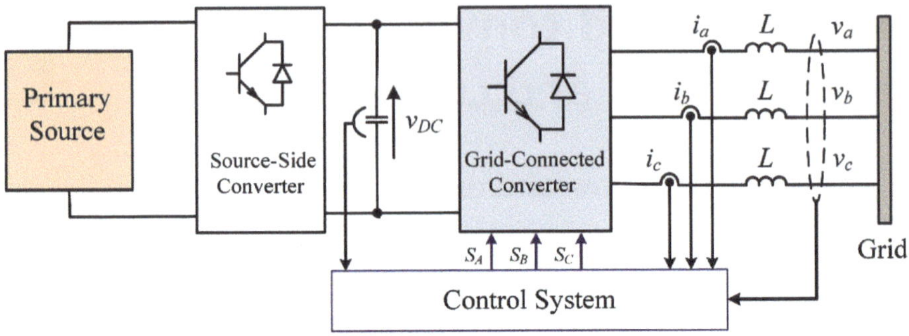

Figure 2.1 A general structure of a GCC.

In conventional power grids, the synchronous generators (SGs) dominate the electrical grid, establishing a nominal voltage amplitude and frequency that allow GCCs (in grid-following mode) to inject the power to the grid and to synchronize with AC voltage at the PCC by the PLL or the zero-crossing detection techniques. The term of *grid-feeding* refers the same concept.

However, with the increase in GCCs, their control capabilities, and the grid needs, this operation mode may change, and it is likely that some GCCs will need to switch from current source mode to a voltage source mode. In this mode, which is called *grid-forming*, the GCCs follow their own control reference signals and can operate/connect like the SGs.

In other words, in the modern power grids which the SGs are not dominant power generators, the grid-following GCCs are not able to inject the power into the grid. For autonomous operation, the control principle of the GCCs must be dramatically changed to no longer follow the grid but instead lead the grid.

The emulation of SGs inertia property makes the GCC as a promise solution to avoid the problems associated with low inertia in the renewable integrated power grids. The GCCs in the grid-forming operation mode are able to provide damping to frequency changes and their behavior is similar to that of SGs. This is possible since the grid-forming GCC creates its own internal voltage reference angle based on the output power of the converter [1].

Since both grid-following and grid-forming converters can be controlled in the available power range to support the grid voltage and supply the scheduled power, they are also called *grid-supporting* voltage/current source converters.

The grid-following and grid-forming converters are more discussed in Section 2.2. In the present book, some general ideas are discussed for the grid-following GCCs, but more focus is made on the GCCs in the grid-forming mode.

2.1.2 Reference frames

Since the GCCs are implemented as three-phase systems, in order to have a full control over the current, it is required to track three sinusoidal current waveforms. However, by the internal model principle, tracking sinusoidal waveforms is complicated and would require a complex controller (e.g., resonant controller), which its design is not so straightforward.

A more common solution for control analysis and synthesis problems is to change the reference system. This enables the synthesis of desirable controllers that are significantly easier than the design in original three phase system. There are two reference systems referred to as *$\alpha\beta$-frame* and *dq-frame*.

2.1.2.1 The αβ-frame

A symmetric and balanced three-phase system can be described as

$$\begin{aligned} v_a(t) &= V\cos(\omega t + \theta_0) \\ v_b(t) &= V\cos\left(\omega t + \theta_0 - \frac{2\pi}{3}\right) \\ v_c(t) &= V\cos\left(\omega t + \theta_0 - \frac{4\pi}{3}\right) \end{aligned} \tag{2.1}$$

where V is the amplitude, θ_0 the initial phase angle, and ω the angular frequency. The functions in Eq. (2.1) can be written as one state phasor, defined as (where j is the imaginary unit that satisfies $j^2 = -1$)

$$\boldsymbol{v}(t) = \frac{2}{3}\left[e^{j0}v_a(t) + e^{j\frac{2\pi}{3}}v_b(t) + e^{j\frac{4\pi}{3}}v_c(t)\right] \tag{2.2}$$

$\boldsymbol{v}(t)$ can be decomposed into its real and imaginary parts

$$\boldsymbol{v}(t) = v_\alpha(t) + jv_\beta(t) \tag{2.3}$$

where α is the real and β is the imaginary component. Equating the real and imaginary parts of (2.2) and (2.3) results in the following signal transformation

$$\begin{bmatrix} v_\alpha(t) \\ v_\beta(t) \end{bmatrix} = \frac{2}{3}\begin{bmatrix} 1 & -\frac{1}{2} & -\frac{1}{2} \\ 0 & \frac{\sqrt{3}}{2} & -\frac{\sqrt{3}}{2} \end{bmatrix}\begin{bmatrix} v_a(t) \\ v_b(t) \\ v_c(t) \end{bmatrix} \tag{2.4}$$

This can be seen as a projection of Eq. (2.1) on the real and imaginary axis. Since $\boldsymbol{v}(t)$ is a space vector rotating with the grid frequency the value of the α- and β-components will change in time. The α-vector will vary along the real-axis and β will vary along the imaginary axis, hence the α- and β- will behave as

$$v_\alpha(t) = \boldsymbol{v}(t)\cos\theta(t)$$
$$v_\beta(t) = \boldsymbol{v}(t)\sin\theta(t) \tag{2.5}$$

where $\theta(t)$ is the phase angle of the grid. This transformation into the $\alpha\beta$-frame has reduced the system of three components into two components. But, the signals still behave as sinusoidal signals, to solve this problem an additional transformation into the rotating *dq*-frame is introduced [2].

2.1.2.2 The dq-frame

The *dq*-frame presents a solution to the problem of oscillating control signals and transforms them into DC-waveforms during steady-state conditions. This is possible by rotating the $\alpha\beta$-reference frame with the frequency of the AC-signals. This can be mathematically formulated as

$$\boldsymbol{v}(t) = v_d(t) + jv_q(t) = \left(v_\alpha + jv_\beta\right)e^{-j\phi(t)} \tag{2.6}$$

and can be seen as a phase shift with an angle $\phi(t)$. Using Euler's identity $e^{-j\phi(t)} = \cos\phi(t) - j\sin\phi(t)$, Eq. (2.6) can be rewritten as

$$\begin{bmatrix} v_d(t) \\ v_q(t) \end{bmatrix} = \begin{bmatrix} \cos\phi(t) & \sin\phi(t) \\ -\sin\phi(t) & \cos\phi(t) \end{bmatrix} \begin{bmatrix} v_\alpha(t) \\ v_\beta(t) \end{bmatrix} \tag{2.7}$$

Combining the two transformations results in a direct transformation from *abc*-frame to *dq*-frame [2].

$$\begin{bmatrix} v_d(t) \\ v_q(t) \end{bmatrix} = \frac{2}{3} \begin{bmatrix} \cos\phi(t) & \cos\left(\phi(t) - \frac{2\pi}{3}\right) & \cos\left(\phi(t) - \frac{4\pi}{3}\right) \\ \sin\phi(t) & \sin\left(\phi(t) - \frac{2\pi}{3}\right) & \sin\left(\phi(t) - \frac{4\pi}{3}\right) \end{bmatrix} \begin{bmatrix} v_a(t) \\ v_b(t) \\ v_c(t) \end{bmatrix} \tag{2.8}$$

Fig. 2.2 shows the result of a three-phase voltage waveform being transformed into the $\alpha\beta$- and *dq*-frame. Note that the $\alpha\beta$- and *dq*-frame are also commonly known as the *stationary* and *rotating* frames [3].

It can be shown that the active and reactive power in the *dq*-frame can be calculated with the following relations

$$P(t) = \frac{3}{2}\left[v_d(t)i_d(t) + v_q(t)i_q(t)\right] \tag{2.9}$$

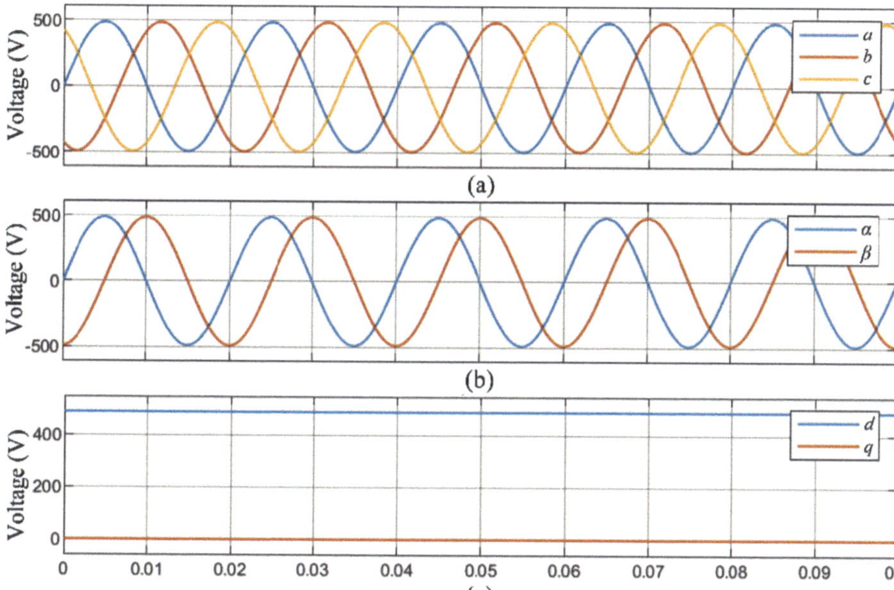

Figure 2.2 Transformation of a three-phase voltage waveform into the $\alpha\beta$- and dq-frames; (A) abc-frame, (B) $\alpha\beta$-frame, and (C) dq-frame.

$$Q(t) = \frac{3}{2}\left[-v_d(t)i_q(t) + v_q(t)i_d(t)\right] \tag{2.10}$$

If $v_q(t)$ is zero, which often is the case for a GCC in steady state, active power is proportional to $i_d(t)$ and the reactive power to $i_q(t)$ [2].

2.2 Configurations and applications

The most grid connected DGs use a back to back structure with two converters, one for the source-side converter and one for the grid-side converter, which we will mainly focus on the later one that is also known as the GCC. In PV generator and wind turbine applications, as explained before, the source-side converters are usually controlled to implement a maximum power point tracking (MPPT), and the GCC is usually a three-phased inverter. The GCCs can provide more flexible operation compared to the direct connection of synchronous and induction generators to the grid.

A practical configuration to interface a primary source (such as solar and wind energy sources) with the utility grid using the GCC is shown in Fig. 2.3. It also includes a source-side converter to extract the maximum power in the given environmental conditions. In order to inject power to the grid, the control structure must first synchronize with the grid using a PLL that measures the angle of the grid voltage. The power drawn

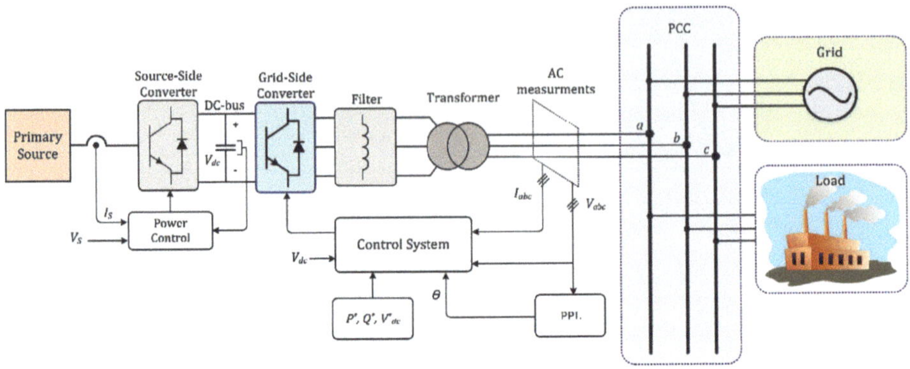

Figure 2.3 A typical DG with primary source and GCC.

from the primary source is injected into the DC bus using the source-side converter. Then it is converted utilizing a three-phase grid-side converter (GCC), which is driven based on the angle of the grid (θ) to synchronize the power electronic-based unit with the main grid.

As discussed in Section 2.1, the GCC systems can be mainly categorized as grid-forming and grid-following systems, which are also likely known as dispatchable and nondispatchable systems, respectively. It means the output power of the grid-forming GCCs can be controlled in the range of no load to full load, while the grid-following GCCs can only be controlled from no load to the maximum power, which is determined by the primary source. Therefore, the grid-forming GCCs can support the frequency and shape the output voltage, while, the grid-following GCCs can only supply the maximum power into grid like a current source. Basically if a GCC loses the voltage/frequency reference, it must be shut down.

The grid-following GCCs may also disconnect and/or stop operation following transmission faults, and in case of islanding in the renewable integrated power system, islanded systems only with the grid-following GCCs cannot be able to operate, autonomously. Moreover, after a blackout, these GCCs may not contribute to the restoration process of the power grid [4].

The grid-following GCCs control the output of real and reactive power by injecting a current at a given phase angle, but they cannot directly regulate system voltage and frequency. The voltage and frequency reference is provided externally either by a grid-forming GCC or the power grid. Simplified control structures of the mentioned GCC classes are shown in Fig. 2.4.

As shown in Fig. 2.4A, the grid-following GCC control system contains two main subsystems: a PLL that estimates the instantaneous angle/frequency of the measured grid bus voltage and a current controller that regulates the AC current injected into the grid. To produce the current reference, the active and reactive powers are calculated/controlled independently. Since the current is the physical quantity that is regulated, the grid-following GCC is often referred to as current controlled source. The PLL provides the angular reference of the current reference signal to follow the assigned output active and reactive power target (for fixed power reference signals,

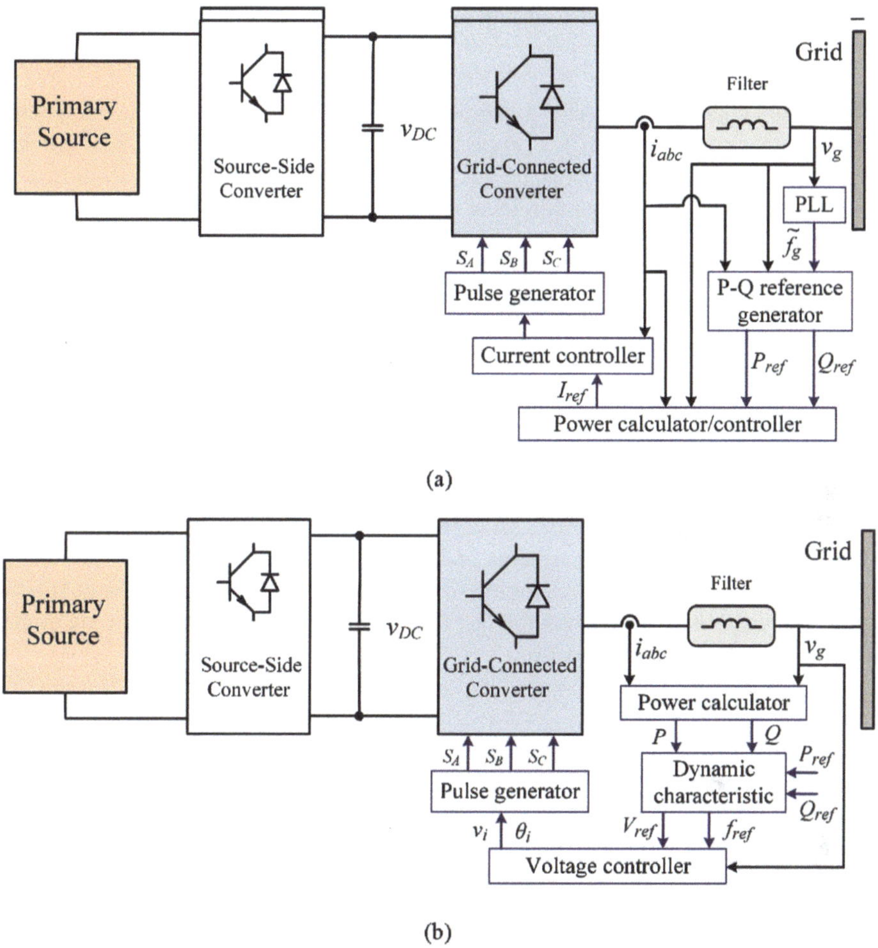

Figure 2.4 Simplified block diagram of two main GCCs categories; (A) grid-following and (B) grid-forming.

the GCC operates as a constant real-reactive power source). In the grid-following GCC, the system voltage and frequency are regulated by the connected external grid.

Fig. 2.4B illustrates a typical structure of a grid-forming GCC. As shown, the control system of this type of GCC contains two main subsystems: a voltage controller and a dynamic characteristic unit which provides the main reference signals for the voltage controller. The dynamic characteristic unit contains a control mechanism to emulate required dynamics.

Therefore, this control unit can act as a droop controller, virtual synchronous generator (VSG), virtual oscillator controller, or any other desirable dynamics. Unlike the grid-following GCC that acts as a current source, requires a PLL, and cannot operate without an externally regulated voltage, the grid-forming GCC may not require a PLL.

As can be seen, the grid-forming GCCs are intrinsically different from grid-following GCCs. A grid-forming converter is a controllable voltage source behind a coupling reactance much like the grid tied SGs. They follow a dynamic characteristic such as droop characteristic, rotor dynamic characteristic, and oscillator dynamic characteristic, which allows them for properly control of output voltage and frequency. Even, in the face of contingencies, the grid-forming GCCs can change their output power to track the loads and maintain local voltage and frequency. There is no considerable delay between the changes of output power and frequency in dynamic characteristic-based grid-forming GCCs. Therefore, they can respond much faster to any contingencies than the response of the SGs. An important advantage of this type of GCC, which is able to change its output much faster than the SGs, is to provide primary frequency regulation particularly for low-inertia renewable integrated power systems. The fast response characteristic causes arresting system's frequency changes before any load shedding is triggered [5].

In fact, the grid-forming GCCs objective is to provide functionalities that are traditionally provided by the SGs, such that they can smoothly operate in the low-inertia renewable integrated power grids. Modern grid-forming GCCs are being considered for deployment in large power systems because of their ability to improve the stability and performance of these grids in regions or during times when loads are largely being served by GCC-based DGs. Over the last few years, a variety of grid-forming control strategies have been developed [4].

2.3 Basic control loops

Over the years, the GCCs have drastically increased their importance on the market of energy conversion for renewable power applications. Two technology breakthroughs enabled these remarkable developments [6]: (1) advances in the field of power electronics, which brings improvements in many directions such as efficiency, power density, power quality, common mode voltage, and electromagnetic interference; and (2) advances in the field of control technology, which currently contains many sophisticated control functions. The basic control loops of the GCCs are directly influenced by both technologies, particularly the second one. Some important control blocks of a typical GCC are shown in Fig. 2.5. The current/voltage/power controls are known as *inner control loops*, while dc-link voltage control, power control (MPPT), reference generator unit and dynamic characteristic emulator (to support the power grid) are known as *outer control loops*. Other control function such as active/reactive power control to support the grid, harmonic compensator, synchronizing block, and PLL system are not shown in Fig. 2.5.

The implemented functions by the control system of GCCs must be distinguished according to the operating mode of the GCC. For the grid-following mode, the ac voltage and frequency are supplied by the grid. Therefore, the GCC can be controlled in a "P/Q mode" to inject the assigned powers. For a grid-forming mode, the GCC has

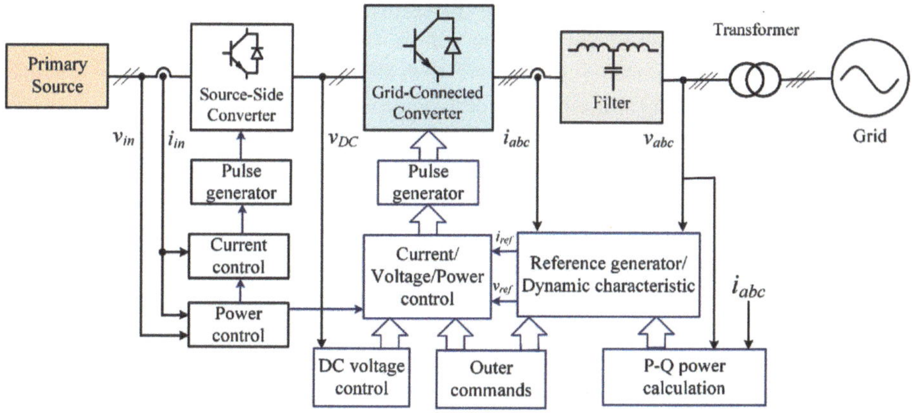

Figure 2.5 A typical GCC with some control blocks.

to control its output voltage and frequency using the local produced references like a voltage source converter (VSC).

2.3.1 Control system for grid-following GCC

A control system example for the grid-following GCC is shown in Fig. 2.6. In this example, the *abc* to *dq* transform is used to convert the three-phase voltage v_{abc} and current i_{abc} into the stationary *dq* frame. The GCC can control the voltage across the DC-link capacitor and control the exchanged power with the grid. This is a useful strategy for the GCC-based DGs/renewable energy sources (RESs), such as wind and PV generators, which due to uncertainty in the primary energy resource, may not be able to perform a precise DC bus voltage control.

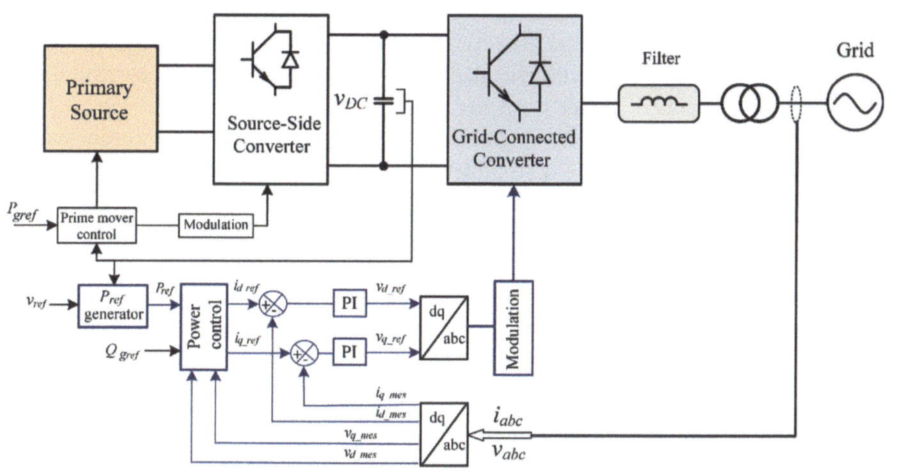

Figure 2.6 An example for grid-following GCC control system.

After regulating the DC-link voltage, different DC voltage values can be used to control the output power of the primary source (e.g., MPPT) [6]. In this example, the GCC can control the exchanged power with the grid.

The same function for regulating of voltage across the DC-link capacitor is done by the source-side converter as depicted in Fig. 2.6. In the grid-following strategy, due to use of a choke filter, the GCC is controlled as a current-controlled voltage source. Direct and quadrature components of currents are calculated by a Park transformation for power calculation. Active power fluctuation of the primary source causes a DC-bus voltage error, which can be corrected via a DC bus control block producing an active power reference (P_{ref}) through a proportional integral (PI) controller to regulate the magnitude of the currents injected to the grid. The reactive power output can be also controlled using a PI controller by adjusting the magnitude of the GCC reactive current output. This GCC can be operated with a unit power factor or can receive a set-point from the outer control system for the output reactive power.

In the GCC-based DGs with possibilities to increase/decrease the amount of primary power, as shown in Fig. 2.6, an additional active power reference (Pg_{ref}) can be used by the outer control system to follow a specified active power pattern. This control system is known as P-Q control, which makes the DG as a dispatchable power source.

The GCCs in the grid-following mode only follow up the grid frequency and phase angle to appropriately supply the desired power into the grid. The DGs using this GCC (e.g., PV and wind turbine) are often controlled to feed the maximum power into the grid. The active power reference is determined by the MPPT mechanism, and the reactive power reference can be chosen to achieve a desired power factor. In order to properly supply the desired active and reactive power into the grid, a synchronization system needs to follow up the grid-phase angle using the PLL or the power control unit [7].

2.3.2 Control system for grid-forming GCC

A control system example for the grid-forming GCC is shown in Fig. 2.7. In the grid-forming control strategy, the voltage of the DC-link capacitor is often regulated by controlling the generated power from the primary source, as shown in Fig. 2.7. The primary source is responsible for responding to this power demand in order to keep a constant DC voltage bus. This is suitable for applications with more controllable power producers such as hydro turbines, gas-microturbines, and fuel cells as the availability of the primary energy is guaranteed. Then, the voltage across the DC bus can be considered as constant and the GCC can control the output active and reactive power based on a control scheme to emulate a specific dynamic characteristic (such as oscillator $f(Q)$ function and control system which is shown in Fig. 2.7). The given GCC example also uses a P-Q control scheme, which consists in an outer power control loop that calculates current references for the current control inner loop. If we consider that the role of DC bus in providing a constant DC voltage as inertia, a similar behavior can be found for the GCC-based DGs with this control strategy and the conventional SGs.

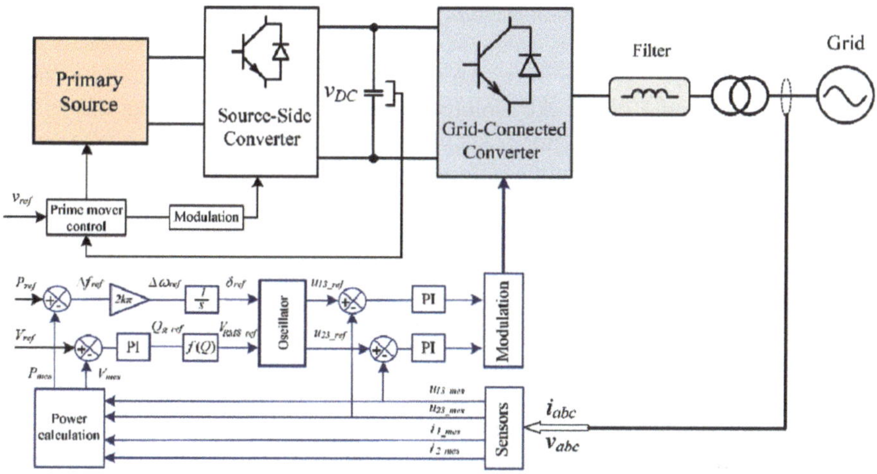

Figure 2.7 An example for grid-forming GCC control system.

The output power of the grid-forming GCCs can be managed based on their output and the corresponding line impedances. However, using droop-based method is more common to control the power sharing among the grid-forming GCCs. This method is inspired from the droop control of the governor system of the conventional SGs [8]. Droop-controlled GCCs, also called grid-supporting GCCs, can support the grid voltage and frequency.

Fig. 2.8 shows another example for grid-forming GCC, with a control system containing the droop and the cascaded PI controls. The advantage of the cascaded PI controllers is in ability to obtain the current reference from the voltage control so that the GCC prevents itself from overcurrent in case of a grid fault. Moreover, the structure of a PI controller is simple and there are numerous tuning procedures for its two parameters. The PI controllers in a cascaded structure are often independently tuned in frequency-domain by setting a faster response time for inner loops and slower one for outer loops [9].

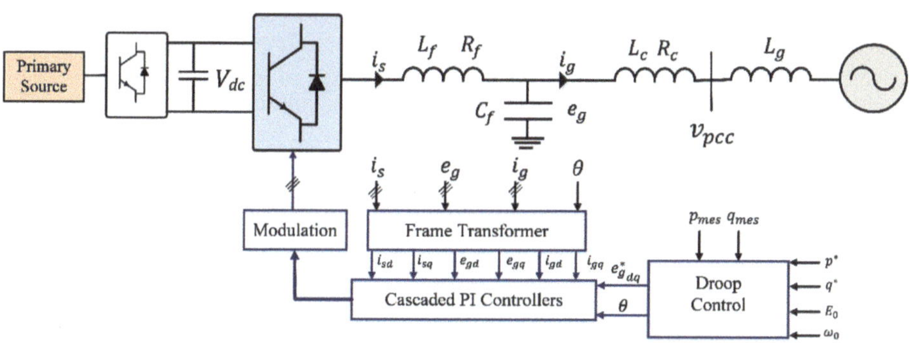

Figure 2.8 Grid-forming GCC example with droop and cascaded PI controls.

The amplitude and frequency of the output voltage (v_{abc}) are determined by a nominal amplitude E_0 and nominal frequency ω_0. The GCC generates a three-phase balanced sinusoidal AC voltage using the cascaded PI controllers in the dq reference frame. The phase angle of the AC voltage is formed by the droop control system. Therefore, the GCC with its primary source can operate as an ideal voltage source, and hence, the power sharing among different grid-forming GCCs is a function of their output impedances as well as line impedances [7].

Parallel operation of the grid-forming GCCs requires a precise synchronization system at the start-up point. In disconnected operation mode, the synchronization system must oscillate at the nominal frequency ω_0 to form the output voltage. During the connection to the grid, the GCCs need to be synchronized with the grid. Therefore, the synchronization system should smoothly change the frequency and the phase angle to synchronize with the grid. The synchronization system also needs to detect the change in condition, when it is disconnected [7].

For contribution to the frequency regulation of the grid, the GCCs need to operate at a power lower than the MPPT power. The remaining power is considered as a regulation power reserve for frequency control support. The supported power can be shared according to the corresponding droop gains.

Fig. 2.9 shows block diagrams of typical droop controls. The control is $P-\omega$ and $Q-E_g$ droop since the nature of the grid is inductive enough as $X_g \gg R_g$. The injected three-phase instantaneous active and reactive power components, p_{mes} and q_{mes}, are given by the following equations

$$p_{mes} = e_{gd}i_{gd} + e_{gq}i_{gq} \tag{2.11}$$

$$q_{mes} = e_{gd}i_{gq} - e_{gq}i_{gd} \tag{2.12}$$

To avoid power noises, a low-pass filter is added on the way of active and reactive power measurements, where ω_c is the cutoff frequency.

$$p = \frac{\omega_c}{\omega_c + s}p_{mes} \tag{2.13}$$

$$q = \frac{\omega_c}{\omega_c + s}q_{mes} \tag{2.14}$$

Here, the droop control determines the amplitude, phase, and frequency of the capacitor voltage reference, thus grid-forming GCC can be synchronized with the grid and can share power without any detection of the grid-phase.

2.4 Dynamic characteristics emulation

As mentioned before, the grid-forming GCCs are able to track any desirable static and dynamic characteristics virtually by designing an appropriate control scheme.

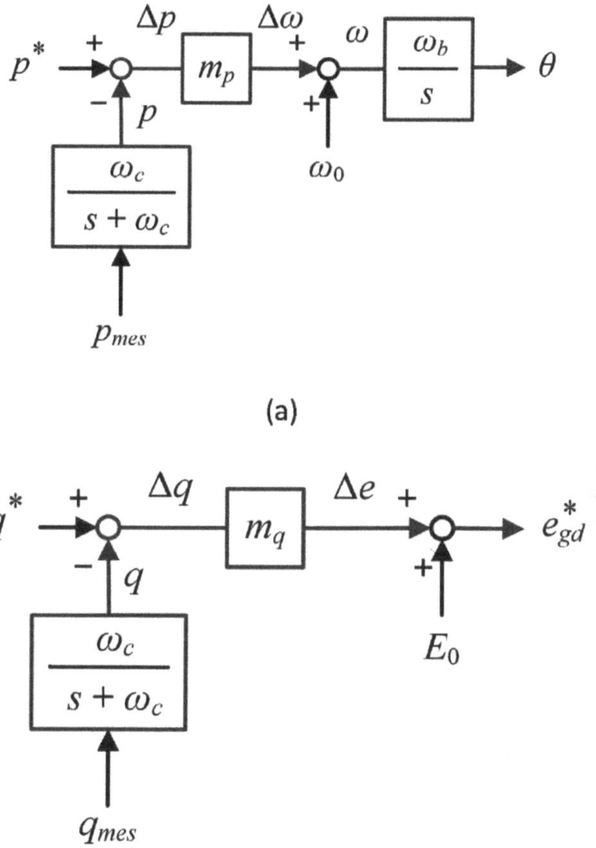

(a)

(b)

Figure 2.9 Block diagram of droop controls; (A) $P - \omega$ droop, and (B) $Q - V$ droop.

Designing virtual resistor and virtual impedance is well known over the last few decades, and recently this concept is enhanced by emulating more complicated dynamics such as SG behaviors in form of VSG. In this section, this concept and its application to the GCCs are briefly explained. Extended discussion is given in Chapters 5 and 9.

2.4.1 Review and applications

It should be noted that the master-slave control structure like that exists in microgrid is not possible to apply in a transmission system and each GCC must operate independently using the local measurements. A grid-forming GCC should be able to generate an AC voltage with a given amplitude and frequency at PCC like the SGs. Three well-known primary control mechanisms for this autonomous operation have been proposed in the literature: the droop control [10–13], synchronverter [14], and the VSG or virtual synchronous machine (VSM) [6,14]. Of course, inertia emulation can be also viable in the grid-following GCCs via mapping the PLL dynamics to a swing equation [15].

Unlike GCCs, the SGs possess inertias and damping factors in terms of their physical structure properties, which introduce a slow dynamic to the frequency that is a result of rotational speed. However, in the grid-forming GCCs, it is possible to produce the specified frequency and perform a virtual inertia and damping factor by its control system. The control system also allows the GCCs to synchronize to the grid without detecting the grid-phase and to share power with other generators [9].

The idea of emulating inertia dynamics for the GCCs has been presented in different control structures using many control methodologies. Firstly, the VSM concept is introduced to allow a grid compatible integration of GCC-based DGs to behave as electromechanical SGs [11]. The idea is presented as a solution for injecting the virtual inertia by the GCC in the form of VSG [16], and it has developed extensively over the last 10 years [14].

In Ref. [17], an internal synchronization mechanism has been proposed for the GCC, where the SG dynamics are mimicked. Finally, the grid connected operation of a GCC with an ideal DC link is realized by emulating the SG dynamics entitled Synchronverter [18], which is developed employing a nonlinear stabilizer [19], mimicking an SG, and improving the dynamic response [6].

Among the virtual inertia control structures, the VSG is more taken into account and developed. The VSG-based GCCs are allowed to be some active units to support the frequency dynamics of the microgrids [6]. By adjusting the VSG inertia and/or damping coefficient, the output power fluctuation and the overall system behavior can be improved significantly, which is demonstrated by the small-signal modeling and sensitivity analysis [10,11]. A virtual capacitor control to enhance the reactive power sharing in VSG-based MGs in Ref. [12] and an adaptive linear quadratic regulator-based VSG in Ref. [13] have also been presented to improve the inertial response of the system. In order to minimize the frequency deviation and output power fluctuation, a selfadaptive VSG control method based on rate of change of frequency (RoCoF) [20] and a selftuning-based algorithm [21] have been employed to continuously optimize the virtual inertia among predefined amounts of inertia moment and damping coefficient. Similar methods are addressed in Ref. [6], where different values of the parameters have been employed to improve the frequency response. The virtual inertia response is improved by tuning droop gain in Ref. [6]. In Ref. [22], a fuzzy control method is used to improve the frequency response of a wind turbine system, which is developed for islanded and grid connected microgrids [23].

Transition from rotating mass-based SGs toward GCCs, which are employed to converse renewable energies in a more controllable way, is becoming more and more important and challengeable. This transition leads to a renewable integrated power system with no or few rotating mass-based SGs, which presents a stochastic and time variant inertial feature. The inertial features for these systems are expected to be a function of the available variable primary power. It means that the system's inertia is going to be a stochastic variable and mainly dependent on environment conditions. In addition, sometimes situations may be emerged in which a required inertial support or an acceptable inertial response cannot be provided properly. Possible solutions to provide a sufficient level to satisfy the inertial requirements are needed not only remodeling through a time-dependent stochastic manner but also considering the primary source type and capacity.

Selfsynchronization of SGs through the grid, with kinetic energy as safeguard against grid-side disturbance and well-known dynamics is going to be replaced by the GCCs with no inherent synchronization through the grid and no (a sufficient) energy storage. In confront, inertial emulation in renewable integrated power systems has not been matured enough.

Table 2.1 sums up technical aspects to be considered for inertial proposes in both planning and control steps [23]. These issues can be employed to find a better solution for the existing concerns like: Which storage solution is better for the weak grids? How much capacity is needed for inertial support? What is the dynamic effect of energy storage systems (ESSs) on the inertial response during contingencies? How can designers find the best location of ESSs for inertial purposes? Which required control and coordination of ESSs can be done for inertial support?

2.4.2 An example: virtual resistor

To suppress unnecessary resonant oscillations in the GCCs, one simple solution is to use a virtual resistor to perform a damping control method. The system structure for a PWM-based GCC is shown in Fig. 2.10. An LCL filter of L_1, C, L_2 is used to reduce the harmonic components sufficiently, and to convert the voltage input v_i to the current i_{L2}. Now consider to have a resistive component R_1 in series with the inductor L_1. It is easy to investigate analytically that this resistor provides a damping impact and may suppress the resonant peak of the filter characteristic. As using an actual resistor causes losses, it can be virtually inserted by a proper control mechanism. The differential equations for the converter and virtual resistor can be obtained as follows [24,25]:

Table 2.1 Challenges for inertia emulating in the GCC-based power grids.

	Challenging field	**Challenging issue**
Planning	Cost evaluation	Capital costs, repair and maintenance costs, power losses
	ESS type	Fly wheels, supercapacitors, Li-Ion batteries, etc.
	Sizing issue	Arbitrary, simulation based, probabilistic based
	Market design	Considering RoCoF and nadir constraints for frequency control in market designs
	Placement issue	Power limit of converters, SoC, and line capacity constraints
Control	Adaptive and robust approaches	
	Time-delay considerations	
	Saturation drawbacks for power injection in inner loops	

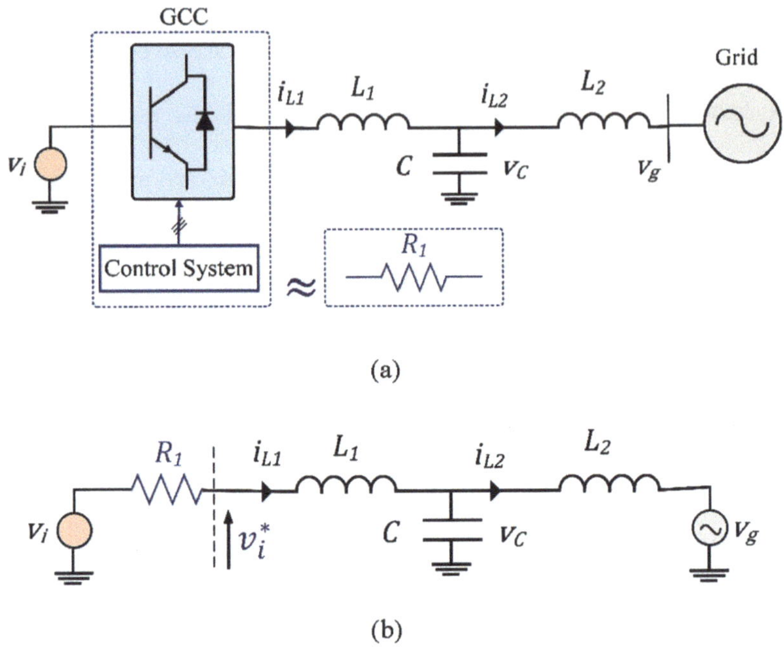

Figure 2.10 (A) A simplified PWM-based GCC, (B) Equivalent circuit with the system with virtual resistor.

$$L_2\frac{di_{L2}}{dt} = v_C - v_g$$

$$L_1\frac{di_{L1}}{dt} = v_i - R_1 i_{L1} - v_C$$

$$C\frac{dv_C}{dt} = i_{L1} - i_{L2} \tag{2.15}$$

Therefore, by an appropriate control scheme, the damping resistor can be emulated and the converter output voltage v_i can be modified to v_i^*:

$$v_i^* = v_i - R_1 i_{L1} \tag{2.16}$$

2.5 Relevant grid codes and standards

Early GCC-based DGs were connected to distribution systems, but, in recent years, more and larger DGs/RESs have been connected to transmission systems. Each of these systems has specific requirements when connecting to the GCC-based DGs.

The connection issues are related to grid stability which is influenced by power flows and GCCs behavior in the presence of grid faults. Therefore, the grid transmission/distribution operators which are often called transmission system operator (TSO) and/or distribution network operator need to develop some technical requirements, known as *grid codes*, for connecting the mentioned GCC-based DGs.

The grid codes have been complemented by specific rules for the GCC-based DGs, and the manufacturers have responded to these requirements by design measures, mainly in the area of DG system and GCC control designs.

Increasing interest on the grid integration of GCC-based DGs leads to continuously update and redefinitions of the grid codes by the system operators. The aim of the requirements is to ensure that GCC-based DGs do not adversely affect the power system operation security, reliability, and power quality.

In power system where GCC-based DGs have become a significant part of the total electrical power production. Disregarding topologies and geographical location, grid codes state how GCCs should behave under both normal and contingency conditions while maintaining a safe and reliable operation. The countries with a high penetration of GCC-based DGs in their national grid have developed detailed technical requirements for interconnection of GCCs. The designed grid codes must focus on the important technical requirements for some issues such as voltage and frequency operation, voltage and reactive power dependencies, active power curtailment, and low voltage ride through requirements [26].

In grid codes, it is specified the steady-state frequency and voltage operation ranges in which the GCCs should operate in normal conditions. For instance, normal condition can be considered for voltages close to 1.0 pu and frequency around the nominal frequency (50/60 Hz), with deviation of about ± 0.05 pu of voltage and ± 0.5 Hz of frequency. In this case, any grid condition outside these values is defined with a minimum operational time, and in some cases, followed by a control action for active or reactive power from the GCCs [27].

As mentioned, the grid code sets out the most important operating procedures and principles governing the relationship between the GCC-based DGs and the grid. It specifies minimum technical, design, and operational criteria which must be complied for connection of the GCC-based DGs to the power system [28]. In other word, grid codes are designed to (1) allow the development, maintenance, and operation of an efficient, coordinated, and economical system for the renewable integrated power grid; (2) facilitate competition in the generation and supply of electricity; and (3) promote the security and efficiency of the whole electricity generation, transmission, and distribution systems in the grid.

2.5.1 Grid code and standards requirements

As mentioned, grid codes are the requirements imposed by the power system operators that the GCC-based DGs must accomplish to be connected to the grid. These requirements are detailed in a document, called grid code that details what are the behavior of the GCC-based DGs both in steady state and transient operation conditions. The most important issues that must be considered in developing the grid codes can be summarized as follows:

(i) *Active power and frequency control*: The GCC-based DGs can be required to curtail or control the active power to a given set-point. This is necessary for participating to the energy dispatch as the conventional SGs and to prevent overloading of lines. Also some grid codes require the GCCs to participate to the frequency control as the conventional SGs. Therefore, the frequency must be kept within acceptable limits to ensure the security of supply, prevent the overloading of electric equipment, and satisfy the power quality standards. In this direction, the relevant grid codes define the frequency operating range, and frequency-power droop characteristic.

(ii) *Reactive power and voltage control*: The GCC-based DGs can be demanded to regulate the reactive power to a given set-point. In this direction, the relevant grid codes define the voltage operating range, reactive power limits depending on active power (according to the specified PQ curves), and voltage limits depending on the given power factor. Furthermore, a set of different requirements may also need from voltage quality point of view (e.g., rapid changes, flicker, and harmonics). Some grid codes require that GCC-based DGs must be equipped with tap-changing grid-side transformer in order to be able to vary the voltage ratio between the GCC and the grid when needed.

(iii) *Low- and high-voltage ride through*: The GCC-based DGs are required to remain connected when there is large voltage deviation during a fault. This issue also may be addressed as fault-ride through (FRT) which satisfies the capability to remain connected to the main grid during temporarily faults, and the voltage must stay above the FRT profile during faults.

(iv) *Ramp control*: The GCC-based DGs need to limit the rate of change of the power injected to the grid.

(v) *Inertia emulation and ancillary services*: The GCC-based DGs can be demanded to emulate the inertia of SG to support the grid dynamic performance and stability enhancement. The GCCs can be required to contribute to control the system synchronism, and to damp oscillations of the system, by adjusting active and/or reactive power injections. They are required to contribute to the voltage and frequency supports with reactive power and active power, respectively.

(vi) *Modeling, verification data, and online measurements*: Some grid codes require the GCC-based DG owners/developers to provide models and system data, to enable the grid system operator to investigate technical issues by simulations the interaction between the GCC and the grid. They also require installation of monitoring equipment to verify the actual behavior of the DG during faults and to check the model. Grid codes may also require that the DG operator provides online measurement of some important variables for the grid operator to enable proper operation of the power system [29].

(vii) *Black-start*: The GCC-based DGs can be required to restore the system voltage and frequency and contribute to restart the system after a blackout [30].

In a wide-area renewable integrated power system, different regulation levels can be identified for the various GCCs. These levels may be classified as local, regional, and global levels. To clarify this issue, the European power grid provides a good example with three different levels of grid codes requirement levels for some locals (TSO), regional (national), and global (wide-area)-specific requirements.

For example, the European Commission regulation 2016/1447 [31] has set a legal framework for the offshore wind farms GCCs via HVDC connections in whole the

European Union (EU). These rules should ensure EU-wide electricity trade possibilities, system security, the integration of renewable electricity sources, increased competition, and a more efficient use of the grid and energy sources for the benefit of society in the whole EU [31]. At the same time, it is needed to adjust a set of grid codes for each country in order to specific regional/national requirements.

For instance, the main objectives of the Technical Connection Rules provided in Germany, mostly for integrating offshore wind energy, are forming the basis for the safe application of new technologies in existing systems with the lowest possible losses, and possibility of providing ancillary services. Finally, the local grid codes are often transferred to grid connection rules that are applicable for certain grid operators. For example, the grid operator TenneT in Germany created a set of grid codes for offshore wind farms that are connected in the area of its operating [32]. The local grid codes usually match the regional requirements [33].

2.5.2 Protection, power quality and synchronization

Satisfy the assigned threshold values in order to system protections is a critical case that should be considered, according to the existing standards (like IEC 50549-1, IEC 62786, and Institute of Electrical and Electronics Engineers [IEEE] 929-2000), when a new GCC is introduced [34–36]. These standards determine the GCC ability and the control procedure for monitoring the grid frequency, control of the active and reactive power, power quality, and the leakage current control between the GCC and grid application sides. For example, a relevant standard given by VDE-126-1-1 and VDE-AR-N 4105 expresses that the leakage currents more than 300 mA (for PV as primary source) should be controlled; otherwise, the GCC should be disconnected from the grid [37].

These standards also address the frequency control procedures for the system. In this context, the GCC must continuously monitor the grid and to be disconnected from the grid if the specified voltage and frequency thresholds are exceeded. The threshold values of voltage and frequency values are evaluated depending on the operating values determined by the grid-code of each country. Some global norms according to IEEE 929-2000, IEC 50549-1:2019, and IEC 62786:2017 for the GCCs are shown in Table 2.2 [34–36]. V_0 and f_0 demonstrate the nominal voltage and nominal

Table 2.2 Threshold values according to IEEE 929–2000 for a reliable grid connection.

Case	CB operating time	Frequency (Hz)	Voltage (V)
1	Normal operation	$f_0 - 0.7 \leq f \leq f_0 + 0.5$	V_0
2	6 cycles	$f < f_0 - 0.7$	V_0
3	6 cycles	$f > f_0 + 0.5$	V_0
4	6 cycles	f_0	$0.5V_0$
5	2 s/120 cycles	f_0	$0.5V_0 < v < 0.88V_0$
6	Normal operation	f_0	$0.88V_0 \leq v \leq 1.10V_0$
7	2 s/120 cycles	f_0	$1.10V_0 < v < 1.37V_0$
8	2 cycles	f_0	$1.37V_0 \leq v$

frequency, respectively. Nominal values of grid voltage and frequency are defined in IEEE 929-2000, and a circuit breaker must be available to be operated when the variables are out of threshold values ($\pm 10\%$) [26].

With the high penetration of GCC-based DGs, the power system operators are facing new challenges in reliability, availability, and power quality point of views. Grid codes ensure that in order to maintain the power quality and reliability of the power grid, the GCC-based DGs should possess low-voltage ride-through (LVRT) capability. Transient voltage sags are very common transient phenomenon in the grid that requires simultaneous injection of active and reactive powers, according to the grid codes. Thus, different TSOs around the world require that the GCCs must remain in connection, as long as the voltage at their terminals are above the specified thresholds, in order to prevent the sudden loss of large amount of energy generation. This requirement is known as LVRT, and both voltage sag duration and variation size are determined by the relevant grid codes, which may differ worldwide from a country to country.

In addition to the LVRT, modern grid codes demand the voltage support by the GCC-based DGs, by means of additional positive sequence reactive current injection (RCI). During the LVRT operation, the grid support throughout the RCI requirement must be in the priority, and the GCC may require to provide its maximum current capacity, even in a low active power production scenario. Moreover, the grid codes require the RCI during the LVRT operation [38].

Since most of GCC-based DGs exhibit voltage source-like behavior, they have distinct behaviors that imply a need for tailored standards. For instance, existing IEEE 1547 standards for distribution-level assets emphasize current harmonics limits, reactive power limits, and antiislanding functions. In contrast, they are predominantly used for voltage regulation instead of current regulation, reactive power can vary for voltage support, and grid-forming GCCs natively provide uninterrupted power during islanded conditions. Ref. [4] suggested the unintentional islanding functions should be reconsidered in the context of ongoing efforts to modernize standards for the GCCs within the IEEE, the International Electrotechnical Commission, Federal Energy Regulatory Commission, and North American Electric Reliability Corporation (NERC). For instance, see NERC 2018c for an example of ongoing modernization efforts. Because a grid-following GCC needs a well-defined voltage at its terminals, it can function in islanded settings only under specific conditions (e.g., GCC power must match load demand before the system is islanded). Because grid-forming GCCs act like voltage sources, they generally continue operating during islanded conditions. This behavior departs from the existing regulatory framework. Future standards must reconcile how can distribution engineers deenergize systems before carrying out maintenance and should islanded subsystems within a larger grid remain energized to enhance reliability and facilitate a system black-start [4]?

The grid codes are often under scrutiny and regularly updated [39,40]. For example, IEEE Std. 1547-2003 and IEEE Std. 929-2000 are required standards that must be strictly complied for ensuring the power quality of the renewable integrated power grids. As mentioned, the grid codes consider factors such as LVRT capability, power regulation, frequency/voltage/power factor control, and power system protection [40]. The grid experiences several kinds of disturbances, such as short-circuit faults, sudden extreme voltage drops, and load-inducing harmonics. Such events might require GCC-based DGs to shut down or disconnect for protection purposes. Again, sudden disconnection of generation units may cause utility network instability and prolonged delay for restart. The LVRT is a major grid code that concerns the manufacturers of the GCC-based DGs. According to LVRT requirements, the DGs should remain connected to the grid at the time of grid fault, regardless of any generator or converter configuration being used. It also applies to the generation unit to supply reactive power in order to help the grid to be able to pull through adverse scenarios [40]. In order to meet the grid codes requirements, especially to ensure the FRT ability of the system, grid side filters are used [41].

Another important issue is the current versus voltage waveform quality. Under the existing regulatory paradigm, the GCCs are controlled to inject sinusoidal currents with minimal harmonic content. This aligns with prevailing practices in which harmonics demanded by nonlinear loads are provided by machines and reactive components, whereas GCCs inject only the components of nominal frequency. If machines are displaced by the GCCs, however, then GCCs will need to provide these current harmonics, which some loads will require. In the future, the primary function of GCCs will be to provide well-regulated system voltages for loads. The challenge is to devise a set of standards that emphasizes voltage control while recognizing the physical current-carrying limits of GCCs [4].

The grid-following GCCs may disconnect and/or stop operation following transmission faults, and this response is contrary to the defined grid codes of connected resources to the bulk power system during a severe event (NERC, 2017b) [4].

The synchronization controller gains are optimized to decrease the inrush current and guarantee a seamless transfer [42], where the threshold values of the voltage, phase, and frequency differences of the PCC and grid are considered as 3%, 1 degree, and 0.05 Hz, respectively. To synchronize DGs less than 500 kVA rated capacity to the utility grid, IEEE Std. 1547-2003 recommends the synchronization criteria as voltage, phase, and frequency differences as $\pm 10\%$, 20°, and 0.3 Hz, respectively [43].

The performance of PLL-based GCCs on the distribution system has also been a major topic of discussion and has resulted in updated equipment standards that are being deployed—for example, the IEEE updated Standard 1547 (IEEE1547-2018) [4]. The grid codes required for the GCCs, not only to tolerate grid disturbances but also to provide ancillary services as conventional SGs do.

2.5.3 Grid voltage support

Conventional SGs are able to provide fast RCI under grid faults, which is important not only for the grid voltage support but also for activation of protective relays. Therefore, RCI from the GCCs may be required by grid operator. For example, for the case of wind turbine, Germany enforces additional RCI in terms of positive-sequence amounting to at least 2% of the rated current for each percent of voltage dip, which should reach 90% of steady-state value within 50 ms.

However, there are still uncertainties in the present grid codes such as how reactive current should be calculated, how rising time is defined. As the correct behavior of

GCCs under grid unbalanced faults is still under open discussion, it is necessary to revise the existing control strategies for short circuit power provision, and update grid codes continuously [44].

In order to facilitate penetrating a more GCC-based DGs, the GCCs must have the characteristics for fulfilling the defined grid codes as the grid interconnection requirements. As the key block of the renewable integrated power systems, the GCCs must inevitably contribute to the tasks of energy trading, FRT capabilities, and the required power quality for the supplying consumers; as well as the connection requirements such as power limitation, and short-circuit current limitation. These requirements also provide the least possible duration of the connection process. Most of available grid codes require GCCs to possess reactive energy injection, inertia, and oscillation damping to support system transience and steady-state stability. Moreover, they need to offer voltage support under different grid faults capabilities, that is, FRT. The FRT requirements given by the relevant grid codes in different European countries are given in Ref. [45]. In accordance with these, the control of the GCC must satisfy two main requirements during the grid faults: (1) keep connection to the grid in the case of grid faults for the specified amount of time and (2) support the voltage at the PCC by injecting the requested amount of reactive energy [45].

Updated standards for Q-V droop functions is a significant need. Industry practice is largely predicated on grid-following GCCs that operate at or near unity power factor. In recent years, this constraint has been relaxed with slow-acting volt-VAR controls that are intended to support system voltages. As machines are gradually displaced by grid-forming GCCs, the burden of satisfying reactive power demanded by loads will shift further toward these GCCs. In this future scenario, grid-forming GCCs must be allowed to respond according to their autonomously executed and fast-acting Q-V droop functions to simultaneously satisfy loads and support system voltages. These Q-V droop functions for grid-forming GCCs are distinct from the volt-VAR standards currently in place [4].

2.5.4 Grid frequency support

Like the voltage-related grid codes, a set of grid codes must be determined for the frequency-related requirements. In most of existing frequency control performance standard over the world, a time window of average frequency, for example, 10 s, is defined to evaluate the frequency performance, and the grid frequency needs to be within a specific range around the nominal frequency (f_0). For example, the European HVDC connections must satisfy $f_0 \pm 2\%$ for 95% of the week. The grid codes also define the frequency-dependent power supply. For instance, the offshore wind farms in the Europe have to feed active power without limitations into the grid within a frequency range of 49.0–50.2 Hz. It is permitted to limit the active power in-feed when the frequency goes below 49 Hz. The maximum active power reduction is 2% of the maximum generator power per Hz [32,33].

The responsible utilities (e.g., TSOs) need also define some requirements for the RoCoF index, as well as for the storage units and DGs which provide a synthetic flywheel mass to limit the frequency gradient in the case under/overfrequency conditions. The RoCoF should not be very steep according to the grid code requirements; otherwise, the GCCs would disconnect from the grid and thus it may magnify the accrued power imbalance.

Based on the rules given by the European Network of Transmission System Operators for Electricity (ENTSO-E), the RoCoF need to be specified regarding the minimum short-circuit capacity at the PCC in pre- and postfault, as well as the active and reactive power operating point of the DG and the voltage at the PCC in prefault [46]. Furthermore, based on the given grid codes, the DGs need to remain connected to the grid at the maximum frequency deviation of ± 2 Hz for 500 ms, ± 1.5 Hz for 1000 ms, and ± 1.25 Hz for 2000 ms.

The requirement of providing synthetic inertia and therefore supporting the frequency control of the grid is specified in the grid codes of many countries. The synthetic inertia provided by the GCCs in case of the grid frequency deviations should be activated by a fast adjustment of the active power supplied to/from the grid to compensate the frequency gradient. The relevant basic parameters of thresholds will be discussed between the grid operator and the owner of the DG system.

The need for synthetic inertia for the grid dynamic improvement and frequency regulation is discussed by the ENTSO-E in Ref. [47]. In the case of a step change in frequency, the initial active power frequency response should be activated in a short time (e.g., 2 s). This activation time can be even shorter for no inertia GCCs. Moreover, the full active power frequency response should be available for a time duration which can be specified by the responsible utility (e.g., up to 15–30 min) [31,33].

Regarding the frequency-related requirements, in addition to synthetic inertia and thus the RoCoF, the frequency nadir and settling time must be also reflected in the designed grid codes. The frequency nadir should not drop below a specific value to ensure that the GCCs remain connected to the grid and follows the grid code limits. The settling time needs to be minimized to reach a stable operating state as quickly as possible [33].

2.6 Summary

In this chapter, the overall structure, main classification, and basic control loops of the GCCs are briefly addressed. The well-known reference frame transformations are explained, and the applications and dynamic characteristics of grid-following and grid-forming converters with some examples are emphasized. Then, the capability of GCCs in dynamic characteristics emulation is discussed, and finally, the most important required relevant grid codes and standards imposed by grid operators are summarized.

References

[1] R. Lasseter, Z. Chen, D. Pattabiraman, Grid-forming inverters: a critical asset for the power grid, IEEE J. Emerg. Select. Topics Power Electr. 8 (2) (2019) 925–935.

[2] A. Yazdani, R. Iravani, In Voltage-Sourced Converters in Power Systems: Modeling, Control, and Applications, IEEE, 2010, ISBN 9780470551561, pp. i−xix, https://doi.org/10.1002/9780470551578.
[3] P. Roos, A Comparison of Grid-Forming and Grid-Following Control of VSCs (Master thesis Report), Uppsala University, June 2020.
[4] Y. Lin, et al., Research Roadmap on Grid-Forming Inverters, NREL Technical Report, NREL/TP-5D00-73476, November 2020.
[5] H. Bevrani, et al., Power system frequency control: an updated review of current solutions and new challenges, Elec. Power Syst. Res. 194 (2021).
[6] H. Bevrani, B. Francois, T. Ise, Microgrid Dynamics and Control, Wiley, NJ, USA, July 2017.
[7] S. Peyghami, H. Mokhtari, F. Blaabjerg, Y. Yang, Grid synchronization for distributed generations, in: Encyclopedia of Sustainable Technologies 4, 2017, pp. 179−194.
[8] H. Bevrani, Robust Power System Frequency Control, second ed., Springer, Gewerbestrasse, Switzerland, 2014.
[9] K. Oue, T. Kato, K. Inoue, Stability analysis of grid-forming inverter in dq frequency domain, in: 20th Workshop on Control and Modeling for Power Electronics (COMPEL), June 2019, https://doi.org/10.1109/COMPEL.2019.8769653.
[10] J. Liu, Y. Miura, H. Bevrani, T. Ise, Enhanced virtual synchronous generator control for parallel inverters in microgrids, IEEE Trans. Smart Grid 8 (5) (September 2017) 2268e−2277.
[11] H. Wu, et al., Small-signal modeling and parameters design for virtual synchronous generators, IEEE Trans. Ind. Electron. 63 (7) (July 2016) 4292−4303.
[12] H. Xu, X. Zhang, F. Liu, R. Shi, C. Yu, R. Cao, A reactive power sharing strategy of VSG based on virtual capacitor algorithm, IEEE Trans. Ind. Electron. 64 (9) (September 2017) 7520−7531.
[13] U. Markovic, Z. Chu, P. Aristidou, G. Hug-Glanzmann, LQR-based adaptive virtual synchronous machine for power systems with high inverter penetration, IEEE Trans. Sustain. Energy 10 (3) (July 2019) 1501−1512.
[14] H. Bevrani, T. Ise, Y. Miura, Virtual synchronous generators: a survey and new perspectives, Int. J. Electr. Power Energy Syst. 54 (2014) 244−254.
[15] Y. Khayat, S. Golestan, J. Guerrero, J.C. Vasquez, H. Bevrani, DC-link voltage control aided for the inertial support during severe faults in weak grids, IEEE J. Emerg. Sel. Topics Power Electron. (2020), https://doi.org/10.1109/JESTPE.2020.3033657.
[16] J. Driesen, K. Visscher, Virtual synchronous generators, in: IEEE Power Energy Soc. General Meeting-Conv. and Del. of Electr. Energy the 21st Century, 2008, pp. 1−3.
[17] L. Zhang, L. Harnefors, H.P. Nee, Power-synchronization control of grid-connected voltage source converters, IEEE Trans. Power Syst. 25 (2) (2009) 809−820.
[18] Q.C. Zhong, G. Weiss, Synchronverters: inverters that mimic synchronous generators, IEEE Trans. Ind. Electron. 58 (4) (2010) 1259−1267.
[19] M. Ashabani, M.Y. Abdel-Rady, A flexible control strategy for grid-connected and islanded microgrids with enhanced stability using nonlinear microgrid stabilizer, IEEE Trans. Smart Grid 3 (3) (2012) 1291−1301.
[20] D. Li, Q. Zhu, S. Lin, X. Bian, A self-adaptive inertia and damping combination control of VSG to support frequency stability, IEEE Trans. Energy Convers. 32 (1) (2017) 397−398.
[21] M.A. Torres L, L.A. Lopes, L.A. Moran T, J.R. Espinoza C, Self-tuning virtual synchronous machine: a control strategy for energy storage systems to support dynamic frequency control, IEEE Trans. Energy Convers. 29 (4) (December 2014) 833−840.

[22] S. Zhang, Y. Mishra, M. Shahidehpour, Fuzzy-logic based frequency controller for wind farms augmented with energy storage systems, IEEE Trans. Power Syst. 31 (2) (March 2015) 1595−1603.
[23] Y. Khayat, M. Naderi, H. Bevrani, F. Blaabjerg, Virtual Inertia Emulating in Power Electronic, Chapter 19 in Control of Power Electronic Converters and Systems, Elsevier, 2021, https://doi.org/10.1016/B978-0-12-819432-4.00019-6.
[24] T. Kato, K. Inoue, M. Ueda, Lyapunov-based digital control of a grid-connected inverter with an LCL filter, IEEE J. Emerg. Select. Topics Power Electr. 2 (4) (2014) 942−948.
[25] T. Kato, K. Inoue, Y. Akiyama, K. Ohashi, Optimum and adjustable damping control of grid-connected inverter with an LCL filter, in: 9th International Conference on Power Electronics and ECCE Asia (ICPE-ECCE Asia), 2015, https://doi.org/10.1109/ICPE.2015.7167775.
[26] D. Ghaderi, G. Bayrak, J.M. Guerrero, Grid code compatibility and real-time performance analysis of an efficient inverter topology for PV-based microgrid applications, Int. J. Electr. Power Energy Syst. 128 (2021).
[27] N. Espinoza, M. Bongiorno, O. Carlson, Grid code testing of full power converter based wind turbines using back-to-back voltage source converter system, in: EWEA Annual Event 2013 Conference Proceedings, 2013. Available on, https://research.chalmers.se/publication/174669.
[28] National Grid Electricity System Operator Limited, The Grid Code, Issue 5, Revision 38, September 2019. Available at: https://www.nationalgrideso.com/document/33821/download.
[29] G.A. Raducu, Control of Grid Side Inverter in a B2B Configuration for WT Applications, Master Thesis Report, Aalborg University, 2008.
[30] O. Gomis-Bellmunt, Grid Code Development for PV System Integration, Lecture Note in International Solar Alliance Expert Training Course, September 2019. Available at: https://cleanenergysolutions.org/sites/default/files/documents/pv3-grid-code-development.pdf.
[31] Commission Regulation (EU) 2016/1447—Establishing a Network Code on Requirements for Grid Connection of High Voltage Direct Current Systems and Direct Current-Connected Power Park Modules, August 2016, p. 65. Available online: https://eur-lex.europa.eu/eli/reg/2016/1447/oj (Accessed on 24 August 2021).
[32] T.T. GmbH, Offshore-Netzanschlussregeln—O-NAR, 2017. Available online: https://docplayer.org/56149134-Offshore-netzanschlussregeln-o-nar-tennet-tso-gmbh.html (Accessed on 24 August 2021).
[33] M. Hoffmann, et al., Grid code-dependent frequency control optimization in multi-terminal DC networks, Energies 13 (2020).
[34] IEEE recommended practice for utility interface of photovoltaic (PV) systems, in: IEEE Std 929-2000, 2000, https://doi.org/10.1109/IEEESTD.2000.91304.
[35] European Committee for Electrotechnical Standardization, Requirements for generating plants to be connected in parallel with distribution networks - Part 1: Connection to a LV distribution network - Generating plants up to and including Type B, EN 50549-1-2019, CEN-CENELEC Management Centre: Rue de la Science 23, B-1040 Brussels.
[36] International Electrotechnical Commission. Distributed energy resources connection with the grid, IEC TS 62786, Edition 1.0 2017-04. ICS 29.240.01, ISBN 978-2-8322-4218-6.
[37] Germany Standards, Automatic Disconnection Device between a Generator and the Public Low-Voltage Grid 0126-1-1, DIN VDE, 2006.
[38] R.M. Silva, et al., Power control strategies for grid connected converters applied to full-scale wind energy conversion systems during LVRT operation, Elec. Power Syst. Res. 184 (2020).

[39] M. Mohseni, S.M. Islam, Review of international grid codes for wind power integration: diversity, technology and a case for global standard, Renew. Sustain. Energy Rev. 16 (6) (2012) 3876–3890.
[40] F. Blaabjerg, R. Teodorescu, M. Liserre, A.V. Timbus, Overview of control and grid synchronization for distributed power generation systems, IEEE Trans. Ind. Electron. 53 (5) (October 2006) 1398–1409.
[41] A.L. Maswood, H.D. Tafi (Eds.), Advanced Multilevel Converters and Applications in Grid Integration, Wiley, 2019.
[42] X. Hou, et al., Distributed hierarchical control of ac microgrid operating in grid-connected, islanded and their transition modes, IEEE Access 6 (2018), pp. 77 388–77 401.
[43] L. Xia, L. Hai, A seamless transfer strategy based on multi-master and multi-slave microgrid, in: 9th IEEE Int. Symp. Power Electron. For Distributed Gen. Syst. (PEDG), IEEE, 2018, pp. 1–5.
[44] J. Jia, G. Yang, A.H. Nielsen, A review on grid-connected converter control for short circuit power provision under grid unbalanced faults, IEEE Trans. Power Deliv. 33 (2) (April 2018) 649–661.
[45] B. Dumnic, et al., Control strategy for a grid connected converter in active unbalanced distribution systems, Energies 12 (2019).
[46] Rate of Change of Frequency (RoCoF) Withstand Capability—ENTSO-E Guidance Document for National Implementation for Network Codes on Grid Connection. Available online: https://eepublicdownloads.entsoe.eu/clean-documents/Network%20codes%20documents/NC%20RfG/IGD_RoCoF_withstand_capability_final.pdf. (Accessed 24 August 2021).
[47] Need for Synthetic Inertia (SI) for Frequency Regulation—ENTSO-E Guidance Document for National Implementation for Network Codes on Grid Connection. Available online: https://eepublicdownloads.entsoe.eu/clean-documents/Network%20codes%20documents/NC%20RfG/IGD_Need_for_Synthetic_Inertia_final.pdf. (Accessed on 24 August 2021).

Modeling and dynamic performance of grid connected converters

3

3.1 A background and overview

To provide desirable dynamics for the grid, GCC with dynamic characteristics feature becomes a focus of worldwide attention. The GCC-based DGs play an important role in the future power grids owing to the development of renewable energy generation. Conventionally, as mentioned in Chapter 2, the grid-following GCCs are controlled as current sources to provide constant or maximum available power generation. This feature, however, is largely dependent on the presence of a stiff voltage source and may not operate during power outage. To fully utilize the GCCs, recently the concept of microgrid (MG) has been proposed [1], which requires one or several GCCs not only capable to operate in grid connected mode but also can operate in an islanded MG during power outage or power quality deterioration of the grid.

A typical control scheme of a GCC is the droop-based control, in which the well-known $P-\omega$ and $Q-V$ droops are generally applied. However, due to the resistive line impedance of the low voltage distribution system, a virtual impedance control needs to be applied to make the output impedance remain inductive [2]. Perform a mathematical model for dynamic and performance analysis of a GCC in the presence of different control loops and network characteristics in the first stage, and use the developed model for synthesis of effective compensators and control systems is a challenging issue.

An effective modeling method must satisfy the following points: (1) it could be easily updated for any new control loop/scheme; (2) the method must provide a small-order model, to simplify the analytical insights and computation requirements; (3) it must be applicable for different single and parallel operation modes of GCCs; and (4) the method must not need to identify the initial value of many variables (since it is difficult to obtain all initial values during the design phase) [3]. The inductive property of the connection lines and grid network makes it possible to study $P-\omega$ and $Q-V$ characteristics of the GCCs, separately.

3.1.1 Active power and frequency response modeling

With a successive growth of power generation using renewable energy sources (RESs), such as photovoltaics and wind turbines, the penetration rate of GCC-based DGs in the power grid is increasing at a rapid pace. Unlike conventional centralized generation using synchronous generators (SGs), the GCC-based DGs do not have a rotating mass to provide inertia support for the grid. Therefore, by growing the number of GCCs, the grid operators are faced with the lack of inertia issue, which intrinsically

Grid Connected Converters. https://doi.org/10.1016/B978-0-323-99902-1.00005-7

leads to a large rate of change of frequency (RoCoF) in the grid. Larger frequency deviations can cause reliability issues such as frequency-based tripping of loads and legacy equipment in the power grid. As a result, the power grid is faced to serious frequency fluctuation, performance degradation, and the design of RoCoF-based relays should be reconfigured [4].

In the past few years, a concept to control the GCC to behave like an SG, called virtual synchronous generator (VSG) control [5], or virtual synchronous machine [6], or synchronverter [7], is proposed. This concept inherits the droop characteristics from the droop control and emulates the swing equation of an SG to provide inertia support for the power grid. Therefore, it performs a new type of grid-forming control with an enhanced $P - \omega$ control loop. Since the emulation of the swing equation results in more oscillatory dynamics, various GCC control schemes with different damping methods have been proposed [2,8].

Practically, it is shown that by having a primary energy source, the GCCs are able to emulate desirable dynamic characteristics such as the inertia of kinetic energy of a rotating mass and damping property of winding of the SG. The GCC with emulated droop characteristic allows for direct control of voltage and frequency. During contingencies the droop-controlled GCCs will increase or decrease their output power instantaneously to balance loads and maintain specified voltage level and nominal frequency. Unlike the conventional SGs, there is no significant delay between the change of output power and the change of output frequency in droop-based GCCs. Therefore, they respond much faster to any contingencies than the response of the SGs. Developing a dynamic model for the GCCs makes it easy to fully understand these characteristics.

As the principle of these concepts is similar, for convenience sake, all these grid-forming GCC topologies are referred to as GCC in this chapter. Despite the similar principle, the basic control method of GCCs is not unique. For instance, the inertia and damping effect can be realized through different approaches [2]. The inertia is mostly emulated using the first-order differential swing equation of the SGs; however, for virtual damping, different methodologies are reported. In the literature, GCC control possessing no dedicated damping unit is adopted in early studies [7] and then improved by dedicated damping methods such as the damper windings emulation method [2,9], the damping correction loop method [10], the state feedback method and its advanced version, and the state feedback with a low-pass filter (LPF) method [8,11]. Besides, other damping methods are also proposed, that is, using a virtual conventional power system stabilizer [12], increasing output reactance through virtual impedance control [13], and using adaptive inertia and/or damping [2,14]. Moreover, various inner loop controls in dq and $\alpha\beta$ frames can be adopted in a GCC. One important advantage of GCC control is its ability to make a dispatchable DG operation. This feature is usually inherited from the droop control [2].

However, although with the same control, a GCC-based DG may behave differently in different operation modes, and this becomes more apparent when the dynamics emulation feature is given [8]. To understand this fact, having a mathematical and physical interpretation model is very important. To address this issue, a modeling method to study the active power and frequency dynamics is introduced in Section 3.2.

Previous works in the literature can be verified and better interpreted using the addressed modeling method, for example, the analytical comparison between different GCC control strategies.

3.1.2 Reactive power and voltage response modeling

The progress in $Q-V$ control loops and relevant studies is more focused on the issue of inaccurate reactive power sharing between parallel GCCs due to the local feature of voltage. In this direction, communication-less solutions like eliminating the mismatch of GCC output impedance [15], compensation of voltage drop [16], and communication-based solutions like secondary control from an upper-level controller [17], consensus control between adjacent GCCs [18], have been proposed.

Regarding the voltage and reactive power dynamics of the GCCs, the inner control loops also show a high diversity. The voltage/current double loop-based (*VI*-based) and reactive power control-based (Q-based) control schemes (in different frames) are arguably the most popular solutions [2,13,19]. Meanwhile, a single voltage magnitude loop is also widely adopted, which behaves differently from the *VI*-based control [20]. Moreover, small-signal modeling is also suggested for studying this issue [21–23]. Besides, large-signal nonlinear modeling is studied in Ref. [24] for better accuracy. However, the improved accuracy is not necessary for the design-oriented modeling of the $Q - V$ response of GCCs, and usually a proper model reduction is needed.

To design $Q - V$ and $P - \omega$ controls of GCCs, it is important to find simple and effective modeling methods that cover the main closed-loop system dynamics for different operation modes. To address the above issues, two state-space dynamic models for $Q-V$ and $P-\omega$ dynamic responses of GCCs are explained in Sections 3.2 and 3.3, which are basically developed in Refs. [3,8]. These methods demonstrate high fidelity by the well-matched analytical and experimental step responses.

3.2 Active power and frequency response model

In this section, a state-space modeling method to study the active power and frequency control response of a GCC is explained. It is assumed that the control system of GCC is capable to emulate desirable dynamics such as virtual inertia, droop loop, and damping property. The introduced model is verified by the step response analyses using simulation and experimental results, which demonstrate the correctness of the proposed modeling method. More details can be found in Ref. [8].

3.2.1 Typical GCC and relevant control system

For the sake of modeling, a typical grid-forming GCC with its control system is considered as shown in Fig. 3.1. The control system uses a root-mean-square (RMS) value-based control scheme without using an additional inner voltage or

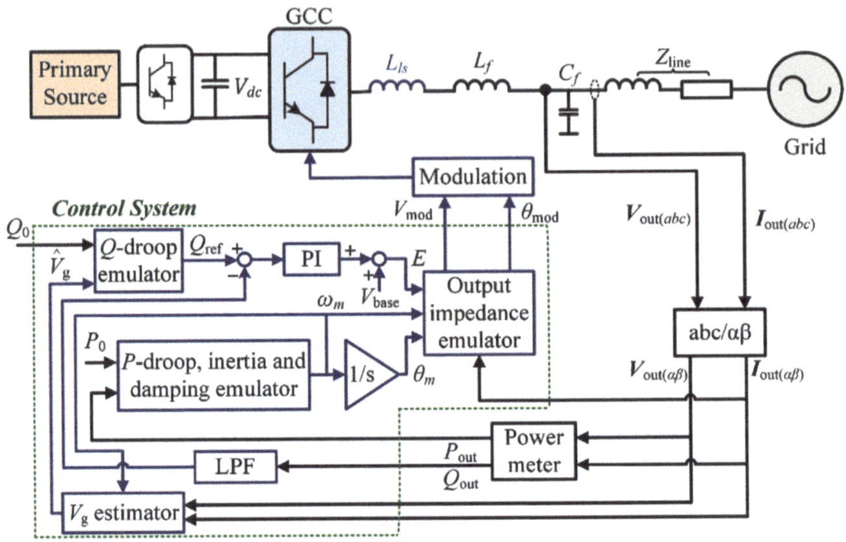

Figure 3.1 Typical control scheme of a grid-forming GCC.

current control loop that regulates the output voltage through the reactive power control loop. Owing to the absence of an inner voltage loop, the active power control loop becomes quite simple, and bus voltage deviation becomes smaller and less-sensitive to the output impedance [13].

For performing the dynamic characteristic emulation system which is discussed in Chapter 2, several blocks are considered. The "P-droop, inertia and damping emulator" block is the core of GCC system control. In the block "Output impedance emulator," a virtual impedance control loop is applied. A voltage drop over virtual inductance L_{ls} is generated to adjust the equivalent output reactance X of the inverter as shown in Eq. (3.1).

$$X = \omega_0 \left(L_{ls} + L_f + L_{\text{line}} \right) \tag{3.1}$$

where L_f and L_{line} are the GCC output filter and connecting line inductances. The block "V_g estimator" estimates the grid voltage from the measurement of output voltage and current, to provide a common reference for the "Q-droop" block, in which the droop relation between voltage and reactive power is applied [2].

To emulate the steady-state operation of an SG, a P (or $P - \omega$) droop model is usually emulated as

$$P_{\text{in}} = P_0 - k_p (\omega_m - \omega_0). \tag{3.2}$$

where P_0 is the active power reference, k_p is the droop coefficient, ω_m is the generated virtual frequency, ω_0 is the nominal frequency, and P_{in} is the virtual input power. To mimic the inertia characteristic, the swing equation of conventional SGs is emulated.

In case of neglecting the effect of damper windings (damping characteristic), the swing equation of an SG can be expressed as

$$P_{\text{in}} - P_{\text{out}} = J\omega_m \frac{\mathrm{d}\omega_m}{\mathrm{d}t} \tag{3.3}$$

Combining Eq. (3.2) and Eq. (3.3) yields

$$P_0 - P_{\text{out}} = J\omega_0 \frac{\mathrm{d}\omega_m}{\mathrm{d}t} + k_p(\omega_m - \omega_0). \tag{3.4}$$

P_{out} is the output power, and J is the moment of inertia. Eq. (3.4) can be realized in form of a block diagram as shown Fig. 3.2. The GCC system control emulating (Eq. 3.4) in the "P-droop, inertia and damping emulator" block. But in Eq. (3.4), the damping effect is not considered.

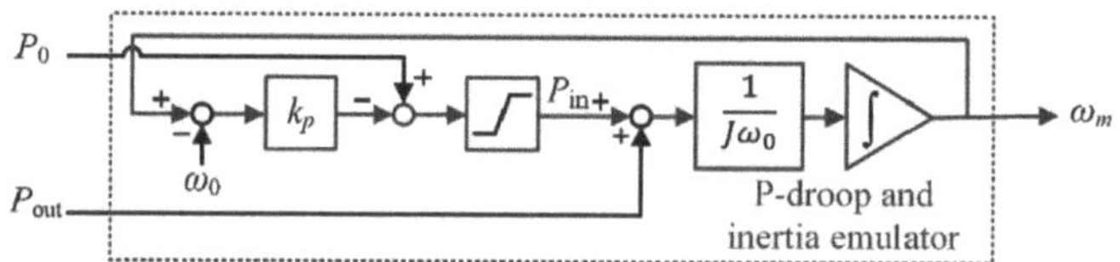

Figure 3.2 The block of "P-droop, inertia and damping emulator block" without damping.

$$P_0 - P_{out} = J\omega_0 \frac{d\omega_m}{dt} + k_p(\omega_m - \omega_0) + D(\omega_m - \omega_g) \tag{3.5}$$

In the literature, for adding the damping characteristic to the GCC control system, various approaches using different damping technologies are proposed. A straightforward approach to generate a dedicated damping term is to add the effect of damper windings into Eq. (3.4), as done in Eq. (3.5).

However, the grid frequency ω_g is difficult to measure directly, thus it is approximated by $\widehat{\omega}_g$, which is the angular frequency measured by a phase-locked loop (PLL) from output voltage $\boldsymbol{V}_{\text{out}}$ [2], as shown in Fig. 3.3.

3.2.2 Modeling methodology

In this section, the modeling method is presented to study the active power-frequency response of a GCC. In the literature, with the assumption of the grid is stiff enough, usually a GCC with the grid is considered as a single DG connected to infinite bus system as shown in Fig. 3.4. In the model of this system, as the PCC voltage is dependent on the grid voltage, ω_g could be considered as a disturbance from the grid side. It is well-known that the synchronizing power coefficient K of the DG can be expressed aswhere

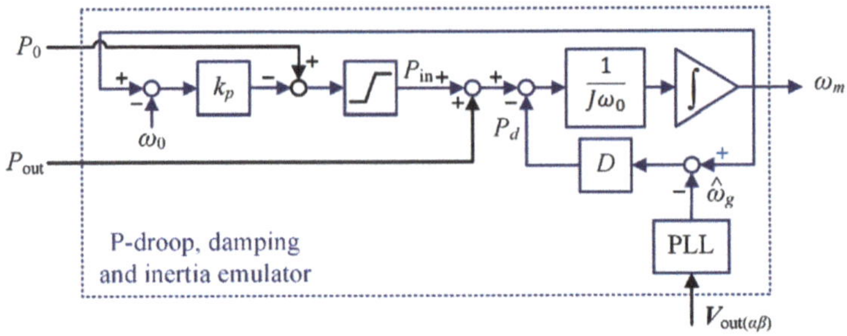

Figure 3.3 The block of "P-droop, inertia and damping emulator" with damping.

$$K = \frac{\partial P_{out}}{\partial \delta} \approx \frac{EV_g cos\delta_0}{X} \approx \frac{V_{base}^2\sqrt{1 - X^{*2}}}{X}, \tag{3.6}$$

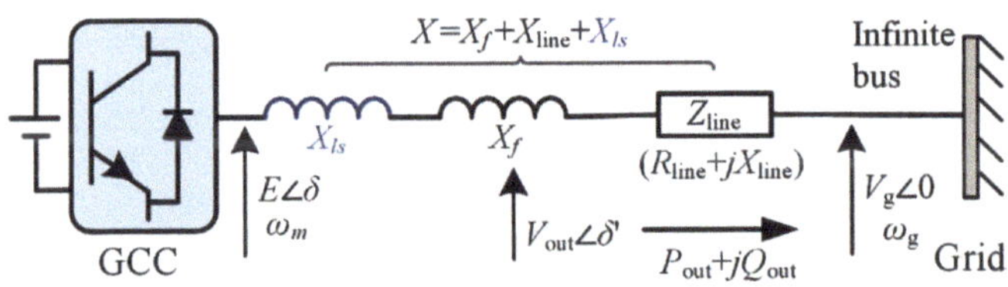

Figure 3.4 The model of GCC connected to an infinite bus.

$$X^* = XS_{\text{base}}/V_{\text{base}}^2. \tag{3.7}$$

δ is the power angle and δ_0 is its operating point. The derivative of power angle can be represented as

$$\frac{\mathrm{d}\delta}{\mathrm{d}t} = \omega_m - \omega_g. \tag{3.8}$$

Here, E is the electromotive force, V_g is the voltage of infinite bus, V_{base} is the rated voltage, S_{base} is the rated power, and X^* is the per unit value of output reactance X.

Thus, for the GCC method, a small-signal state-space model can be derived from Eq. (3.6), (3.8) and respective equations in the "P-droop, inertia and damping emulator" block, as given in Eq. (3.9).

$$\begin{cases} \dot{\boldsymbol{x}} = \boldsymbol{A}x + \boldsymbol{B}u + \boldsymbol{E}_w w \\ \boldsymbol{y} = \boldsymbol{C}x \end{cases} \tag{3.9}$$

where $\boldsymbol{x}$, u, $\boldsymbol{y}$, and w are system states, control input, outputs, and disturbances respectively; and

$$u = \Delta P_0 \tag{3.10}$$

$$w = \Delta \omega_g \tag{3.11}$$

$$\boldsymbol{y} = \begin{bmatrix} \Delta \omega_m & \Delta P_{\text{out}} \end{bmatrix}^{\text{T}} \tag{3.12}$$

$$\boldsymbol{x} = \begin{bmatrix} \Delta \omega_m & \Delta P_{\text{out}} & \boldsymbol{x}_a^{\text{T}} \end{bmatrix}^{\text{T}} \tag{3.13}$$

Here, the vector $\boldsymbol{x}_a$ includes additional state variables besides $\Delta\omega_m$ and ΔP_{out}. The $\boldsymbol{x}_a$ comes from additional terms such as a LPF or an integrator, thus it is not related to the basic GCC control system. Also, it can include state variables of the turbine model of an SG or inner control loops of a multi-loop GCC. In several basic GCC control systems, $\boldsymbol{x}_a$ not exist.

In Eq. (3.9), the state variables $\boldsymbol{x}$, and state-space matrix coefficients $\boldsymbol{A}$, $\boldsymbol{B}$, $\boldsymbol{E}_w$ and $\boldsymbol{C}$ depend on the structure and its dynamics in the "P-droop, inertia and damping emulator" block (In this section, the subscript w is used for the matrix $\boldsymbol{E}$ in the above state-space model to avoid confusing with the output voltage of GCC in Fig. 3.4). For instance, for case of without damping characteristic (Fig. 3.2),

$$\boldsymbol{x} = \begin{bmatrix} \Delta \omega_m & \Delta P_{\text{out}} \end{bmatrix}^{\text{T}} \tag{3.14}$$

$$\boldsymbol{A} = \begin{bmatrix} -\dfrac{k_p}{J\omega_0} & -\dfrac{1}{J\omega_0} \\ K & 0 \end{bmatrix} \tag{3.15}$$

$$\boldsymbol{B} = \begin{bmatrix} \dfrac{1}{J\omega_0} & 0 \end{bmatrix}^{\text{T}} \tag{3.16}$$

$$\boldsymbol{E}_w = \begin{bmatrix} 0 & -K \end{bmatrix}^{\text{T}} \tag{3.17}$$

$$\boldsymbol{C} = \boldsymbol{I}_2. \tag{3.18}$$

From Eq. (3.5), ignoring the PLL dynamics, the model of an ideal damper winding emulation method can be derived as

$$\boldsymbol{A} = \begin{bmatrix} -\dfrac{k_p + D}{J\omega_0} & -\dfrac{1}{J\omega_0} \\ K & 0 \end{bmatrix} \tag{3.19}$$

$$\boldsymbol{E}_w = \begin{bmatrix} \dfrac{D}{J\omega_0} & -K \end{bmatrix}^{\text{T}}, \tag{3.20}$$

where $\boldsymbol{x}$, $\boldsymbol{B}$, and $\boldsymbol{C}$ are the same as those of the no-damping case.

However, in practice, the fast dynamic response of the PLL requires a large loop gain, which makes $\widehat{\omega}_g$ sensitive to the ripples in $\boldsymbol{V}_{\text{out}}$. Therefore, the PLL in damper winding emulation method is usually tuned to have a moderate response, thus it cannot be neglected. If a typical PLL using *dq* transformation and a proportional integral (PI) controller is applied, the model of the damper winding emulation method becomes

$$\boldsymbol{x} = \begin{bmatrix} \Delta\omega_m & \Delta P_{\text{out}} & \Delta x_{1_{pll}} & \Delta x_{2_{pll}} \end{bmatrix}^{\text{T}} \tag{3.21}$$

$$\boldsymbol{A} = \begin{bmatrix} -\dfrac{k_p}{J\omega_0} & -\dfrac{1}{J\omega_0} & \dfrac{DK^*_{p_{pll}}}{J} & \dfrac{D}{J\omega_0 T_{i_{pll}}} \\ K & 0 & 0 & 0 \\ 0 & 0 & -\omega_0 K^*_{p_{pll}} & -\dfrac{1}{T_{i_{pll}}} \\ 0 & 0 & \omega_0 K^*_{p_{pll}} & 0 \end{bmatrix} \tag{3.22}$$

$$\boldsymbol{B} = \begin{bmatrix} \dfrac{1}{J\omega_0} & 0 & 0 & 0 \end{bmatrix}^{\text{T}} \tag{3.23}$$

$$\boldsymbol{E}_w = \begin{bmatrix} 0 & -K & 1 & 0 \end{bmatrix}^{\text{T}} \tag{3.24}$$

$$\boldsymbol{C} = \begin{bmatrix} \boldsymbol{I}_2 & 0_{2\times 2} \end{bmatrix}. \tag{3.25}$$

This PLL model is often not considered [2], and in this case the damping control loop acts ideally. The proposed model which is realized in Fig. 3.5 can be easily implemented in mathematical computing software environment such as MATLAB. The most important issue in this modeling is in existence of an infinite bus, whose frequency is independent of the state variables, thus $\Delta\omega_g$ could be considered as an external disturbance. Meanwhile, as there is no external power generation, load power deviation determines a power disturbance in the system, thus ΔP_{load} could be considered as a disturbance, and included in w in the performed model, (Eq. 3.9) and (Eq. 3.11) [8].

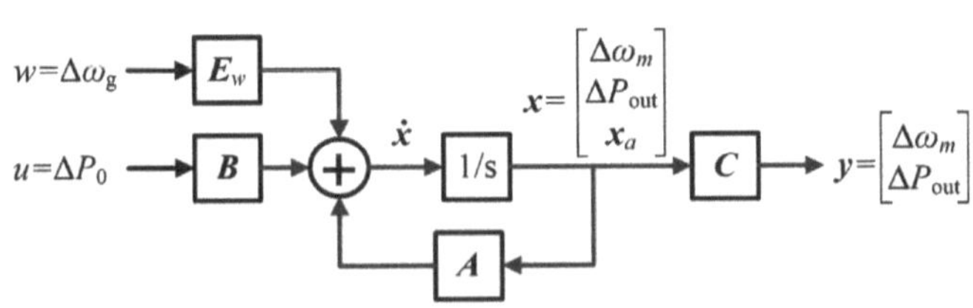

Figure 3.5 Block diagrams of the performed GCC state-space model.

It is noteworthy that the presented model does not include the detailed differential equations of the LCL filter(s), and Eq. (3.6) is a linearized quasi-static approximation for analyzing relatively slow dynamics [25]. Besides, although it is not the case in this section that some GCC control schemes have inner voltage and current control loops. Including these complete equations will add new state variables in $\boldsymbol{x}_a$; however, the resulted poles are generally much faster than the interested range of dominant poles.

Therefore, including these equations will only complicate the model and introduce some nondominant poles, which may marginally influence the GCC dominant poles. Nevertheless, if the proposed modeling method is applied to a detailed stability study, it could be better to include all these equations to make a more complete model.

It is also noteworthy that the presented linearized model is an approximation of the real nonlinear system. Although the system nonlinearities, for example, caused by PLL, affect the accuracy of the analytical prediction; the resulted errors are generally ignorable.

3.2.3 Step-response analysis and evaluation

From the obtained state-space model (Eq. 3.9), the transfer function matrices from the control inputs and the disturbance input shown as $\boldsymbol{G_c}(s)$ and $\boldsymbol{G_d}(s)$, can be easily calculated as follows:

$$\boldsymbol{G_c}(s) = \boldsymbol{C}(s\boldsymbol{I} - \boldsymbol{A})^{-1}\boldsymbol{B} \tag{3.26}$$

$$\boldsymbol{G_d}(s) = \boldsymbol{C}(s\boldsymbol{I} - \boldsymbol{A})^{-1}\boldsymbol{E} \tag{3.27}$$

Thus, it is straightforward to analyze the responses of outputs during a given step change of an input. The most important advantage of using the damping characteristic in the GCC control system is to assign the dominant closed-loop poles of the single machine infinite bus model. As it is discussed in Ref. [2], the closed-loop poles of a GCC can be represented as

$$\lambda_{1,2} = \omega_n e^{j(\pi \pm \cos^{-1}\xi)} \tag{3.28}$$

$$\omega_n = \sqrt{\frac{K}{J\omega_0}} \tag{3.29}$$

where ω_n and ξ are undamped natural frequency and damping ratio of the desired poles, and K is the synchronizing power coefficient determined by output reactance and rated voltage as shown in Eq. (3.6).

From Eq. (3.28), it can be concluded that the dominant poles $\lambda_{1,2}$ is fixed once the desired moment of inertia J and damping ratio ξ are given. To assign the closed-loop pole system with damping loop, D should be designed as

$$D = 2\xi\sqrt{KJ\omega_0} - k_p \tag{3.30}$$

Based on the above discussion, the control parameters of GCC are designed and considered as given in Table 3.1. It is noteworthy that parameters are given in per unit value such as

$$M^* = J\omega_0^2/S_{\text{base}} \tag{3.31}$$

$$k_p^* = k_p\omega_0/S_{\text{base}} \tag{3.32}$$

$$D^* = D\omega_0/S_{\text{base}} \tag{3.33}$$

where M^* is the inertial constant.

3.2.3.1 Response to active power reference change

The GCC output power P_{out} is expected to follow its reference P_0 as fast as possible. To compare this dynamic response, small-signal transfer functions $\frac{\Delta P_{\text{out}}}{\Delta P_0}$ are derived for all the discussed methods.

From Eqs. (3.6 and 3.8), and the control law in the "P-droop, inertia and damping emulator" block, $\frac{\Delta P_{\text{out}}}{\Delta P_0}$ can be obtained by eliminating δ and ω_m. The resulted transfer function without and with damping loop are shown in Eqs. (3.34 and 3.35), respectively.

$$\frac{\Delta P_{\text{out}}}{\Delta P_0} = \frac{K}{J\omega_0 s^2 + k_p s + K} \tag{3.34}$$

$$\frac{\Delta P_{\text{out}}}{\Delta P_0} = \frac{K}{J\omega_0 s^2 + (k_p + D)s + K} \tag{3.35}$$

From the above transfer functions, the zero-state responses are obtained as Fig. 3.6. It can be noticed that the damping loop with dedicated damping shows a good damping effect compared to the case of no damping loop.

Table 3.1 Parameters of GCC.

Common Parameters	Parameter	Value	Parameter	Value	Parameter	Value
	V_{base}	200 V	ω_0	377 rad/s	X^*	0.3 pu
	S_{base}	5 kVA	P_0^*	1 pu	M^*	8 s
			k_p^*	20 pu	ζ	0.9
With damping loop	D^*	156 pu	$K_{p\_pll}^*$	0.1 pu	T_{i_pll}	0.5 s

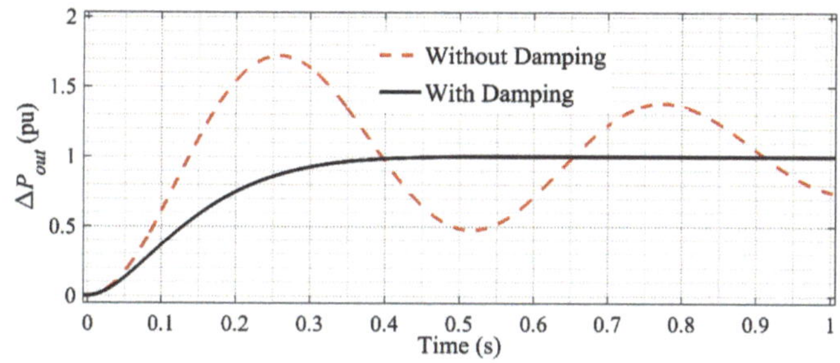

Figure 3.6 Dynamic response of GCC to step change in active power reference (ΔP_0).

3.2.3.2 Ripple attenuation ability

The input of the "P-droop, inertia and damping emulator" block, that is, the measured output power P_{out} may contain ripples due to unbalance phenomena and/or harmonics. As a result, the attenuation of transfer function $\frac{\Delta\omega_m}{\Delta P_{\text{out}}}$ at high-frequency area is important. Therefore, the ripple attenuation ability can be compared by Bode plot of $\frac{\Delta\omega_m}{\Delta P_{\text{out}}}$ obtained in previous section. It is noteworthy that in case of using damping control loop, one can obtain:

$$\Delta\omega_m = -\frac{1}{J\omega_0 s + k_p + D}\Delta P_{\text{out}} + \frac{D}{J\omega_0 s + k_p + D}\Delta\widehat{\omega}_g \tag{3.36}$$

Eq. (3.36) implies that ripples in ω_m of damping loop is not only affected by P_{out}, but also by $\widehat{\omega}_g$, which is obtained by the PLL. Therefore, ripples in V_{out} may penetrate in $\widehat{\omega}_g$, thus affect ω_m. However, as this effect is difficult to analyze, here it is assumed that $\Delta\widehat{\omega}_g = 0$.

Bode plot of $\frac{\Delta\omega_m}{\Delta P_{\text{out}}}$ of the studied GCC methods is shown in Fig. 3.7, and it can be concluded that at low frequency, both with and without damping loops provide a desirable ripple attenuation ability. At high frequencies, they provide same disturbance attenuation. Of course, it is noteworthy that in case of using damping loop, the influence of $\widehat{\omega}_g$ is ignored; therefore, practically, the ripple attenuation ability is worse than that shown in Fig. 3.7.

3.2.3.3 Experimental verification

To verify the analytical comparison, a laboratory experiment is conducted for a GCC with the same parameters as Table 3.1. An overview of laboratory system and the relevant experimental circuit is shown in Fig. 3.8.

In the performed experiment, the GCC-based DG with a two-phase RL load is already connected to the grid, and the two-phase RL load is connected. At $t = 0$ sec, active power reference P_0 is changed from 0.5 pu to 1 pu. The dynamic response of P_{out} for the cases of with and without damping loop is shown in Fig. 3.9, which are similar to the simulation results for the same system (Fig. 3.6).

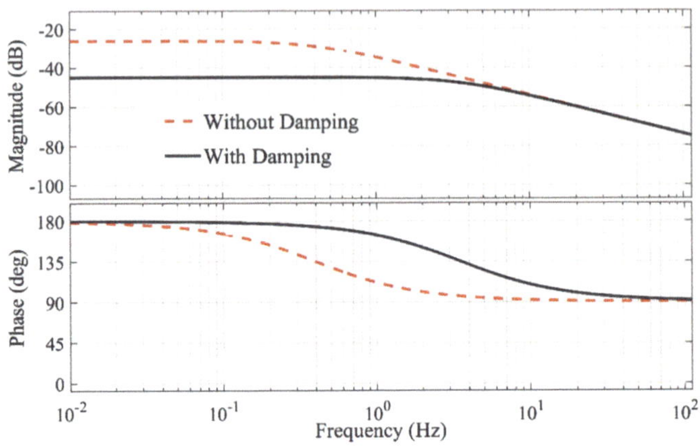

Figure 3.7 Bode plot of the transfer function $\frac{\Delta\omega_m}{\Delta P_{\text{out}}}$.

(a)

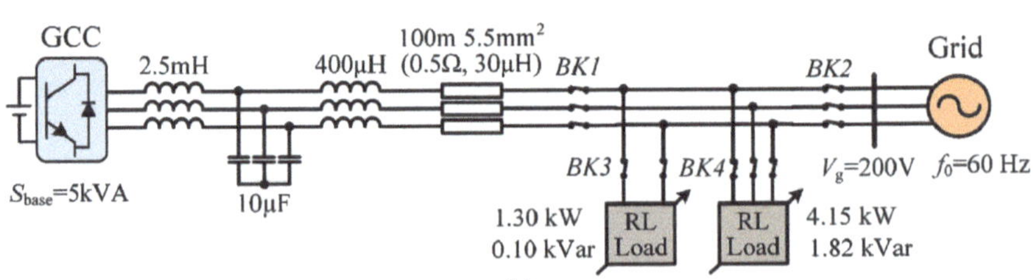

(b)

Figure 3.8 (A) Laboratory system overview, and (B) experiment circuit.

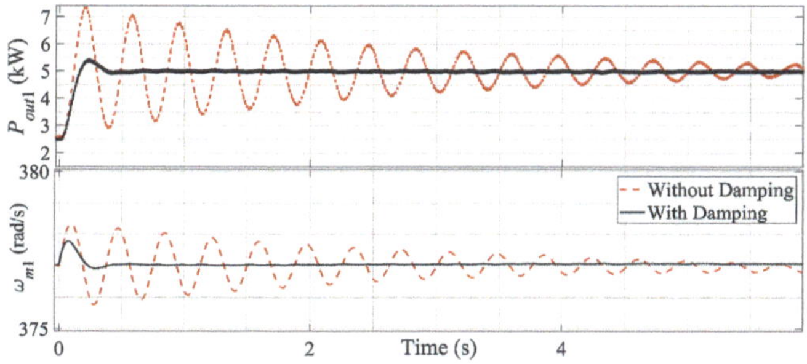

Figure 3.9 Experimental results.

A small overshoot can be seen in the response of system with damping loop, because the PLL is also influenced by V_{out} fluctuation caused by loading transition. This fact is not considered in simulation shown in Fig. 3.6. Furthermore, it can be seen that the frequency of V_{out} does not exactly equal to ω_g during transients. Fig. 3.9 verifies the dynamic response of ω_m in the employed transient condition.

3.3 Reactive power and voltage response model

A modeling method for grid-forming GCCs to study the $P-\omega$ dynamic response and control loops is introduced in Section 3.2. In this Section, a modeling method to study the $Q-V$ dynamic response control system is addressed. More details can be found in Ref. [8]. Compared to the perturbation and/or Kron based reduced-order models [26,27], the present modeling framework intrinsically excludes the fast network states, which are known as nondominant dynamic states.

3.3.1 Typical GCC and relevant control system

Fig. 3.10 shows a typical GCC and relevant control scheme useful for $Q-V$ dynamic study. It is connected to the grid through an *LCL* filter and a line. For simplicity of explanation, the outer inductor of the *LCL* filter is aggregated into the line inductance in the impedance model.

The control scheme of a grid-forming GCC can be divided into two major parts, the $P-\omega$ control part, and the $Q-V$ control part including inner loops and virtual impedance control. Such division helps us to study the GCC dynamics in an easier way owing to the following facts: (1) as long as the $P-\omega$ and $Q-V$ control parts are well decoupled, the interaction between the two control parts is negligible. Therefore, modeling them separately can significantly simplify the relevant studies; (2) the design of inner control loops can greatly affect the $Q-V$ response, whereas it makes a little influence on the $P-\omega$ response; and (3) the developed virtual inertia concept tends to slow down the $P-\omega$ response. This makes different time constants orders of the $P-\omega$ and $Q-V$ responses.

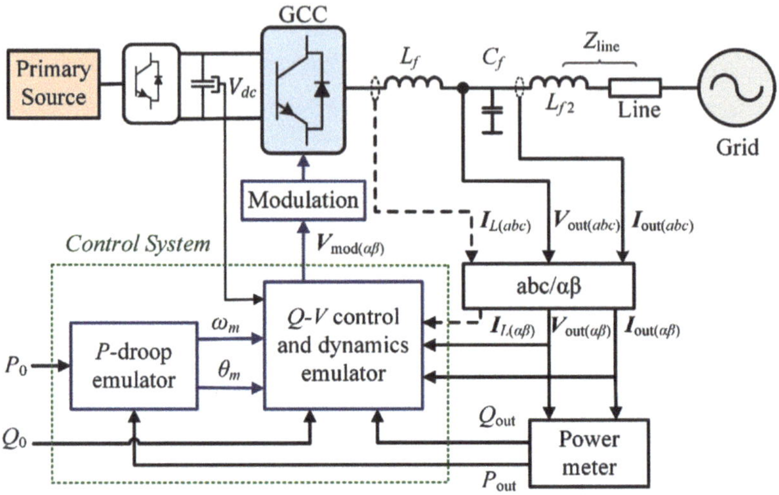

Figure 3.10 Typical control scheme of a grid-forming GCC.

As a modeling method for the $P-\omega$ control loop has been already presented in Section 3.2, we only focus on the $Q-V$ control and relevant dynamic response in this section. A complete model consisting of both models is given in the next section. The P (P-ω) droop control adopted in this scheme is performed by the dynamic characteristic emulator with damping that discussed in previous section.

The "Q−V control and dynamics emulator" block in Fig. 3.10 can be mainly realized using two different schemes of inner loops in *dq* frame, that is, *VI*- based and *Q*-based methods, which are depicted in Figs. 3.11 and 3.12, respectively.

3.3.1.1 Inner VI-based control

The inner VI-based control (Fig. 3.11) [28] requires feedbacks of $\boldsymbol{V}_{out}$, $\boldsymbol{I}_{out}$, and $\boldsymbol{I}_L$, which are the vectors of output voltage, output current, and inductor current, respectively, as shown in Fig. 3.11. It begins with a $Q-V$ droop control as follows.

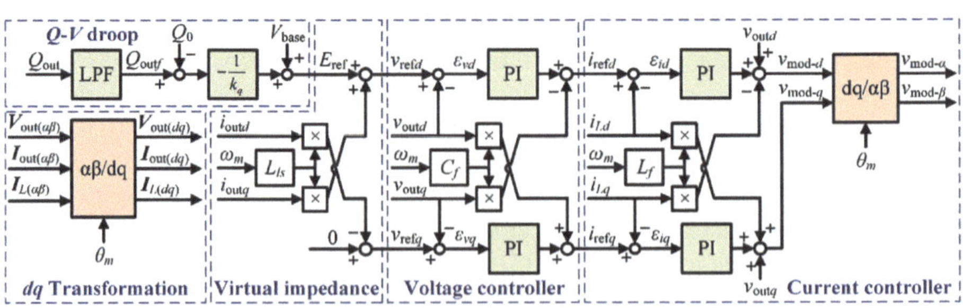

Figure 3.11 The "$Q-V$ control and dynamics emulator" block with inner *VI*-based control.

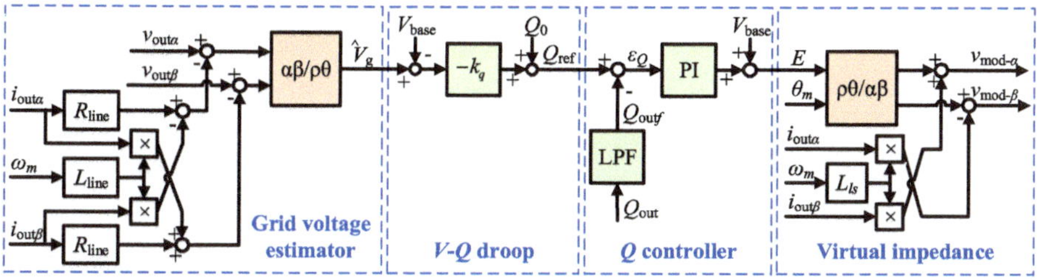

Figure 3.12 The "$Q-V$ control and dynamics emulator" block with inner Q-based control.

$$Q_{outf} = \frac{1}{1 + T_{fq}s} Q_{out} \tag{3.37}$$

$$E_{ref} = V_{base} - (1/k_q)(Q_{outf} - Q_0) \tag{3.38}$$

where Q_{out} is the output reactive power and Q_{outf} is its value after the LPF, Q_0 is the reactive power command, E_{ref} is the reference of virtual electromotive force (*emf*), V_{base} is the rated voltage, T_{fq} is the LPF time constant, and k_q is the $Q-V$ droop coefficient.

With the obtained E_{ref}, the voltage loop references v_{refd} and v_{refq} are obtained through the dq frame-based virtual impedance control as

$$v_{refd} = E_{ref} + \omega_m L_{ls} i_{outq} \tag{3.39}$$

$$v_{refq} = -\omega_m L_{ls} i_{outd} \tag{3.40}$$

where L_{ls} is the virtual inductance.

The voltage controller applies PI controls for both v_{outd} and v_{outq}, which can be expressed as

$$\varepsilon_{vd} = v_{refd} - v_{outd} \tag{3.41}$$

$$i_{refd} = K_{pv}[1 + 1/(T_{iv}s)]\varepsilon_{vd} - \omega_m C_f v_{outq} \tag{3.42}$$

$$\varepsilon_{vq} = v_{refq} - v_{outq} \tag{3.43}$$

$$i_{refq} = K_{pv}[1 + 1/(T_{iv}s)]\varepsilon_{vq} + \omega_m C_f v_{outd} \tag{3.44}$$

where ε_{vd} and ε_{vq} are the control errors of v_{outd} and v_{outq}, respectively, K_{pv} and T_{iv} are the proportional gain and integral time constant of voltage PI control, respectively.

Similarly, the current controller can be expressed as

$$\varepsilon_{id} = i_{\text{ref}d} - i_{Ld} \tag{3.45}$$

$$v_{\text{pwm}d} = K_{pi}[1 + 1/(T_{ii}s)]\varepsilon_{id} - \omega_m L_f i_{Lq} + v_{\text{out}d} \tag{3.46}$$

$$\varepsilon_{iq} = i_{\text{ref}q} - i_{Lq} \tag{3.47}$$

$$v_{\text{pwm}q} = K_{pi}[1 + 1/(T_{ii}s)]\varepsilon_{iq} + \omega_m L_f i_{Ld} + v_{\text{out}q} \tag{3.48}$$

where ε_{id} and ε_{iq} are the control errors of i_{Ld} and i_{Lq}, respectively, K_{pi} and T_{ii} are the proportional gain and integral time constant of current PI control, respectively.

3.3.1.2 Inner Q-based control

Fig. 3.12 shows the control scheme of an inner *Q*-based control [13]. It uses an $\alpha\beta$ frame-based grid voltage estimator to estimate the RMS value of bus line voltage $\widehat{V}_{\text{g}}$, and applies a *V-Q* droop control shown in Eq. (3.50), which reverses the input and output of the *Q*–*V* droop control given in Eq. (3.38).

$$Q_{\text{ref}} = Q_0 - k_q\left(\widehat{V}_{\text{g}} - V_{\text{base}}\right) \tag{3.49}$$

Afterward, a PI controller for the reactive power is applied to generate the virtual *emf E* as

$$\varepsilon_Q = Q_{\text{ref}} - Q_{\text{out}f} \tag{3.50}$$

$$E = V_{\text{base}} + K_{pq}\left[1 + 1/\left(T_{iq}s\right)\right]\varepsilon_Q \tag{3.51}$$

where ε_Q is the control error of $Q_{\text{out}f}$, K_{pq}, and T_{iq} are the proportional gain and integral time constant of reactive power PI control, respectively. For a fair comparison, the LPFs for Q_{out} in both control schemes are the same in the presented work.

With the obtained *E* and the phase angle θ_m from the *P*–ω control loop, an $\alpha\beta$ frame-based virtual impedance control is applied to determine the final output voltage. It can be noticed that the *Q*-based control is much simpler than the *VI*-based control. Only one PI controller is applied, and *dq* transformation is not needed. Besides, the sensor for $\boldsymbol{I}_L$ can be omitted in the *Q*-based control.

3.3.1.3 Output impedance models

As will be seen in the next section, the proposed modeling method is based on a quasi-stationary approximation of the power flow equation. Therefore, it is important to correctly model the equivalent output impedance of each control scheme.

The output impedance model of the *VI*-based control is shown in Fig. 3.13. It indicates that L_f and C_f are merged into the voltage control loop, thus they do not contribute to the output impedance. The equivalent output reactance *X* between the

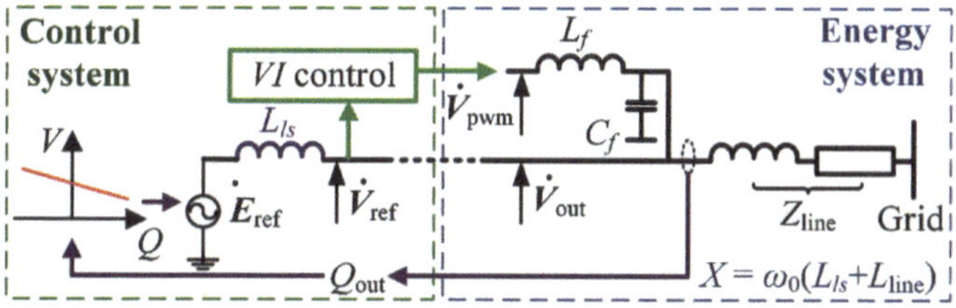

Figure 3.13 Output impedance model of inner *VI*-based control method.

emf reference $\dot{\boldsymbol{E}}_{ref}$ and the grid bus is made up of the virtual and line impedances, as illustrated in Fig. 3.13.

On the other hand, the *Q*-based control does not have a voltage loop; therefore, its equivalent output reactance *X* also includes L_f and C_f, as shown in Fig. 3.14. Practically, the susceptance of C_f is quite small at fundamental frequency; therefore, it is negligible when applying the quasi-stationary approximation of the power flow equation. Moreover, since the virtual impedance is designed to make the output impedance significantly inductive ($X/R > 5$, where $R \approx R_{line}$) to assure the decoupling between $P-\omega$ and $Q-V$ control loop, the output resistance *R* can also be neglected.

Another important point shown in Figs. 3.13 and 3.14 is how the $Q-V$ droop is performed in each control scheme. Both methods measure the reactive power in the same location; however, according to Eqs. (3.38) and (3.50), it is used to determine the *emf* reference E_{ref} in the *VI*-based control, but the grid voltage V_g in the *Q*-based control. This difference results in significantly different sensitivity to the voltage drop across *X*, which will be further discussed in Chapter 4.

It is noteworthy that for the *Q*-based control, a significant voltage drop on L_{ls} will occur if the inverter outputs a large amount of reactive power. However, this voltage drop will only affect $\dot{\boldsymbol{E}}$, which is a virtual value. The dc-link voltage requirement, which is proportional to $\dot{\boldsymbol{V}}_{pwm}$, will be barely affected by the voltage drop on L_{ls}.

3.3.2 Modeling methodology

As shown in Fig. 3.15, the adjacent grid bus is considered as an infinite bus, as a stiff bus voltage is determined by the grid. Therefore, the infinite bus voltage V_g is

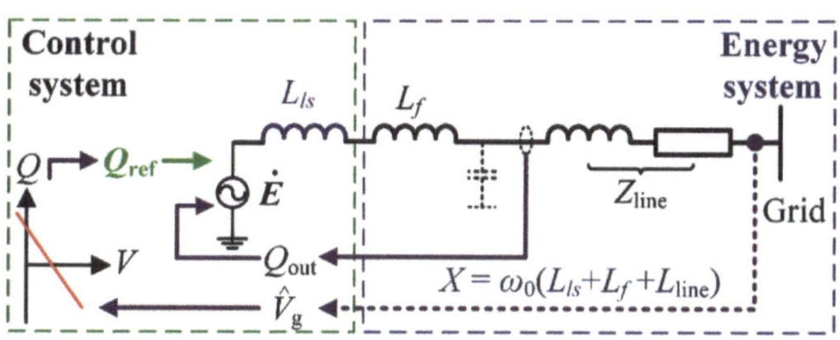

Figure 3.14 Output impedance model of inner *Q*-based control.

independent of the GCC, and thus its deviation should be considered as a disturbance input. Meanwhile, as illustrated in Fig. 3.10, the reactive power command Q_0 can be used to regulate the $Q-V$ control loop, either by the user or an upper-level control. Therefore, Q_0 should be modeled as a control input. To study the response of E and Q_{out}, the following small-signal state-space model is expected.

$$\begin{cases} \dot{\boldsymbol{x}} = \boldsymbol{A}x + \boldsymbol{B}u + \boldsymbol{E}_w w \\ \boldsymbol{y} = \boldsymbol{C}x + \boldsymbol{D}u + \boldsymbol{F}w \end{cases} \tag{3.52}$$

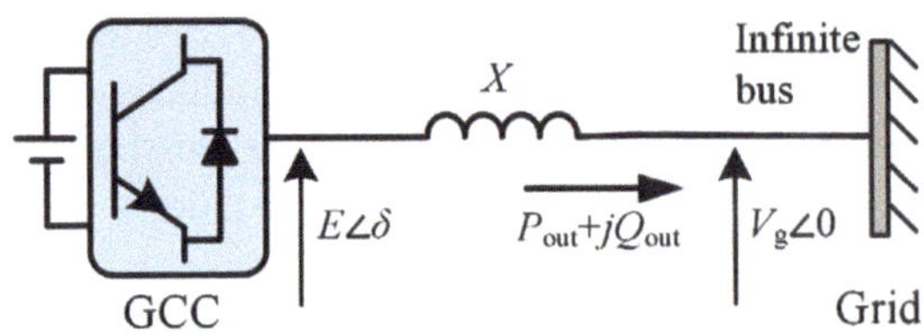

Figure 3.15 The model of GCC connected to an infinite bus.

$$u = \Delta Q_0 \tag{3.53}$$

$$w = \Delta V_g \tag{3.54}$$

$$\boldsymbol{y} = [\Delta E \quad \Delta Q_{out}]^T \tag{3.55}$$

According to the previous discussion, the control input u, the disturbance input w, and the output vector $\boldsymbol{y}$ should be as follows.where Δ denotes a small-signal variation.

To find the expression of Eq. (3.53) for each control scheme, we start from the quasi-stationary approximation of the power flow equation for a significantly inductive line, which is known as

$$P_{out} \approx EV_g \sin\delta / X \tag{3.56}$$

$$Q_{out} \approx \left(EV_g \cos\delta - V_g^2\right) / X. \tag{3.57}$$

The complete models usually include the full electromagnetic model of output inductance, which means a differential term is also considered [25]. However, the quasi-stationary approximation is correct enough for a virtual impedance-based GCC, since the virtual impedance controls applied in both Figs. 3.11 and 3.12 are based on quasi-stationary models, and they usually contribute to a large proportion of X. Linearizing (3.58) yields

$$\Delta Q_{out} = K_{QE}\Delta E + K_{QV}\Delta V_g + K_{Q\delta}\Delta\delta \tag{3.58}$$

where

$$K_{QE} \approx V_{\text{base}}\cos\delta_0/X \tag{3.59}$$

$$K_{QV} \approx V_{\text{base}}(\cos\delta_0 - 2)/X \tag{3.60}$$

$$K_{Q\delta} \approx -V_{\text{base}}^2\sin\delta_0/X \tag{3.61}$$

where δ_0 is the equilibrium point of the power angle δ, and its value at rated active power can be assumed as

$$\cos\delta_0 \approx \sqrt{1 - X^{*2}} \tag{3.62}$$

$$\sin\delta_0 \approx X^* = XS_{\text{base}}/V_{\text{base}}^2 \tag{3.63}$$

Here, superscript $*$ denotes a per-unit value, and S_{base} is the rated power. It is noteworthy that $K_{Q\delta}$ introduces coupling with the $P-\omega$ control loop. Nevertheless, since δ_0 is generally a small value close to 0, $\sin\delta_0$ is much smaller than $\cos\delta_0$. Therefore, $K_{Q\delta}$ is much smaller than K_{QE} and K_{QV}. Consequently, here, $K_{Q\delta}$ is neglected. The case of considering the coupling between $P-\omega$ and $Q-V$ is discussed in Section 3.4.1.

First, the reduced-order model of inner *VI*-based control is studied. As can be seen later, the voltage loop parameters can significantly influence the performance of the $Q-V$ control loop, thus should not be neglected as it is done in the reported reduced-order models [26,27]. However, the current loop, which is generally designed to have a 10 times wider bandwidth than the voltage loop, can be neglected.

$$\Delta i_{\text{ref}d} \approx \Delta i_{Ld} \tag{3.64}$$

$$\Delta i_{\text{ref}q} \approx \Delta i_{Lq} \tag{3.65}$$

The *LCL* filter and feeder line can be modeled as

$$0 \approx sC_f\Delta v_{\text{out}d} = \Delta i_{Ld} - \Delta i_{\text{out}d} + \omega_0 C_f \Delta v_{\text{out}q} \tag{3.66}$$

$$0 \approx sC_f\Delta v_{\text{out}q} = \Delta i_{Lq} - \Delta i_{\text{out}q} - \omega_0 C_f \Delta v_{\text{out}d} \tag{3.67}$$

$$0 \approx sL_{\text{line}}\Delta i_{\text{out}d} = \Delta v_{\text{out}d} - \Delta v_{gd} - R_{\text{line}}\Delta i_{\text{out}d} + \omega_0 L_{\text{line}}\Delta i_{\text{out}q} \tag{3.68}$$

$$0 \approx sL_{\text{line}}\Delta i_{\text{out}q} = \Delta v_{\text{out}q} - \Delta v_{gq} - R_{\text{line}}\Delta i_{\text{out}q} - \omega_0 L_{\text{line}}\Delta i_{\text{out}d} \tag{3.69}$$

where ω_0 is the nominal frequency. The differential terms in Eqs. (3.67–3.70) are neglected to reduce four orders of the system. It is reasonable because the values of C_f and L_{line} are generally small, and we have confirmed that including these orders barely changes the dominant poles and step responses. On the other hand, although resistance is neglected in Eqs. (3.57 and 3.58) because R_{line} is quite smaller than X; it is not proper to neglect R_{line} in Eqs. (3.69 and 3.70) because it is considerable comparing to $\omega_0 L_{\text{line}}$. It is also noteworthy that the Kirchhoff's voltage law related to L_f is not included, because these equations are a part of the current loop, which has already been neglected.

According to Fig. 3.11, the d-axis is fixed in the direction of $\dot{\boldsymbol{E}}_{\text{ref}}$. Therefore, from Fig. 3.13:

$$\Delta E \approx \Delta e_d = \Delta v_{\text{out}d} - \omega_0 L_{ls} \Delta i_{\text{out}q}. \tag{3.70}$$

From Fig. 3.15, the grid bus voltage can be expressed as

$$\Delta v_{gd} = \cos\delta_0 \Delta V_g \tag{3.71}$$

$$\Delta v_{gq} = -\sin\delta_0 \Delta V_g \tag{3.72}$$

The control part Eqs. (3.2)−(3.45) are almost all linear, except that ω_m in these equations should be replaced by ω_0 after linearization. Afterward, by combining them with the linearized power flow Eq. (3.59), the approximation of current loop (Eqs. 3.65 and 3.66), and the plant model (Eqs. 3.67−3.73), we can finally solve Eq. (3.53) for inner VI-based control as follows.

$$\boldsymbol{x} = \begin{bmatrix} \Delta Q_{\text{out}f} \\ K_{pv}\Delta\varepsilon_{vd} - \Delta i_{\text{out}d} \\ K_{pv}\Delta\varepsilon_{vq} - \Delta i_{\text{out}q} \end{bmatrix} \tag{3.73}$$

$$\boldsymbol{A} = \boldsymbol{A}_o + \boldsymbol{H}_o \boldsymbol{H}_x^{-1} \boldsymbol{A}_x \tag{3.74}$$

$$\boldsymbol{B} = \boldsymbol{H}_o \boldsymbol{H}_x^{-1} \boldsymbol{B}_x \tag{3.75}$$

$$\boldsymbol{E}_w = \boldsymbol{H}_o \boldsymbol{H}_x^{-1} \boldsymbol{E}_x \tag{3.76}$$

$$\boldsymbol{C} = \begin{bmatrix} \left(\boldsymbol{H}_x^{-1}\boldsymbol{A}_x\right)_{(1)} / K_{QE} \\ \left(\boldsymbol{H}_x^{-1}\boldsymbol{A}_x\right)_{(1)} \end{bmatrix} \tag{3.77}$$

$$\boldsymbol{D} = \begin{bmatrix} \left(\boldsymbol{H}_x^{-1}\boldsymbol{B}_x\right)_{(1)} / K_{QE} \\ \left(\boldsymbol{H}_x^{-1}\boldsymbol{B}_x\right)_{(1)} \end{bmatrix} \tag{3.78}$$

$$\boldsymbol{F}=\begin{bmatrix}\left(\left(\boldsymbol{H}_x^{-1}\boldsymbol{E}_x\right)_{(1)}-K_{QV}\right)/K_{QE}\\ \left(\boldsymbol{H}_x^{-1}\boldsymbol{E}_x\right)_{(1)}\end{bmatrix} \tag{3.79}$$

Here, the expression $\boldsymbol{a}_{(k)}$ indicates the kth row of the matrix/vector $\boldsymbol{a}$, and

$$\boldsymbol{A}_o=\begin{bmatrix}-1/T_{fq} & 0 & 0\\ 0 & -1/T_{iv} & 0\\ 0 & 0 & -1/T_{iv}\end{bmatrix} \tag{3.80}$$

$$\boldsymbol{H}_o=\begin{bmatrix}1/T_{fq} & 0 & 0\\ 0 & -1/T_{iv} & 0\\ 0 & 0 & -1/T_{iv}\end{bmatrix} \tag{3.81}$$

$$\boldsymbol{H}_x=\begin{bmatrix}1 & 0 & K_{QE}\omega_0 L_{ls} & -K_{QE} & 0\\ 0 & 1/K_{pv} & -\omega_0 L_{ls} & 1 & 0\\ 0 & \omega_0 L_{ls} & 1/K_{pv} & 0 & 1\\ 0 & R_{\text{line}} & -\omega_0 L_{\text{line}} & -1 & 0\\ 0 & \omega_0 L_{\text{line}} & R_{\text{line}} & 0 & -1\end{bmatrix} \tag{3.82}$$

$$\boldsymbol{A}_x=\begin{bmatrix}0 & 0 & 0\\ -1/k_q & -1/K_{pv} & 0\\ 0 & 0 & -1/K_{pv}\\ 0 & 0 & 0\\ 0 & 0 & 0\end{bmatrix} \tag{3.83}$$

$$\boldsymbol{B}_x=\begin{bmatrix}0 & 1/k_q & 0 & 0 & 0\end{bmatrix}^{\mathrm{T}} \tag{3.84}$$

$$\boldsymbol{E}_x=\begin{bmatrix}K_{QV} & 0 & 0 & -\cos\delta_0 & \sin\delta_0\end{bmatrix}^{\mathrm{T}} \tag{3.85}$$

From Eq. (3.74), it can be noticed that the $Q-V$ control loop of inner VI-based control is modeled as a third-order system. One order comes from the LPF of Q_{out}, and the other two come from the voltage-loop PI controllers. This is significantly simplified compared to the detailed complete models [21–23]. It can be noticed that C_f does not appear in this model, because the term $\omega_0 C_f$ in the plant model Eqs. (3.67–3.68) is canceled by the decoupling control shown in Eqs. (3.43 and 3.45).

Second, the reduced-order model of inner Q-based control is studied. Since it does not regulate voltage, it is not necessary to include the plant model. Therefore, the GCC model (Eq. 3.53) can be simply obtained from the linear control part Eqs. (3.2), (3.50)–(3.52), and the linearized power flow Eq. (3.59) as follows.

$$\boldsymbol{x} = \begin{bmatrix} \Delta Q_{\text{outf}} \\ K_{pq}\Delta\varepsilon_Q - \Delta E \end{bmatrix} \tag{3.86}$$

$$\boldsymbol{A} = \begin{bmatrix} -\left(1 + K_{QE}K_{pq}\right)/T_{fq} & -K_{QE}/T_{fq} \\ K_{pq}/T_{iq} & 0 \end{bmatrix} \tag{3.87}$$

$$\boldsymbol{B} = \begin{bmatrix} K_{QE}K_{pq}/T_{fq} \\ -K_{pq}/T_{iq} \end{bmatrix} \tag{3.88}$$

$$\boldsymbol{E}_w = \begin{bmatrix} \left(K_{QV} - K_{QE}k_qK_{pq}\right)/T_{fq} \\ k_qK_{pq}/T_{iq} \end{bmatrix} \tag{3.89}$$

$$\boldsymbol{C} = \begin{bmatrix} -K_{pq} & -1 \\ -K_{QE}K_{pq} & -K_{QE} \end{bmatrix} \tag{3.90}$$

$$\boldsymbol{D} = \begin{bmatrix} K_{pq} \\ K_{QE}K_{pq} \end{bmatrix} \tag{3.91}$$

$$\boldsymbol{F} = \begin{bmatrix} -k_qK_{pq} \\ K_{QV} - K_{QE}k_qK_{pq} \end{bmatrix} \tag{3.92}$$

Eq. (3.87) implies that the proposed reduced-order model of the Q-based control is a second-order system. One order comes from the LPF of Q_{out}, and the other comes from the reactive power PI controller.

Including the coupling terms between $P-\omega$ and $Q-V$ controls will significantly complete the unified formulas, as discussed in Section 3.4. Therefore, it is not practical to use this method for a strong-coupled system. Nevertheless, for the design purpose, it is not necessary to consider the coupling terms as long as a decoupling strategy like the virtual impedance method has already been adopted.

3.3.3 Step-response analysis and evaluation

To verify the fidelity of the presented model, and the previous given discussion, experimental step responses are obtained using the testbed shown in Fig. 3.8, and compared with the analytical ones calculated from the reduced-order models obtained through the proposed design-oriented modeling method. The testbed uses the configuration as Fig. 3.8B, but this time the two-phase RL load is removed. The system parameters are given in Table 3.2. A 360-V dc-link voltage and an 18-kHz switching frequencies were used for the GCC. The inverter was controlled by a digital signal processor system, and the resulted data were collected using an appropriate function, which transferred the real-time variable values from the controller to the interfaced PC. It is worthwhile to mention that in this work, instead of connecting GCC to the real power system, it was connected to an AC power supply, which behaves as an infinite bus.

The dynamics of the systems following a Q_0 step change are studied, since Q_0 is generally used to regulate the reactive power and/or the bus voltage of a MG. It is noteworthy that 20 Hz LPFs are applied to the experimental results for ripple filtering; therefore, the same LPFs are also applied to the analytical results for a rigorous comparison.

The breakers BK1 and BK2 are closed, and other breakers are opened. The infinite bus voltage is set to 200 V by the ac power supply, and the Q_{out} response to a step change of Q_0 from 0 pu (0 var) to 0.5 pu (2500 var) is investigated.

The model-based simulation and experimental results are compared in Fig. 3.16. As shown in this figure, the analytical results based on the models are well verified by the experimental results, despite a slight difference due to the linearization and assumptions in Eqs. (3.59−3.64).

Table 3.2 System parameters.

GCC parameters					
Parameter	**Value**	**Parameter**	**Value**	**Parameter**	**Value**
S_{base}	5 kVA	ω_0	377 rad/s	T_{fq}	7.96×10^{-3} s
		X^*	0.3 pu	k_q^*	10 pu
V_{base}	200 V	Q_0^*	0 pu		
Parameters of the inner *VI*-based control scheme					
K_{pv}^* (K_{pv0}^*)	1.0 pu	T_{iv}	2.5×10^{-3} s	L_{ls}	5.94×10^{-3} mH
K_{pi}^*	1.5 pu	T_{ii}	3.0×10^{-3} s		
Parameters of the inner *Q*-based control scheme					
K_{pq}^* (K_{pq0}^*)	5.0×10^{-3} pu	T_{iq}	1.25×10^{-4} s	L_{ls}	3.44×10^{-3} mH

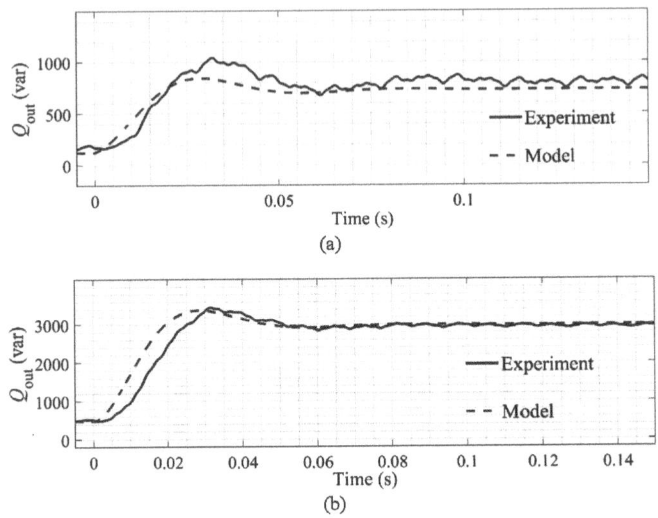

Figure 3.16 Q_{out} responses to Q_0 step change in the system: (A) inner *VI*-based control; (B) inner *Q*-based control.

Another interesting point is that the final variation of Q_{out} is about 2500 var with the *Q*-based control, which follows the change of command Q_0. However, the final variation of Q_{out} is much smaller with the *VI*-based control. This is because the *Q*-based control regulates Q_{out} based on V_g, whereas the *VI*-based control regulates Q_{out} based on E_{ref}, as shown in Figs. 3.13 and 3.14. The former is fixed in the grid connected mode, whereas the latter is affected by the voltage drop across X. An increase in Q_{out} will increase this voltage drop, and thus increases E_{ref}. According to the $Q-V$ droop, this will finally lead to a decrease in Q_{out}, and thus cancel the command following reactive power. This defect of the *VI*-based control will tend to be amplified when a large virtual impedance is applied, since it will increase the voltage drop across X.

3.4 Comprehensive and reduced GCC models

3.4.1 A comprehensive model

The models of GCC that are obtained as given in Eq. (3.9) for $P-\omega$ dynamic response and in Eq. (3.53) for $Q-V$ dynamics response can be rewritten in form of Eq. (3.94 and 3.95), respectively.

$$\begin{cases} \dot{\boldsymbol{x}}_{P\omega} = \boldsymbol{A}_{P\omega} x_{P\omega} + \boldsymbol{B}_{P\omega} u_{P\omega} + \boldsymbol{E}_{P\omega} w_{P\omega} \\ \boldsymbol{y}_{P\omega} = \boldsymbol{C}_{P\omega} x_{P\omega} + \boldsymbol{D}_{P\omega} u_{P\omega} + \boldsymbol{F}_{P\omega} w_{P\omega} \end{cases} \tag{3.93}$$

$$\begin{cases} \dot{x}_{QV} = A_{QV}x_{QV} + B_{QV}u_{QV} + E_{QV}w_{QV} \\ y_{QV} = C_{QV}x_{QV} + D_{QV}u_{QV} + F_{QV}w_{QV} \end{cases} \tag{3.94}$$

The complete model of grid-forming GCCs can be obtained considering above $P-\omega$ and $Q-V$ state-space models as follows

$$\begin{cases} \begin{bmatrix} \dot{x}_{P\omega} \\ \dot{x}_{QV} \end{bmatrix} = \begin{bmatrix} A_{P\omega} & 0 \\ 0 & A_{QV} \end{bmatrix}\begin{bmatrix} x_{P\omega} \\ x_{QV} \end{bmatrix} + \begin{bmatrix} B_{P\omega} & 0 \\ 0 & B_{QV} \end{bmatrix}\begin{bmatrix} u_{P\omega} \\ u_{QV} \end{bmatrix} + \begin{bmatrix} E_{P\omega} & 0 \\ 0 & E_{QV} \end{bmatrix}\begin{bmatrix} w_{P\omega} \\ w_{QV} \end{bmatrix} \\ \begin{bmatrix} y_{P\omega} \\ y_{QV} \end{bmatrix} = \begin{bmatrix} C_{P\omega} & 0 \\ 0 & C_{QV} \end{bmatrix}\begin{bmatrix} x_{P\omega} \\ x_{QV} \end{bmatrix} + \begin{bmatrix} D_{P\omega} & 0 \\ 0 & D_{QV} \end{bmatrix}\begin{bmatrix} u_{P\omega} \\ u_{QV} \end{bmatrix} + \begin{bmatrix} F_{P\omega} & 0 \\ 0 & F_{QV} \end{bmatrix}\begin{bmatrix} w_{P\omega} \\ w_{QV} \end{bmatrix} \end{cases} \tag{3.95}$$

where subscripts $P\omega$ and QV denote the $P-\omega$ model matrices in Eqs. (3.94) and the $Q-V$ model matrices in Eq. (3.95), respectively.

Eq. (3.96) implies that the $P-\omega$ and $Q-V$ control loops are independent. As discussed before, this is based on the assumptions that $X \gg R$ and $\sin\delta_0$ is small; otherwise, the zero coupling matrices in Eq. (3.96) will become nonzero. Therefore, despite the explained advantages, this method is not appropriate for accurate modeling of a strong-coupled system. However, as *PQ* decoupling is a general design requirement for the stable operation of grid-forming GCCs, the proposed method can be applied in most of the cases.

3.4.2 Reduced dynamic models

For many application, complex and detailed dynamic model of GCC are not necessary. For example for the case of linear control synthesis problems, only reduced linear model of a GCC is enough. Here, few simple model examples are described.

3.4.2.1 Example 1: focusing on LFC filter dynamics

The system structure for a GCC and its equivalent circuit is shown in Fig. 3.17. An LCL filter of L1, C, L2 is used to sufficiently reduce the harmonic components, and to convert the voltage input v_i to the current i_{L2} as the output of GCC. The linear differential equations for the GCC focusing on the LCL filter and neglecting the dynamics of switching elements and control system can be obtained as follows:

$$L_2\frac{di_{L2}}{dt} = v_C - v_g$$

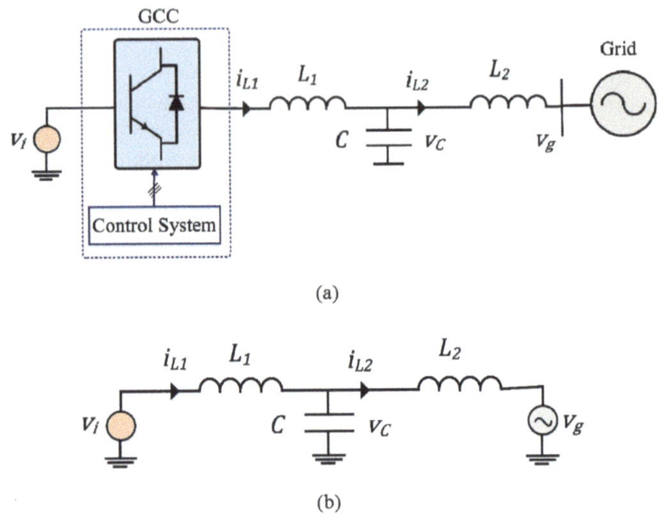

Figure 3.17 (A) A GCC with LCL filter, (B) Equivalent circuit.

$$L_1\frac{di_{L1}}{dt}=v_i-v_C \tag{3.96}$$

$$C\frac{dv_C}{dt}=i_{L1}-i_{L2}$$

Assuming states vector, measured output, control input, and disturbance input as follows:

$$\boldsymbol{x}^T=[i_{L1}\quad v_C\quad i_{L2}];\ y=i_{L2}=x_3;\ u=v_i;\ w=v_g$$

The linear state-space model of the GCC can be obtained as Eq. (3.97).

$$\begin{cases}\dot{\boldsymbol{x}}=\boldsymbol{A}x+\boldsymbol{B}u+\boldsymbol{E}w\\ \boldsymbol{y}=\boldsymbol{C}x+\boldsymbol{D}u+\boldsymbol{F}w\end{cases} \tag{3.97}$$

where,

$$\boldsymbol{A}=\begin{bmatrix}0 & -\frac{1}{L_1} & 0\\ \frac{1}{C} & 0 & -\frac{1}{C}\\ 0 & \frac{1}{L_2} & 0\end{bmatrix};\ \boldsymbol{B}=\begin{bmatrix}\frac{1}{L_1}\\ 0\\ 0\end{bmatrix};\ \boldsymbol{E}=\begin{bmatrix}0\\ 0\\ -\frac{1}{L_2}\end{bmatrix};$$

$$\boldsymbol{C}=[0\quad 0\quad 1];\ \boldsymbol{D}=[0];\ \boldsymbol{F}=[0].$$

3.4.2.2 Example 2: LFC filter and a virtual resistor

Now consider that the GCC control system to be designed such that a resistive component R_I is virtually created in series with the inductor L_I. As discussed in Chapter 2, this virtual resistor realizes a damping characteristic to suppress the resonant peak of the output filter. The equivalent circuit for this example is shown in Fig. 2.10. In this case, only the second differential equation of (3.96) needs to be changed as

$$L_1\frac{di_{L1}}{dt}=v_i-R_1 i_{L1}-v_C. \tag{3.98}$$

The linear state-space model of the modified GCC can be obtained as

$$\begin{cases}\dot{x}=\boldsymbol{A}_R x+\boldsymbol{B}u+\boldsymbol{E}w\\ y=\boldsymbol{C}x+\boldsymbol{D}u+\boldsymbol{F}w\end{cases} \tag{3.99}$$

The virtual resistor only updates system matrix A to A_R, where

$$\boldsymbol{A}_R=\begin{bmatrix}0 & -\dfrac{R_1}{L_1} & 0\\ \dfrac{1}{C} & 0 & -\dfrac{1}{C}\\ 0 & \dfrac{1}{L_2} & 0\end{bmatrix}.$$

3.4.2.3 Example 3: LFC filter and three virtual resistors

Assume that the GCC control system to be designed such that in addition to R_1 in series with the inductor L_1, two more resistors need to be virtually created by the GCC control system as shown in the equivalent circuit in Fig. 3.18. In this case, the system matrix A will be changed to A_{3R} as follows:

$$\boldsymbol{A}_{3R}=\begin{bmatrix}-\dfrac{R_1}{L_1} & -\dfrac{1}{L_1} & 0\\ \dfrac{1}{C} & -\dfrac{G_C}{C} & -\dfrac{1}{C}\\ 0 & \dfrac{1}{L_2} & -\dfrac{R_2}{L_2}\end{bmatrix}.$$

3.4.2.4 Example 4: using dq-frame

Fig. 3.19 shows the three-phase GCC connected with an AC grid through LCL filters and supplied by an ideal DC source. As shown in the figure, the control system uses

Figure 3.18 Equivalent circuit of a GCC with three virtual resistors.

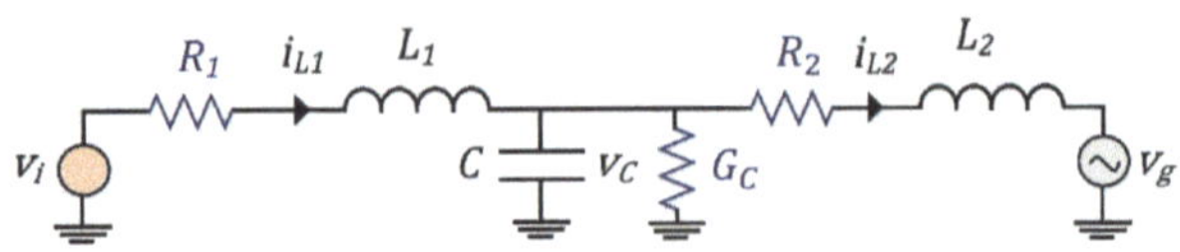

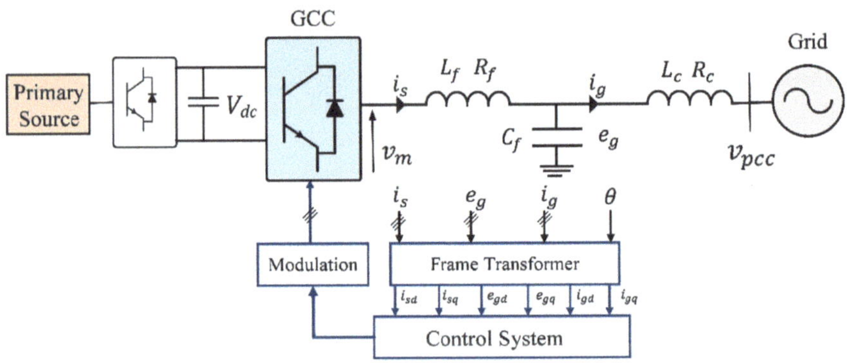

Figure 3.19 A GCC example with *dq*-frame based control system.

the *dq*-frame characteristics; therefore, it is reasonable to have a dynamic model in the same frame. Assume the state variables are the inverter current i_s through the inductor L_f, the voltage e_g across the filter capacitor C_f, and the current i_g through the transformer inductance L_c. V_{dc}, v_m, and v_{pcc} are the ideal DC bus voltage, the output voltage and the grid-voltage, respectively. The state variables are transformed in *dq*-frame where θ is the instantaneous phase of the capacitor voltage deduced by the droop control.

Thus, the state-space model for the three-phase GCC system in *dq*-frame can be obtained as follows:

$$\begin{cases} \dot{\boldsymbol{x}} = \boldsymbol{A}\boldsymbol{x} + \boldsymbol{B}\boldsymbol{u} + \boldsymbol{E}\boldsymbol{w} \\ \boldsymbol{y} = \boldsymbol{C}\boldsymbol{x} \end{cases} \tag{3.100}$$

where,

$$\boldsymbol{x}^T = \begin{bmatrix} i_{sd} & i_{sq} & e_{gd} & e_{gq} & i_{gd} & i_{gq} \end{bmatrix}$$

$$\boldsymbol{u}^T = \begin{bmatrix} v_{md} & v_{mq} \end{bmatrix}$$

$$\boldsymbol{w}^T = \begin{bmatrix} v_{pccd} & v_{pccq} \end{bmatrix}$$

$$A=\begin{bmatrix} -\dfrac{R_f\omega_b}{L_f} & \omega\omega_b & -\dfrac{\omega_b}{L_f} & 0 & 0 & 0 \\ -\omega\omega_b & -\dfrac{R_f\omega_b}{L_f} & 0 & -\dfrac{\omega_b}{L_f} & 0 & 0 \\ \dfrac{\omega_b}{C_f} & 0 & 0 & \omega\omega_b & -\dfrac{\omega_b}{C_f} & 0 \\ 0 & \dfrac{\omega_b}{C_f} & -\omega\omega_b & 0 & 0 & -\dfrac{\omega_b}{C_f} \\ 0 & 0 & \dfrac{\omega_b}{L_c} & 0 & -\dfrac{R_c\omega_b}{L_c} & \omega\omega_b \\ 0 & 0 & 0 & \dfrac{\omega_b}{L_c} & -\omega\omega_b & -\dfrac{R_c\omega_b}{L_c} \end{bmatrix}$$

$$B^T=\begin{bmatrix} \dfrac{\omega_b}{L_f} & 0 & 0 & 0 & 0 & 0 \\ 0 & \dfrac{\omega_b}{L_f} & 0 & 0 & 0 & 0 \end{bmatrix}$$

$$C=\begin{bmatrix} 0 & 0 & 1 & 0 & 0 & 0 \\ 0 & 0 & 0 & 1 & 0 & 0 \end{bmatrix}$$

$$E^T=\begin{bmatrix} 0 & 0 & 0 & 0 & -\dfrac{\omega_b}{L_c} & 0 \\ 0 & 0 & 0 & 0 & 0 & -\dfrac{\omega_b}{L_c} \end{bmatrix}$$

The vectors $\boldsymbol{x}$ and $\boldsymbol{u}$ are the state variables and the system inputs, respectively, and $\boldsymbol{w}$ is the grid-voltage considered as perturbations in this case. The output $\boldsymbol{y}$ is the voltage across the capacitor. The term of ω_b is the base value for the angular frequency in rad/s, corresponding to the rated frequency of the grid.

3.5 Summary

In this Chapter, different state-space models for the GCC are introduced. The proposed models despite simplicity and generality provide mathematical and physical interpretations for the different behaviors of GCCs in different applications and operation modes. The detail procedures of two Q–V and P-ω dynamic response modeling are presented, and the accuracy of obtained models is evaluated using some simulation and experimental tests. Then, a comprehensive model by combination of the former two developed models is introduced. Finally, several examples for simple reduced-order dynamic models useful for linear control synthesis and dynamics analysis are explained.

References

[1] R.H. Lasseter, MicroGrids, in: Proc. IEEE Power Eng. Soc. Winter Meeting, 2002, pp. 305–308.
[2] H. Bevrani, B. Francois, T. Ise, Microgrid Dynamics and Control, Wiley, NJ, USA, July 2017.
[3] J. Liu, H. Bevrani, T. Ise, A design-oriented QV response modeling approach for grid-forming distributed generators considering different operation modes, IEEE J. Emerg. Sel. Topics Power Electron. (2021), https://doi.org/10.1109/JESTPE.2021.3057517.
[4] J. Fang, H. Li, Y. Tang, F. Blaabjerg, On the inertia of future more-electronics power systems, IEEE J. Emerg. Sel. Topics Power Electron. 7 (4) (2019) 2130–2146.
[5] J. Driesen, K. Visscher, Virtual synchronous generators, in: Proc. IEEE Power Energy Soc. Gen. Meeting—Convers. Del. Elect. Energy 21st Century, Pittsburgh, PA, USA, 2008, pp. 1–3.
[6] H.-P. Beck, R. Hesse, Virtual synchronous machine, in: Proc. 9th Int. Conf. Elect. Power Quality Util., Barcelona, Spain, 2007, pp. 1–6.
[7] Q.-C. Zhong, G. Weiss, Synchronverters: inverters that mimic synchronous generators, IEEE Trans. Ind. Electron. 58 (4) (April 2011) 1259–1267.
[8] J. Liu, Y. Miura, H. Bevrani, T. Ise, A unified modeling method of virtual synchronous generator for multi-operation-mode analyses, IEEE J. Emerg. Sel. Topics Power Electron. (2021), https://doi.org/10.1109/JESTPE.2020.2970025.
[9] T. Shintai, Y. Miura, T. Ise, Oscillation damping of a distributed generator using a virtual synchronous generator, IEEE Trans. Power Deliv. 29 (2) (April 2014) 668–676.
[10] S. Dong, Y.C. Chen, Adjusting synchronverter dynamic response speed via damping correction loop, IEEE Trans. Energy Convers. 32 (2) (June 2017) 608–619.
[11] J. Liu, Y. Miura, T. Ise, Fixed-parameter damping methods of virtual synchronous generator control using state feedback, IEEE Access 7 (2019) 99177–99190.
[12] Y. Ma, W. Cao, L. Yang, F. Wang, L.M. Tolbert, Virtual synchronous generator control of full converter wind turbines with short-term energy storage, IEEE Trans. Ind. Electron. 64 (11) (Nov. 2017) 8821–8831.
[13] J. Liu, Y. Miura, H. Bevrani, T. Ise, Enhanced virtual synchronous generator control for parallel inverters in microgrids, IEEE Trans. Smart Grid 8 (5) (2017) 2268–2277.
[14] Y. Khayat, M. Naderi, H. Bevrani, F. Blaabjerg, Virtual Inertia Emulating in Power Electronic, Chapter 19 in Control of Power Electronic Converters and Systems, Elsevier, 2021, https://doi.org/10.1016/B978-0-12-819432-4.00019-6.
[15] Y. Zhu, Z. Fang, F. Wang, B. Liu, Y. Zhao, A wireless load sharing strategy for islanded microgrid based on feeder current sensing, IEEE Trans. Power Electron. 30 (12) (2015) 6706–6719.

[16] Y.W. Li, C.-N. Kao, An accurate power control strategy for power-electronics-interfaced distributed generation units operating in a low-voltage multibus microgrid, IEEE Trans. Power Electron. 24 (12) (2009) 2977–2988.
[17] H. Han, Y. Liu, Y. Sun, M. Su, J.M. Guerrero, An improved droop control strategy for reactive power sharing in islanded microgrid, IEEE Trans. Power Electron. 30 (6) (2015) 3133–3141.
[18] J. Schiffer, T. Seel, J. Raisch, T. Sezi, Voltage stability and reactive power sharing in inverter-based microgrids with consensus-based distributed voltage control, IEEE Trans. Control Syst. Technol. 24 (1) (2016) 96–109.
[19] J. C Vasquez, J.M. Guerrero, M. Savaghebi, J. Eloy-Garcia, R. Teodorescu, Modeling, analysis, and design of stationary-reference-frame droop-controlled parallel three-phase voltage source inverters, IEEE Trans. Ind. Electron. 60 (4) (2013) 1271–1280.
[20] P. Piagi, R.H. Lasseter, Autonomous control of microgrids, In Proc. IEEE Power Eng. Soc. Gen. Meeting (2006) 8.
[21] K. Yu, Q. Ai, S. Wang, J. Ni, T. Lv, Analysis and optimization of droop controller for microgrid system based on small-signal dynamic model, IEEE Trans. Smart Grid 7 (2) (2016) 695–705.
[22] M. Rasheduzzaman, J.A. Mueller, J.W. Kimball, An accurate small-signal model of inverter-dominated islanded microgrids using dq reference frame, IEEE J. Emerg. Sel. Topics Power Electron. 2 (4) (2014) 1070–1080.
[23] S. Leitner, M. Yazdanian, A. Mehrizi-Sani, A. Muetze, Small-signal stability analysis of an inverter-based microgrid with internal model-based controllers, IEEE Trans. Smart Grid 9 (5) (2018) 5393–5402.
[24] M. Kabalan, P. Singh, D. Niebur, Nonlinear Lyapunov stability analysis of seven models of a dc/ac droop controlled inverter connected to an infinite bus, IEEE Trans. Smart Grid 10 (1) (2019) 772–781.
[25] L. Zhang, L. Harnefors, H. Nee, Power-synchronization control of grid connected voltage-source converters, IEEE Trans. Power Syst. 25 (2) (2010) 809–820.
[26] V. Mariani, F. Vasca, J.C. Vásquez, J.M. Guerrero, Model order reductions for stability analysis of islanded microgrids with droop control, IEEE Trans. Ind. Electron. 62 (7) (2015) 4344–4354.
[27] M. Rasheduzzaman, J.A. Mueller, J.W. Kimball, Reduced-order small-signal model of microgrid systems, IEEE Trans. Sustain. Energy 6 (4) (2015) 1292–1305.
[28] S. D'Arco, J.A. Suul, O.B. Fosso, Automatic tuning of cascaded controllers for power converters using eigenvalue parametric sensitivities, IEEE Trans. Ind. Appl. 51 (2) (2015) 1743–1753.

Grid connected converters: stability assessment and sensitivity analysis

4

4.1 Stability analysis methods: an overview

A grid connected converter (GCC) must be designed to assure a stable operation; therefore, stability analysis of this system is a vital issue, which can be handled using different methods in *time* and *frequency* domains. This section provides a short review on some stability analysis methods and their applications.

4.1.1 Frequency domain

In frequency domain, the *impedance* (*admittance*) *method* is a simple, popular, and powerful method to analyze the GCC stability, which is based on the frequency characteristics of the system [1–5]. In this method, first the frequency characteristics of the GCC output impedance are determined. Then they are investigated to satisfy a stability criterion, for example, the Nyquist and Middlebrook criteria. According to the impedance method, when the phase of the output impedance of the GCC is passive, or the phase variation range is within ±90 (180) degrees, the resistive component (the real part of the impedance/admittance) is always positive and it satisfies an impedance-based stability criterion to ensure stability. Three types of passivity-based design examples are well discussed in Refs. [6–12].

The impedance method is based on the study of system frequency characteristics. There are mainly three methods to determine the frequency characteristics [13–15]: (1) *analytical method* (by the state variable approach), (2) *numerical method* using a systematic formulation (such as the modified nodal analysis) [16–18], and (3) *simulator-based frequency scan method* (which may calculate numerical responses for a given excitation signal by the Fast Fourier Transformation-FFT) [19].

Analytical method uses an analytical approach based on the GCC dynamic model (e.g., state-space model) to obtain the output impedance/admittance, and its frequency characteristics. Unlike the analytical method, the numerical method is more useful for a large and complex system, because it is systematic and applicable to any linear system [20]. The numerical method procedure consists of three steps: (1) decomposition of the system into components of circuit and control elements, (2) modeling of each component for its frequency characteristics, and (3) formulation and solution for obtaining the component characteristics.

Analytical and numerical methods can calculate frequency characteristics of the system precisely up to the switching frequency. However, they have difficulties to process a digitally controlled system. The characteristics of such a system can be easily determined using a general-purpose circuit simulator (third method) [20]. In this

Grid Connected Converters. https://doi.org/10.1016/B978-0-323-99902-1.00007-0

method, a perturbed excitation is applied at a desired frequency and the response waveform is analyzed by the FFT. The numerical methods mostly use general-purpose simulation algorithms [16−18,21]. An extension of the method is used for stability analysis of a GCC in next section.

In addition to impedance methods which are explained in Section 4.1.3, the frequency domain-based stability analysis class uses several powerful graphical tools such as the *closed-loop eigenvalues/poles graph* or *Root Locus*, *Bode*, and *Nyquist* diagrams.

4.1.2 Time domain

Since almost all GCCs are nonlinear, it is difficult to analyze the frequency characteristics of the output impedance/admittance for the stability analysis. It is noteworthy that the impedance method is only applicable to a linear GCC system.

In time domain, it is usually desirable to analyze the system transient stability. An important time-domain stability analysis approach is based on the *Poincaré map* method, which is reduced to analysis of the system transition matrix or *sensitivity matrix* for perturbations along an operating trajectory. This method provides a useful alternative, even for nonlinear systems [22−25]. However, it cannot be considered as a measurement-based stability analysis method.

In time domain, the system is analyzed to get a sort of transition matrix or a sensitivity matrix for the state variables during one periodic cycle. The matrix includes complete information about interactions between state variables. This time domain method does not need application of the averaging method and the system can be considered in its original nonlinear form. According to the eigenvalue analysis, it is possible to investigate stability which is related to the nonlinear phenomena including harmonics.

4.1.3 Impedance-based stability analysis methods

For easier understanding the impedance-based stability analysis methods, consider a single-phase inverter with an LCL filter shown in Fig. 4.1 which is controlled with

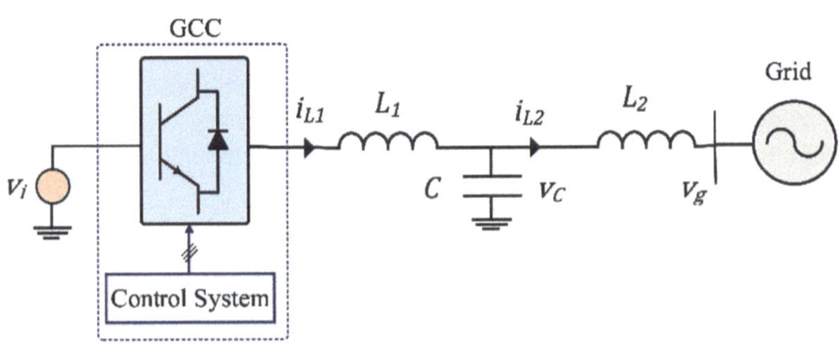

Figure 4.1 A GCC system with an LCL filter.

a digital control scheme. In frequency domain-based stability analysis, the frequency characteristics of the output admittance of the system must be calculated.

Here, the basic principle of the impedance method for a current type is briefly reviewed for stability analyzing the shown GCC in Fig. 4.1 [26]. The linearized equivalent circuit of the GCC is shown in Fig. 4.2A. The GCC is expressed with the

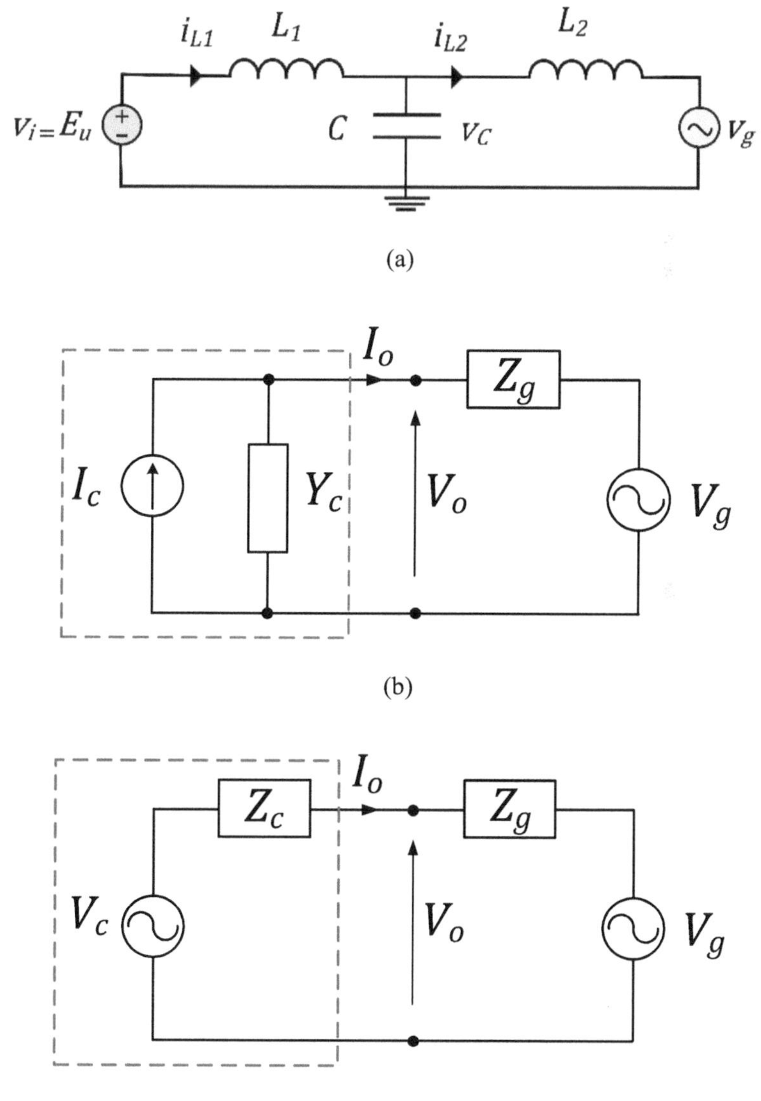

Figure 4.2 Equivalent linear model of the GCC; (A) electric circuit, and (B) Norton model for current-type, and (C) Thevenin model for voltage-type.

Norton's circuit of the output admittance $Y_c(s)$ and the GCC control current $I_C(s)$. The grid system is expressed with the Thevenin's circuit of the equivalent system voltage $V_g(s)$ and the line impedance $Z_g(s)$. Therefore, the GCC can be expressed with an equivalent model as shown in Fig. 4.2B. The output terminal voltage to the grid system $V_o(s)$ becomes

$$V_o(s) = \frac{V_g(s) + Z_g(s)I_c(s)}{1 + Z_g(s)Y_c(s)}. \tag{4.1}$$

By applying the Nyquist criterion to the denominator $1 + Z_g(s)Y_c(s)$, it can be seen that the system is stable when the phase of $Z_g(s)Y_c(s)$ is less than 180° or $|Z_g(s)Y_c(s)| < 1$ at the cross frequency of 180°. It is equivalent with that the phase difference of $Y_c(s)$ and $Y_g(s) = 1/Z_g(s)$ is less than 180° or $|Y_c| < |Y_g|$ at the cross frequency.

For stability analysis based on the impedance method for voltage-type model, frequency characteristics of the output impedance (or admittance) are also needed. For instance, consider the Thevenin's equivalent circuit of the grid side (Fig. 4.2C). For small-signal analysis, the current $I_o(s)$ flowing from the GCC to the grid is

$$I_o(s) = \frac{V_o(s) - V_g(s)}{Z_c(s) + Z_g(s)} = \frac{V_o(s) - V_g(s)}{Z_g(s)} \frac{1}{1 + \dfrac{Z_c(s)}{Z_g(s)}} \tag{4.2}$$

For system stability analysis, it can be assumed that the grid voltage is stable when unloaded and the current is stable when powered from an ideal source. In this case both $V_c(s) - V_g(s)$ and $1/Z_g(s)$ are stable, such that stability of the current depends on the stability of the second term on the right hand side of Eq. (4.2). The system is stable when $Z_c(s)Y_g(s)$ satisfies the Nyquist criterion.

4.1.3.1 Analytical analysis method

For the given GCC, frequency characteristics of the output admittance of the system can be derived analytically. An LCL filter of L_1, C, L_2 is used at the GCC output to convert the voltage input v_i to the current output i_{L_2} and to reduce the harmonic components. The main circuit is nonlinear and of high-order dynamics, while the averaged linearized circuit is a low-order system. The state vector is considered as $\boldsymbol{x} = [\, i_{L_1} \quad i_{L_2} \quad v_C \,]^T$ and its sampled form at $t = iT$ is presented as $\boldsymbol{x}[i]$, where T is a sampling time interval. Its input voltage $v_i[i] = Eu[i]$ is generated according to its input $u[i]$ where E is DC voltage of the inverter source. Discrete state equations for the output current of L_2 $(y[i] = i_{L_2}[i])$ are as follows.

$$\begin{aligned} \boldsymbol{x}[i+1] &= \boldsymbol{A}_d \boldsymbol{x}[i] + \boldsymbol{b}_d\, u[i] + \boldsymbol{h}_d\, v_g[i] \\ y[i] &= \boldsymbol{c}_d \boldsymbol{x}[i] \end{aligned} \tag{4.3}$$

In Eq. (4.2), "d" shows the discrete type. A digital control scheme is used for the GCC as shown in Fig. 4.3. The feedback (FB) controller $(\boldsymbol{k}_{fb})$ and compensator $(G[z])$ in Fig. 4.3 are often designed to track the output current in a desired sinusoidal waveform, and GCC stabilizing using the control theory. According to Fig. 4.3, in frequency domain, the state equations of the system are as follows.

$$\begin{aligned} s\boldsymbol{X}(s) &= \boldsymbol{A}\boldsymbol{X}(s) + \boldsymbol{b}U(s) + \boldsymbol{h}V_g(s) \\ U(s) &= -\boldsymbol{k}_{fb}\, S_m\boldsymbol{X}(s) + G\left(i_r^* - \boldsymbol{c}\, S_m\boldsymbol{X}(s)\right) \end{aligned} \tag{4.4}$$

where S_m is the sample and hold (S/H) function, $\boldsymbol{k}_{fb}$ is the FB gain vector, and i_r^* is the input reference signal.

From Eq. (4.4), the following equation can be obtained.

$$\boldsymbol{X}(s) = \left(s\boldsymbol{I} - \boldsymbol{A} + \boldsymbol{b}\boldsymbol{k}_{fb}S_m + \boldsymbol{b}G\boldsymbol{c}S_m\right)^{-1}\left(\boldsymbol{b}Gi_r^* + \boldsymbol{h}V_g(s)\right) \tag{4.5}$$

Then, the output admittance Y_c is derived as follows.

$$Y_c = -\frac{I_{L_2}}{V_g} = -\boldsymbol{c}\left(s\boldsymbol{I} - \boldsymbol{A} + \boldsymbol{b}\boldsymbol{k}_{fb}\, S_m + \boldsymbol{b}G\boldsymbol{c}\, S_m\right)^{-1}\boldsymbol{h} \tag{4.6}$$

Using Eq. (4.5), the admittance frequency characteristics of the given GCC can be analytically studied via the magnitude and phase plots of the relevant Bode diagrams.

4.1.3.2 Frequency-analysis simulator

As the GCC system model becomes more complex, the analytical method is difficult to apply, not only because of complexity of handling the system equations but also because of difficulty in selection of state variables due to inductor cutset and capacitor

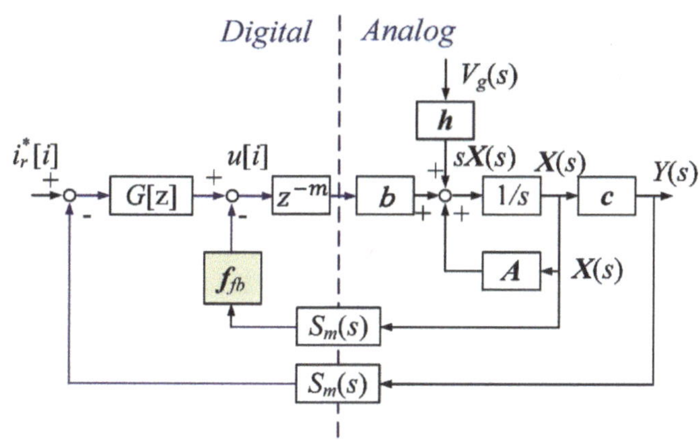

Figure 4.3 Feed-back control block diagram of a GCC.

loop restrictions. To overcome this challenge, a generalized frequency analysis simulator (method) is addressed in Section 4.3, which provides an alternative solution for analyzing the frequency characteristics of GCCs.

4.1.3.3 Time-domain simulation-based frequency-scan method

Output admittances of a GCC system can be computed also by a time-domain simulator. For this purpose, a small voltage perturbation (ΔV_p) is directly added to the output of the GCC and then the associated current response (ΔI_p) is computed. This method can be summarized in the following steps:

1. The given GCC circuit is analyzed for the steady-state response without purturbation (I_p).
2. The perturbed circuit with a small excitation signal ΔV_p at a frequency f_i is analyzed for the complete steady-state response $(I_p + \Delta i_p)$, where its period is $T_i = 1/f_i$.
3. The admittance $Y_c(f_i)$ for the frequency f_i is calculated from dividing the difference signal $\Delta I_p(f_i)$ by the perturbation voltage $\Delta V(f_i) = \Delta V_p$ using the FFT.

$$Y_c(f_i) = \frac{\Delta I_p(f_i)}{\Delta V(f_i)} \tag{4.7}$$

4. The above steps are repeated for necessary f_i samples, because the system is nonlinear and its operating point may slightly change.

A detailed application example for this method is given in Section 4.3.

4.1.3.4 Passivity-based stability analysis

The Nyquist criterion in the impedance method essentially checks if the cross point of eigenvalues of $Z_g Y_c$ at the negative real axis exceeds -1 or not. The line impedance Z_g is passive and its phase stays between $\pm 90°$. If the phase of the output admittance Y_c of the positive sequence (and the negative sequence) can be kept between $\pm 90°$, there is no cross point at the negative real axis and the system is always stable even if the admittance becomes large. It is sufficient when the real part of the admittance of the GCC is positive and passive to ensure stability.

$$\mathrm{Re}\{Y_c(s)\} > 0 \tag{4.8}$$

4.1.4 Poincaré map-based stability analysis method

Consider a nonlinear model of a dynamic system as Eq. (4.9), where $\boldsymbol{x}$ is a state-space variable vector and t is time.

$$\dot{\boldsymbol{x}} = \boldsymbol{f}(t, \boldsymbol{x}) \tag{4.9}$$

The $\boldsymbol{x}_T$ ($\boldsymbol{x}$ at $t = T$) shows the integrated value for one period T from an initial value $\boldsymbol{x}_0$ at $t = 0$ to $t = T$:

$$\boldsymbol{x}_T = \int_0^T \boldsymbol{f}(\tau, \boldsymbol{x}(\tau))d\tau + \boldsymbol{x}_0 \tag{4.10}$$

The dynamic behavior of state vector can be described in a Poincaré map, as depicted in Fig. 4.4, where $\boldsymbol{x}_0$ starts from the Poincaré section at $t = 0$ and comes back to $\boldsymbol{x}_T$ in the section at $t = T$. The periodic steady-state condition is that the two boundary values at $t = 0$ and $t = T$ are equal and can be described as follows

$$\boldsymbol{x}_T = \boldsymbol{P}(x_0). \tag{4.11}$$

where $\boldsymbol{P}(\cdot)$ is mapping relation from $\boldsymbol{x}_0$ to $\boldsymbol{x}_T$, and generally it is a nonlinear function with respect to $\boldsymbol{x}_0$. It is possible to analyze stability of the system by this mapping relation. When all absolute eigenvalues of $\frac{\partial x_T}{\partial x_0}$ at a periodic steady-state condition are less than one, the system is stable because variations of the initial values are not magnified.

4.2 Stability analysis using closed-loop eigenvalues/ poles graph

The graph of eigenvalues (or poles) provides a simple graphical tool, which is widely used for stability and sensitivity analysis of the linearized GCC systems. To clarify the usefulness and importance of this graphical tool, let to recall the proposed $P - \omega$ and $Q - V$ models of the GCC control systems in Chapter 3 and analyze their stability.

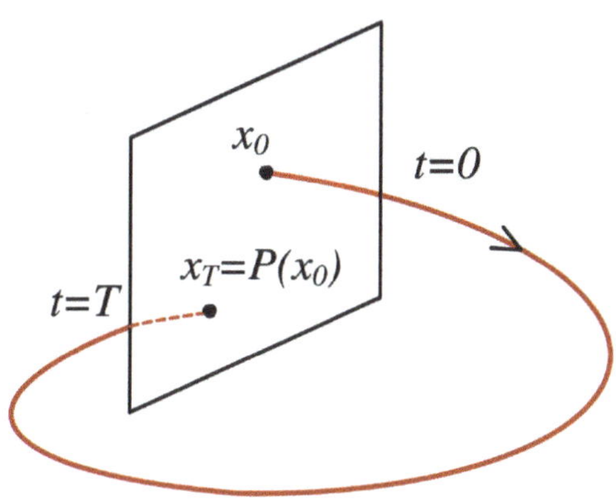

Figure 4.4 Poincaré map and boundary value condition.

4.2.1 Stability analysis of $P - \omega$ model

In Chapter 3 (Section 3.2), a modeling method to study the active power and frequency response behavior of a typical GCC (Fig. 3.1) is proposed. This method uses unified formulas to obtain the state-space model, and this helps us to interpret the relevant dynamics, mathematically. Here, the closed-loop poles analyses of the GCC based on the developed model using Root-Locus graph is presented. The results illustrate the movement and correlation of dominant poles and their impact on the dynamic performance. These common features are valid for most investigated GCCs, and the exceptional conditions are well specified. The revealed analytical solutions of the dominant poles and time constants help us to understand how these dominant poles are determined by the key parameters, and thus this facilitates the parameters design.

For simplicity, the proposed state-space model of the mention GCC with and without damping is rewritten below:

$$\begin{cases} \dot{\boldsymbol{x}} = \boldsymbol{A}\boldsymbol{x} + \boldsymbol{B}u + \boldsymbol{E}w \\ \boldsymbol{y} = \boldsymbol{C}\boldsymbol{x} \end{cases} \tag{4.12}$$

$$u = \Delta P_0;\ w = \Delta\omega_g;\ \boldsymbol{y} = [\Delta\omega_m \quad \Delta P_{\text{out}}]^T$$

where the parameters are defined in Chapter 3, and the state variables and coefficient parameters for case of no damping are

$$\boldsymbol{x} = [\Delta\omega_m \quad \Delta P_{\text{out}}]^T$$

$$\boldsymbol{A} = \begin{bmatrix} -\dfrac{k_p}{J\omega_0} & -\dfrac{1}{J\omega_0} \\ K & 0 \end{bmatrix};\ \boldsymbol{B} = \begin{bmatrix} \dfrac{1}{J\omega_0} & 0 \end{bmatrix}^{\mathrm{T}};\ \boldsymbol{E} = [0 \quad -K]^{\mathrm{T}};\ \boldsymbol{C} = \boldsymbol{I}_2.$$

The state-space model coefficient parameters for case of using damping control loop, and ignoring PLL dynamics are

$$\boldsymbol{x} = [\Delta\omega_m \quad \Delta P_{\text{out}}]^T$$

$$\boldsymbol{A} = \begin{bmatrix} -\dfrac{k_p + D}{J\omega_0} & -\dfrac{1}{J\omega_0} \\ K & 0 \end{bmatrix};\ \boldsymbol{B} = \begin{bmatrix} \dfrac{1}{J\omega_0} & 0 \end{bmatrix}^{\mathrm{T}};\ \boldsymbol{E} = \begin{bmatrix} \dfrac{D}{J\omega_0} & -K \end{bmatrix}^{\mathrm{T}};\ \boldsymbol{C} = \boldsymbol{I}_2.$$

Finally, for case of using damping control, considering PLL dynamics, the following states vector and matrix/vector coefficients can be obtained:

$$\boldsymbol{x} = \begin{bmatrix} \Delta\omega_m & \Delta P_{\text{out}} & \Delta x_{1_{pll}} & \Delta x_{2_{pll}} \end{bmatrix}^T$$

$$\boldsymbol{A} = \begin{bmatrix} -\dfrac{k_p}{J\omega_0} & -\dfrac{1}{J\omega_0} & \dfrac{DK^*_{P_{pll}}}{J} & \dfrac{D}{J\omega_0 T_{i_{pll}}} \\ K & \mathbf{0} & \mathbf{0} & \mathbf{0} \\ \mathbf{0} & \mathbf{0} & -\omega_0 K^*_{P_{pll}} & -\dfrac{1}{T_{i_{pll}}} \\ \mathbf{0} & \mathbf{0} & \omega_0 K^*_{P_{pll}} & \mathbf{0} \end{bmatrix}; \ \boldsymbol{B} = \begin{bmatrix} \dfrac{1}{J\omega_0} & 0 & 0 & 0 \end{bmatrix}^T$$

$$\boldsymbol{E} = \begin{bmatrix} 0 & -K & 1 & 0 \end{bmatrix}^T; \ \boldsymbol{C} = \begin{bmatrix} \boldsymbol{I}_2 & \mathbf{0}_{2\times 2} \end{bmatrix}.$$

Let to assign the dominant poles in desired locations as follows:

$$\lambda_{1,2} = \omega_n e^{j(\pi \pm \cos^{-1}\zeta)} = \sqrt{\frac{K}{J\omega_0}} e^{j(\pi \pm \cos^{-1} 0.9)} \tag{4.13}$$

The resulted parameters are shown in Table 3.1, which are normalized with individual power ratings to facilitate parameter design of the GCC as follows.

$$M^* = J\omega_0^2/S_{\text{base}}, \ k_p^* = k_p\omega_0/S_{\text{base}}, \ D^* = D\omega_0/S_{\text{base}} \tag{4.14}$$

The closed-loop poles of the model are shown in Fig. 4.5. It is clear that $\lambda_{1,2}$ of the dedicated damping methods (with and without PLL), which reflect the dynamic control feature of GCC, are placed to the desired location according to Eq. (4.13), whereas those of the without damping method have small damping ratio. This indicates that the damping method provides better dynamic performance than the without damping method. It is noteworthy that with damping method using PLL, the model have additional poles at λ_3 and λ_4. The additional poles in case of damping method, which introduced by the PLL, are quite close to the origin, thus they may affect the dynamics of the proposed model of GCC.

In [27], it is claimed that assuming K is fixed, as long as ω_n of $\lambda_{1,2}$ is kept the same, the equivalent inertia remains constant as indicated by Eq. (4.13). However, as long as the additional poles in the GCC model are nondominant poles, their impact is quite slight. As long as the nondominant poles of the model are negligible, no matter which

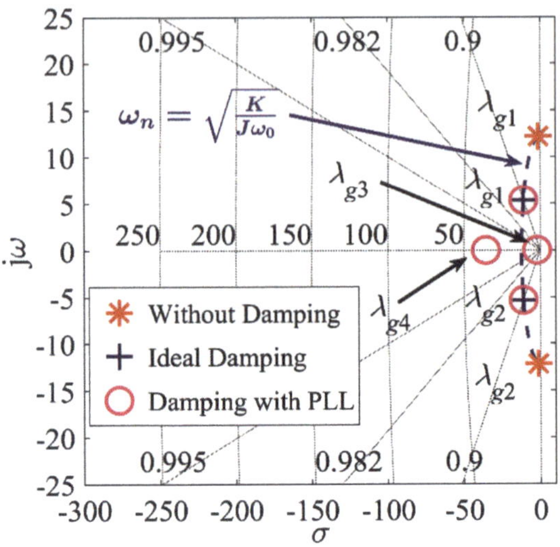

Figure 4.5 All closed-loop poles.

damping control method is applied, the model is of second order, and its respective time constant τ can be quite different as shown in Eq. (4.15). The reason for this reduced order is discussed in Ref. [28].

$$\tau = \frac{1}{\left|\mathrm{Re}(\lambda_{1,2})\right|} = \frac{1}{\zeta}\sqrt{\frac{J\omega_0}{K}} \tag{4.15}$$

Generally, small τ and large ζ are preferred for fast and nonoscillatory power response in the GCC. From Eq. (4.15) it can be seen that large J, ζ, and small X are preferable. As shown in Fig. 4.5, large ζ is available through a proper closed-loop assignment. It is noteworthy that the output impedance of GCC, X, should not be too small, and it is preferred to be inductive in order to ensure the decoupling between active and reactive powers. As shown in Fig. 4.5, this important conclusion is valid for both dominant and nondominant poles. The GCC poles illustrate how the GCC is synchronized to the grid utility (an infinite bus). As it is discussed in the sensitivity analysis section, small τ_g is still preferred for fast power response.

4.2.2 Stability analysis of Q-V model

Here, stability of the obtained two addressed *Q-V* based GCC models (i.e., *VI*- and *Q*-based control schemes) are studied using the closed-loop poles graph. The analyses presented here are aiming at (1) concluding generalized intermode laws of the *Q-V* control loop of GCCs and (2) comparing the different pole distribution of the two control schemes.

Here for simplicity, the proposed state-space model of the GCC for both VI-based and Q-based control schemes (Chapter 3) is rewritten:

$$\begin{cases} \dot{x} = \boldsymbol{A}\boldsymbol{x} + \boldsymbol{B}u + \boldsymbol{E}w \\ \boldsymbol{y} = \boldsymbol{C}\boldsymbol{x} + \boldsymbol{D}u + \boldsymbol{F}w \end{cases} \tag{4.16}$$

$$u = \Delta Q_0;\ w = \Delta V_g;\ \boldsymbol{y} = \begin{bmatrix} \Delta E & \Delta Q_{\text{out}} \end{bmatrix}^T \tag{4.17}$$

where all parameters are defined in Chapter 3. For inner VI-based control scheme, the remained parameters are determined as follows

$$\boldsymbol{x} = \begin{bmatrix} \Delta Q_{\text{out}f} \\ K_{pv}\Delta\varepsilon_{vd} - \Delta i_{\text{out}d} \\ K_{pv}\Delta\varepsilon_{vq} - \Delta i_{\text{out}q} \end{bmatrix};\ \boldsymbol{A} = \boldsymbol{A}_o + \boldsymbol{H}_o\boldsymbol{H}_x^{-1}\boldsymbol{A}_x;\ \boldsymbol{B} = \boldsymbol{H}_o\boldsymbol{H}_x^{-1}\boldsymbol{B}_x;\ \boldsymbol{E} = \boldsymbol{H}_o\boldsymbol{H}_x^{-1}\boldsymbol{E}_x$$

$$\boldsymbol{C} = \begin{bmatrix} \dfrac{\left(\boldsymbol{H}_x^{-1}\boldsymbol{A}_x\right)_{(1)}}{K_{QE}} \\ \left(\boldsymbol{H}_x^{-1}\boldsymbol{A}_x\right)_{(1)} \end{bmatrix};\ \boldsymbol{D} = \begin{bmatrix} \dfrac{\left(\boldsymbol{H}_x^{-1}\boldsymbol{B}_x\right)_{(1)}}{K_{QE}} \\ \left(\boldsymbol{H}_x^{-1}\boldsymbol{B}_x\right)_{(1)} \end{bmatrix};\ \boldsymbol{F} = \begin{bmatrix} \left(\left(\boldsymbol{H}_x^{-1}\boldsymbol{E}_x\right)_{(1)} - K_{QV}\right)/K_{QE} \\ \left(\boldsymbol{H}_x^{-1}\boldsymbol{E}_x\right)_{(1)} \end{bmatrix}$$

$$\boldsymbol{A}_o = \begin{bmatrix} -\dfrac{1}{T_{fq}} & 0 & 0 \\ 0 & -\dfrac{1}{T_{iv}} & 0 \\ 0 & 0 & -\dfrac{1}{T_{iv}} \end{bmatrix};\ \boldsymbol{H}_o = \begin{bmatrix} \dfrac{1}{T_{fq}} & 0 & 0 \\ 0 & -\dfrac{1}{T_{iv}} & 0 \\ 0 & 0 & -\dfrac{1}{T_{iv}} \end{bmatrix};$$

$$\boldsymbol{H}_x = \begin{bmatrix} 1 & 0 & K_{QE}\omega_0 L_{ls} & -K_{QE} & 0 \\ 0 & 1/K_{pv} & -\omega_0 L_{ls} & 1 & 0 \\ 0 & \omega_0 L_{ls} & 1/K_{pv} & 0 & 1 \\ 0 & R_{\text{line}} & -\omega_0 L_{\text{line}} & -1 & 0 \\ 0 & \omega_0 L_{\text{line}} & R_{\text{line}} & 0 & -1 \end{bmatrix};\ \boldsymbol{A}_x = \begin{bmatrix} 0 & 0 & 0 \\ -1/k_q & -1/K_{pv} & 0 \\ 0 & 0 & -1/K_{pv} \\ 0 & 0 & 0 \\ 0 & 0 & 0 \end{bmatrix}$$

$$\boldsymbol{B}_x = \begin{bmatrix} 0 & 1/k_q & 0 & 0 & 0 \end{bmatrix}^T; \ \boldsymbol{E}_x = \begin{bmatrix} K_{QV} & 0 & 0 & -\cos\delta_0 & \sin\delta_0 \end{bmatrix}^T$$

and for the Q-based control scheme:

$$\boldsymbol{x} = \begin{bmatrix} \Delta Q_{\text{outf}} \\ K_{pq}\Delta\varepsilon_Q - \Delta E \end{bmatrix}; \ \boldsymbol{A} = \begin{bmatrix} -\dfrac{(1 + K_{QE}K_{pq})}{T_{fq}} & -\dfrac{K_{QE}}{T_{fq}} \\ \dfrac{K_{pq}}{T_{iq}} & 0 \end{bmatrix}; \ \boldsymbol{B} = \begin{bmatrix} \dfrac{K_{QE}K_{pq}}{T_{fq}} \\ -\dfrac{K_{pq}}{T_{iq}} \end{bmatrix}$$

$$\boldsymbol{E} = \begin{bmatrix} \dfrac{(K_{QV} - K_{QE}k_qK_{pq})}{T_{fq}} \\ \dfrac{k_qK_{pq}}{T_{iq}} \end{bmatrix}; \ \boldsymbol{C} = \begin{bmatrix} -K_{pq} & -1 \\ -K_{QE}K_{pq} & -K_{QE} \end{bmatrix}; \ \boldsymbol{D} = \begin{bmatrix} K_{pq} \\ K_{QE}K_{pq} \end{bmatrix}$$

$$\boldsymbol{F} = \begin{bmatrix} -k_qK_{pq} \\ K_{QV} - K_{QE}k_qK_{pq} \end{bmatrix}$$

For the given case study example in Section 3.2, a GCC-based DG in connected operation mode is considered [29,30]. The equivalent output reactance (X^*) is adjusted to 0.3 pu. Fig. 4.6 illustrates the dominant pole locations of Q-based control scheme, with the nominal control parameters in Table 3.2 (Fig. 4.6A) and changing the gain of K_{pv} (Fig. 4.6B). Poles are denoted by λ_k ($k = 1-2$). By changing the parameters of the system, the poles may move and the root locus can be plotted on the same plane as shown in Fig. 4.6B.

4.3 A frequency characteristics-based stability assessment

This section addresses a frequency characteristics-based stability assessment method for a GCC with a digital control system considering parameter tolerances. The frequency characteristics are computed numerically using the modified nodal analysis [16]. The proposed method which is applicable for any averaged linear time-invariant (LTI) GCC system is supplemented by simulations and experiments.

Most of so far frequency characteristics-based stability analysis methods have difficulties to process a digitally controlled GCC system [13–15]. The characteristics of such a system can be computed using a generalized circuit simulator. A perturbed excitation is applied at a desired frequency, and the response waveform is analyzed by the FFT [19].

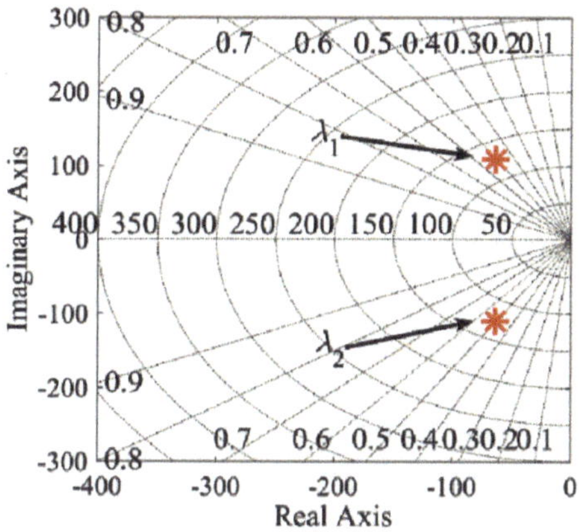

(a)

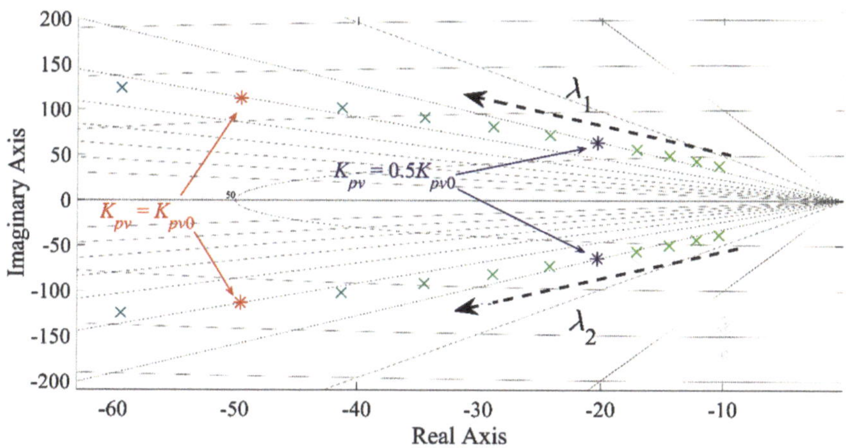

(b)

Figure 4.6 Closed-loop pole locations of inner Q-based control; (A) dominant poles for nominal parameters and (B) root locus graph for changing the gain K_{pv}.

Here, an ideal digital sampler is approximately modeled with a transfer function, which is applicable to a linear equivalent GCC circuit that is averaged for switching operations. In this case, an efficient frequency analysis is applicable; while it is an approximate, but precise enough up to the switching frequency. On the other hand, a digital control system often has a control time delay that affects frequency characteristics of the system and thus stabilities [31−33]. Such delay effects are easily expressed using a time delay transfer function and can be included in the frequency characteristics analysis method.

A GCC often has inductor cutsets and capacitor loops that need a special reduction process to adjust the number of state variables. Additionally, when elements parameters between three phases are asymmetrical, the state equations become too complex to handle, even if the circuit structure is symmetrical. Here, a generic formulation and simulation method is used for frequency analysis of the averaged LTI grid system [16,18,21]. A *stamp* is defined for any circuit element, which only needs node-connection and element-model information. This method is general for numerical processing with the objective system being expressed using a prepared library of element models. When the simulator library does not have a model for a new element, it can be easily generated and to be added to the library. This also proposes a new frequency analysis method that considers parameter purturbation. The tolerance analysis can be done based on the Monte Carlo method [34–36].

4.3.1 Impedance matrix-based formulation

Consider a general structure for a GCC system as shown in Fig. 4.7. An equivalent model can be derived as shown in Fig. 4.2B. It is assumed that the GCC is stable when it is connected to an ideal voltage source. The GCC is expressed using Norton's circuit of the GCC output admittance $\boldsymbol{Y}_c(s)$ and the converter control current $\boldsymbol{I}_c(s)$ [2]. The grid side is expressed using the Thevenin's circuit of the equivalent system voltage $\boldsymbol{V}_g(s)$ and the line impedance $\boldsymbol{Z}_g(s) = Z_g(s)\,\boldsymbol{I}$ that is assumed to be symmetrical, where $\boldsymbol{I}$ is the identity matrix. The GCC output voltage $\boldsymbol{V_o}(s)$ becomes as:

$$\boldsymbol{V_o}(s) = \{\boldsymbol{I} + \boldsymbol{Z}_g(s)\boldsymbol{Y_c}(s)\}^{-1}\{\boldsymbol{V}_g(s) + \boldsymbol{Z}_g(s)\boldsymbol{I}_c(s)\}, \tag{4.18}$$

where

$$\boldsymbol{Y_c} = \begin{bmatrix} Y_{uu} & Y_{vu} & Y_{wu} \\ Y_{uv} & Y_{vv} & Y_{wv} \\ Y_{uw} & Y_{vw} & Y_{ww} \end{bmatrix}; \boldsymbol{Z}_g = \begin{bmatrix} Z_g & 0 & 0 \\ 0 & Z_g & 0 \\ 0 & 0 & Z_g \end{bmatrix} \tag{4.19}$$

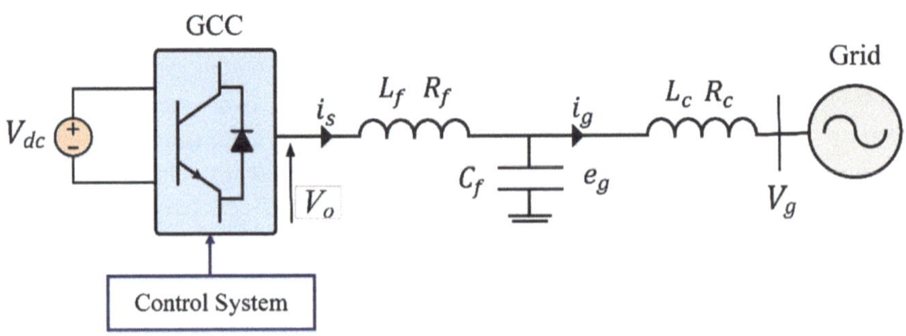

Figure 4.7 A typical GCC.

Symmetry conditions of the main circuit in three phases u, v, and w give

$$Y_{uu} = Y_{vv} = Y_{ww} = Y_g;\ Y_{uv} = Y_{vu} = Y_{uw} = Y_{wu} = Y_{vw} = Y_{wv} = Y_c \tag{4.20}$$

Eq. (4.18) can be diagonalized by similarity transform based on eigenvectors of the admittance matrix $\boldsymbol{Y}_c(s)$ and it is decoupled into three independent components, which correspond to the symmetrical coordinate components when the main circuit is symmetrical. When any phase of the eigenvalues of $\boldsymbol{Z}_g(s)\boldsymbol{Y}_c(s)$ in Eq. (4.18) becomes 180°, its stability is assessed to determine if its magnitude is less than unit according to the Nyquist stability criterion.

4.3.2 Computation of frequency characteristics

4.3.2.1 Preliminaries

In practice, the frequency characteristic can be obtained by measurement of the response current for a small-signal excitation voltage v_u^p using a frequency analyzer, as shown in Fig. 4.8 where the excitation signal is inserted in series by a transformer. Similarly, it is possible to calculate the frequency characteristic by repetitive numerical computation of the response current i_u^p for a small-signal excitation voltage v_u^p at every sampled frequency using a simulator. However, much effort and time are needed for the above repetitive manual operations, and an alternative simple and efficient method for a linear system is required.

Here, a systematic formulation method is introduced for frequency analysis of the *linearized GCC*. For this purpose, only node-connection and element-model information of the circuit elements is required. In order to easily understanding the basic idea, consider the frequency analysis of a trivial RLC circuit shown in Fig. 4.9A, consists of four elements as listed in Table 4.1. The circuit equations are formulated from these

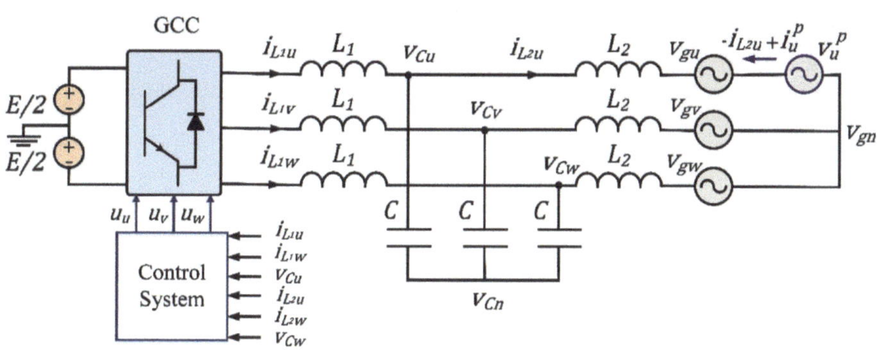

Figure 4.8 Measurement of frequency characteristics with perturbation.

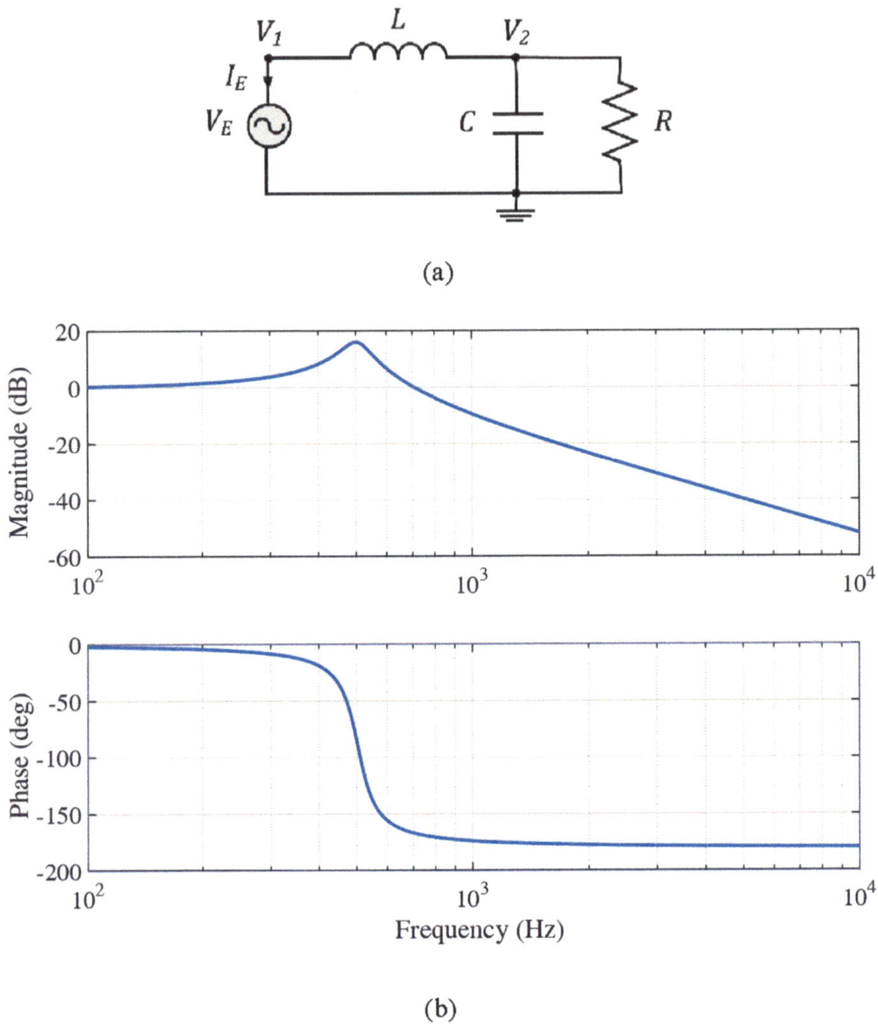

Figure 4.9 Frequency analysis example of a trivial RLC circuit; (A) RLC circuit and (B) Frequency characteristics of the output voltage.

Table 4.1 Element type and node-connection information.

Element type	Node-from	Node-to	Value
L	1	2	10 mH
C	2	0	100 μF
R	2	0	1 Ω
V_E	1	0	1.0 p.u.

element type and node connection information by the modified nodal analysis where V_1 and V_2 are node voltages and I_E is current of the voltage source.

$$\begin{bmatrix} \frac{1}{j\omega L} & -\frac{1}{j\omega L} & 1 \\ -\frac{1}{j\omega L} & \frac{1}{j\omega L}+j\omega C+\frac{1}{R} & 0 \\ 1 & 0 & 0 \end{bmatrix} \begin{bmatrix} V_1 \\ V_2 \\ I_E \end{bmatrix} = \begin{bmatrix} 0 \\ 0 \\ 1 \end{bmatrix} \tag{4.21}$$

This example is trivial and the equations can be formulated manually. However, they can be formulated systematically from type and node connection information of each element according to the *stamp* patterns of the modified nodal analysis [16–18].

To get a transfer function characteristic for the RLC circuit (Fig. 4.9A) from input terminal to the output voltage, a unit voltage source is applied at the input. The computed frequency characteristic curve is shown in Fig. 4.9B. Contribution of the voltage source to a circuit equation can be expressed with a stamp. The RLC circuit contains two Kirchhoff's current law (KCL) equations at the two nodes and one equation of the voltage source characteristic. They can be formulated from the given element and connection node data, according to the stamps of the elements.

4.3.2.2 Proposed method

The proposed analysis method consists of three processing steps: (1) decomposition of the system into components of circuit and control elements; (2) modeling of each component for its frequency characteristics; (3) computation process which includes a consequence of formulation and solution of the system equations from the component characteristics point of view, then the desired characteristics are computed.

In the second processing step, all components are modeled for their frequency characteristics, and their formulation stamps must be prepared. In third step, first circuit information of an object system is inputted for element connection, type, and parameter data. Then, circuit equations for the system are formulated according to the element stamps at a sampled frequency. Finally, they are solved for the desired characteristics. This formulation and solution steps are repeated at all sampled frequencies.

Consider the block diagram of a GCC consisting an analog/digital control system shown in Fig. 4.3. The main power part of GCC may include inductors, capacitors, resistors, and dependent/independent sources. The analog control circuit may include adders, subtractors, differentiators, and integrators. The digital control circuit may include gain, add, and subtract blocks. The digital control is processed only discretely at sampled times in the time domain after processing voltages and currents of the main circuit through the S/H unit. It may have a control time delay $T_d = mT = z^{-m}$. When the switching and S/H operations are averaged and an LTI version of the GCC is given, the proposed frequency analysis method can be easily applied.

According to the model description, symbols are denoted as the node voltage vector $\boldsymbol{v}(s)$, the control signal vector $\boldsymbol{u}(s)$, and the current vector $\boldsymbol{i}\ (s)$. The KCL, signal-flow (SF) between control signal nodes, and circuit characteristic equations (CE) including current variables are formulated as Eq. (4.22) with $\boldsymbol{Y}_v, \cdots, \boldsymbol{C}_i, \boldsymbol{R}_u, \boldsymbol{R}_i,$ which are constant or dependent only on the differential operator s.

$$\begin{array}{l} \text{KCL:} \\ \\ \text{SF:} \\ \\ \text{CE:} \end{array} \begin{bmatrix} \boldsymbol{Y}_v & \boldsymbol{Y}_u & \boldsymbol{Y}_i \\ & & \\ \boldsymbol{F}_v & \boldsymbol{F}_u & \boldsymbol{F}_i \\ & & \\ \boldsymbol{C}_v & \boldsymbol{C}_u & \boldsymbol{C}_i \end{bmatrix} \begin{bmatrix} \boldsymbol{v} \\ \\ \boldsymbol{u} \\ \\ \boldsymbol{i} \end{bmatrix} = \begin{bmatrix} \boldsymbol{0} \\ \\ \boldsymbol{R}_u \\ \\ \boldsymbol{R}_i \end{bmatrix} \tag{4.22}$$

Formulation of the circuit equations is finally reduced to repeated combinations of all model stamps. For example, the inductor is connected from node i to node j. This contribution pattern to the circuit equations is expressed with the stamp in Fig. 4.11A where $Y(s) = \frac{1}{j\omega L}$, and the angular frequency is ω, and the differential operator is $s = j\omega$. For simple RLC circuit in Fig. 4.9A, $i = 1, j = 2$. It expresses the KCL equations at the two nodes. According to this stamp, elements $Y(s) = 1/(j\omega L)$ are added to two element locations and subtracted from two locations in the admittance matrix. Those of capacitors and resistors are the same when an appropriate $Y(s)$ is calculated for each one.

The stamp of the voltage source has to be slightly modified because its admittance $Y(s)$ is infinity and it cannot be expressed only with KCL equations. Its current is then introduced as a new variable and its element CE $\left(v_i - v_j = E\right)$ is added as shown in Fig. 4.10B where RHS means the right-hand side vector.

As another example for control elements, the stamp of a gain component $G(s)$ from input i to output j is shown in Fig. 4.10C which expresses the relationship of $x_j - G(s)x_i = 0$. For example, the stamps for the S/H and for the unit delay z^{-1} of one sample time T, using this stamp are unified into the transfer functions $G_1(s)$ and $G_2(s)$ as follows:

$$\begin{aligned} &G_1(s):\ S_m(s) = \frac{1 - \varepsilon^{-sT}}{sT} \\ &G_2(s):\ z^{-1} = \varepsilon^{-sT} \end{aligned} \tag{4.23}$$

The stamp of a signal source with output signal a at node i is expressed with its CE of $x_i = a$ as in Fig. 4.10D. The stamp of an adder with two inputs from node i and node j and one output to node k is shown in Fig. 4.10E which expresses the relationship of $x_k - x_i - x_j = 0$. The stamp of a subtractor is similar except for the sign of the subtract coefficient.

A simulator based on the above discussion can be developed in the MATLAB environment. It is also possible to utilize a simulator that has frequency analysis

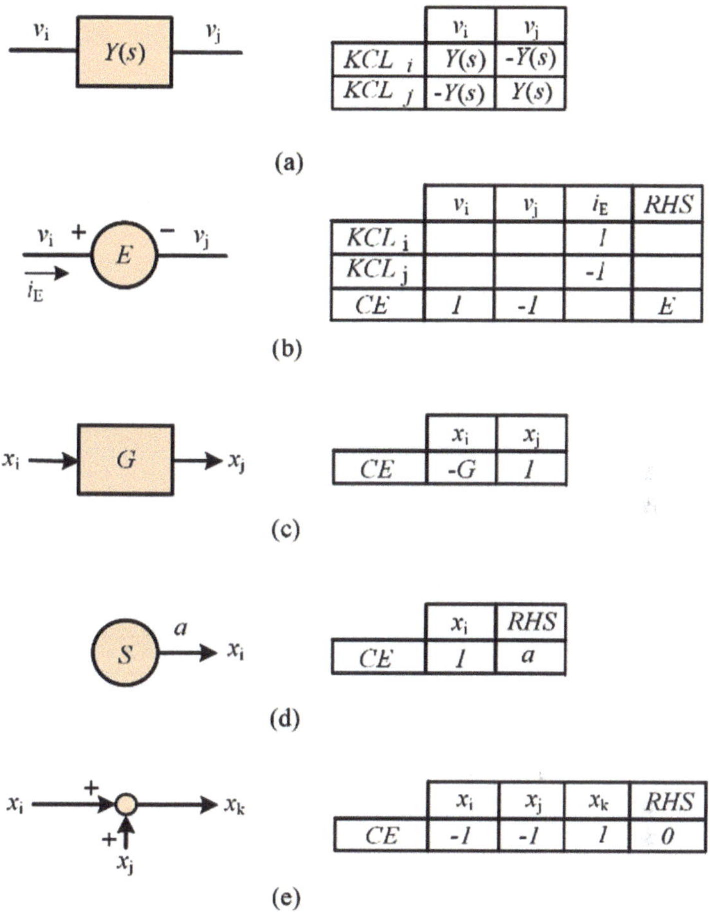

Figure 4.10 Stamp examples in power and control parts of a GCC; (A) admittance, (B) voltage source, (C) gain, (D) signal source, and (E) adder.

capabilities and can model inductors, capacitors, S/H, and delay blocks as frequency-dependent elements. Digital control elements like adders and subtractors do not need any special models and the correspondent analog elements are utilized because their digital behaviors are expressed with the S/H blocks.

4.3.2.3 Application example

As an example of the proposed efficient computation method, a three-phase GCC which is shown in Fig. 4.11A is considered for the sake of frequency characteristics analysis of the output admittance and thus the stability of the system. The GCC is controlled digitally by a digital signal processor. An overview of laboratory setup is also shown in Fig. 4.11B. The GCC and associated control parameters are given in Ref. [20]. The system consists of the power and the control circuits. The power circuit

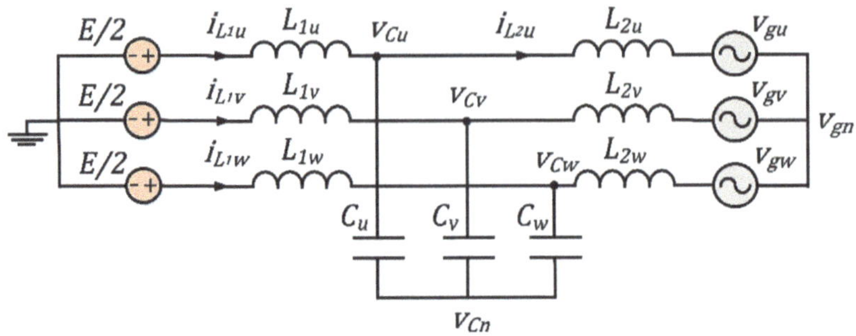

(a)

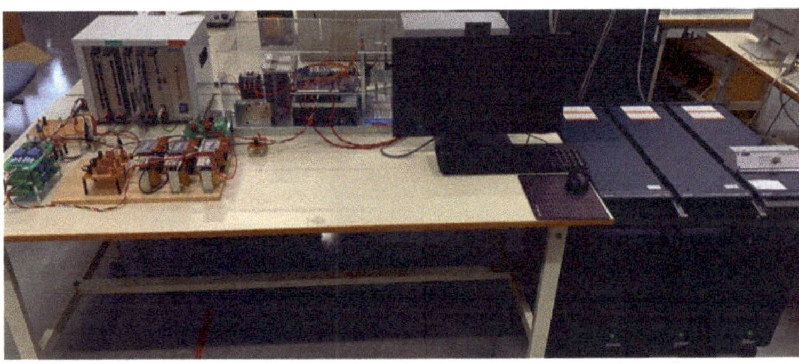

(b)

Figure 4.11 The three-phase GCC example; (A) single line main circuit, and (B) laboratory setup.

consists of three controlled voltage sources, six inductors, three capacitors, nine resistors, three voltage sources, four current sensors, and two voltage sensors. The control circuit, as shown in Fig. 4.12, consists of 6 S/H, six gains, 13 adders, seven delays, and two signal sources. In total, 74 equations are formulated according to the modified nodal analysis, and there are KCL equations at 30 electrical nodes except the ground, signal equations at 31 control nodes, and 10 element CEs for the voltage sources.

The computed Y_{uu}, Y_{uv} and positive sequence Y_p $\left(Y_p = Y_{uu} - Y_{uv}\right)$ of the output admittance are shown as solid curves in Fig. 4.13. This is compared with the result that is based on the accurate PWM model by the perturbation simulation method and is denoted as circle-curves in Fig. 4.13. The two results are nearly identical.

The experimental results, which are measured using a gain-phase analyzer where 5% excitation voltage is injected with a transformer, are shown as dashed curves. This is slightly different from those obtained by a simulation near the resonant frequency of 2.7 kHz. The experimental result is close to those of the proposed and perturbation methods. The proposed method is accurate almost up to the switching frequency, and it is efficient and practical.

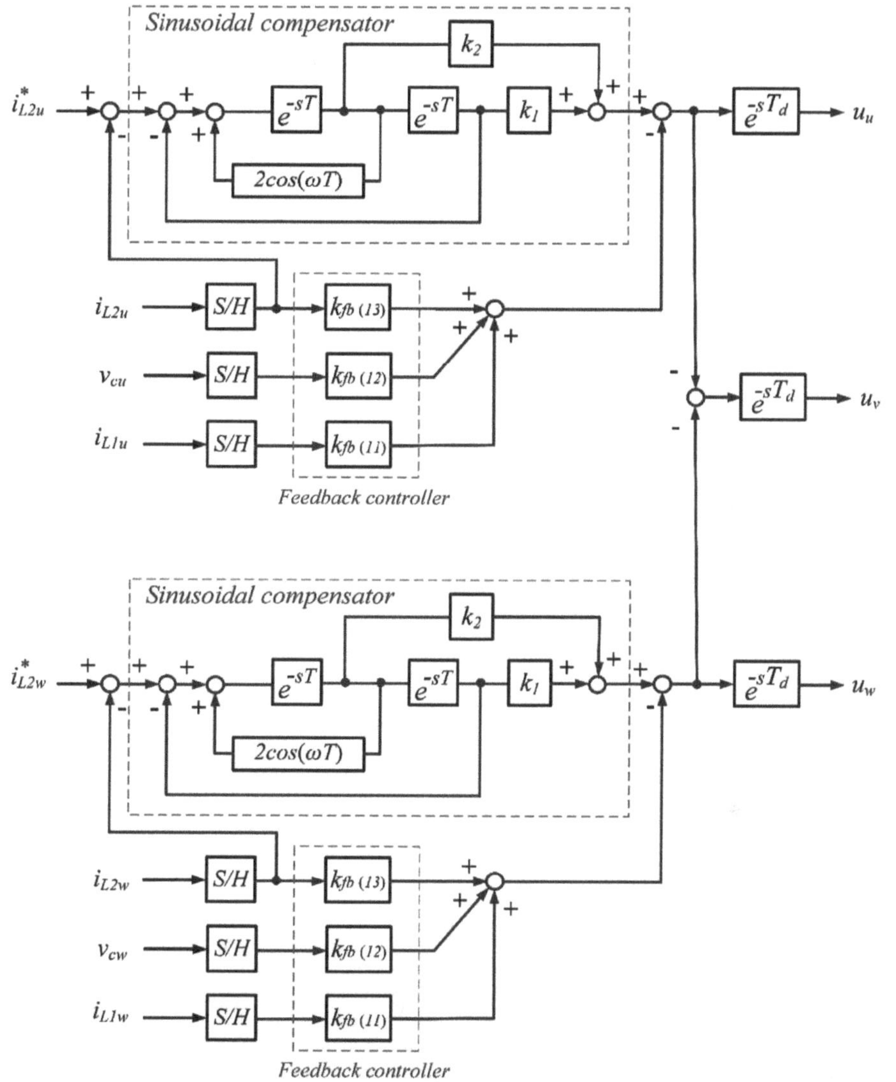

Figure 4.12 The digital control scheme of the three-phase GCC example.

4.3.2.3.1 Stability analysis

According to the impedance method, the stability condition is that the positive sequence $|Y_p|$ of the output admittance must be less than $|1/(\omega L_s)|$, where their phase difference is 180°. This means that the critical grid line inductance is estimated by $L_g = 1/(\omega|Y_p|)$ when the phase of Y_p is 90°, because the phase of the line impedance is 90°.

According to the proposed results in Fig. 4.13, the cross frequency where the phase of Y_p is 90° is 250 Hz and the critical inductance is $L_g = 3.2$ mH. A simulated result

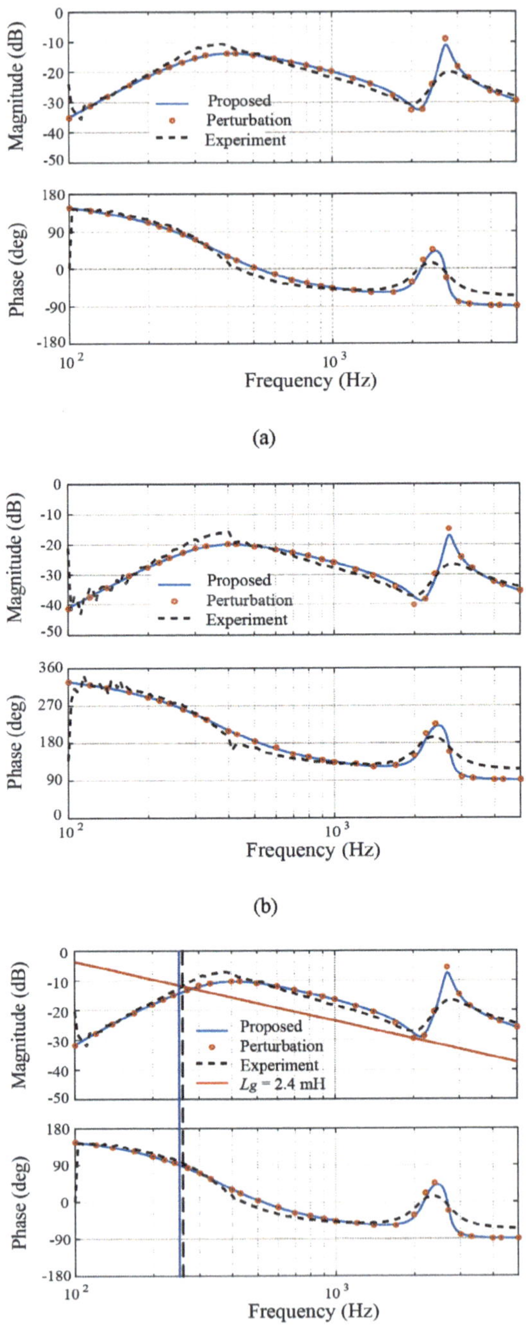

(c)

Figure 4.13 Computed and experimental results for positive sequence admittance Y_p; (A) Y_{uu}, (B) Y_{uv}, and (C) $Y_{uu} - Y_{uv} = Y_p$.

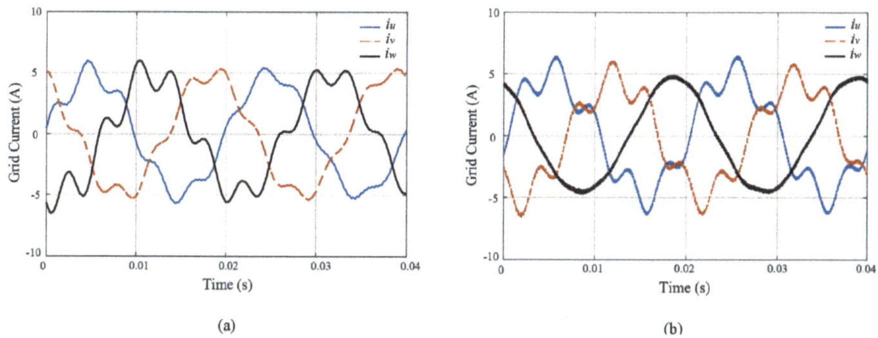

Figure 4.14 Output currents of GCC with line inductor L_g; (A) Simulation for $L_g = 3.0$ mH, and (B) Experiment for $L_g = 2.9$ mH.

for $L_g = 3.0$ mH is shown in Fig. 4.14A that is an oscillation waveform with the fifth harmonic which corresponds to the cross frequency. Because of switching harmonics and the output voltage limit, it was not possible to obtain a stable waveform for $L_g \geq$ 3.1 mH by simulation.

The above process was also verified experimentally according to the characteristics of Fig. 4.13. The cross frequency is 260 Hz and the critical line inductance is estimated to be $L_g = 2.4$ mH from the measured output admittance and the pure inductive L_g characteristics. However, it was possible to obtain a stable waveform up to $L_g = 2.9$ mH in the experiment. A measured result for $L_g = 2.9$ mH is shown in Fig. 4.13B which is a waveform with smaller oscillations that corresponds to the cross frequency. For the parameter design of the GCC control system, the estimated critical condition should be satisfied with some margins to consider switching harmonics and the output voltage limit.

4.3.2.3.2 Perturbation impact analysis

A GCC in practice has variations in circuit parameters which should be considered. For a given set of parameters, regions of eigenvalues of the matrix $Z_g Y_c$ is then computed to check the stability. If any eigenvalue region does not include the point of $(-1, 0)$ as in Fig. 4.15A, the system is stable. Otherwise, the system is unstable, as shown in Fig. 4.15B.

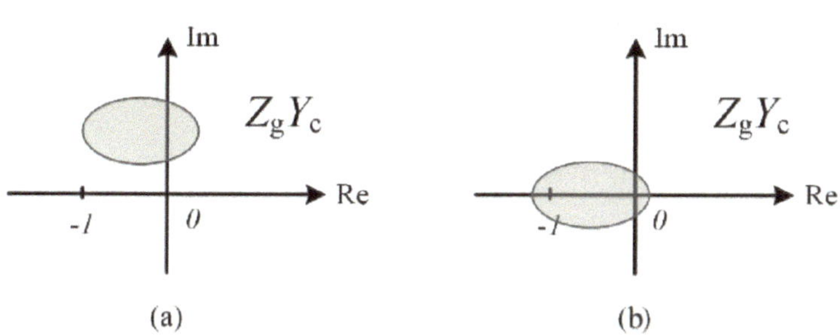

Figure 4.15 Tolerance analysis using the impedance method; (A) stable and (B) unstable.

Here, the projected admittance area is computed practically from randomly sampled sets within the given parameter area using the Monte Carlo method which can be summarized as follows:

(1) The number n of trials for the parameter sample sets and their variation ranges are determined.
(2) For each trial, a set of parameter values is generated randomly within the ranges. Admittance characteristics are computed for the parameter set by the proposed frequency analysis method. This step is repeated for n trial times.
(3) The maximum and minimum admittance characteristics, or tolerance bands, are computed from the n trial sets of the characteristics.

For the example at hand, the maximum and minimum bands of the GCC output admittances within the 5% parameter variations are computed for $n = 200$ using the proposed tolerance analysis method. The circuit becomes asymmetrical with the variations; however, the admittance matrix is diagonalized by similarity transform based on the eigenvectors. The frequency responses are shown in Fig. 4.16 for positive and negative sequences *related* components.

There are tolerances in all phases, both the positive and negative related sequences are affected by the parameter purturbations. The tolerance bandwidth is narrow in the low-frequency region and it does not affect stability seriously, because the analysis only considers the filter parameter tolerances.

According to the nominal parameter values, the positive related component of $Z_g(s)Y_c(s)$ for $L_g = 3.2$ mH is shown as a black solid curve in Fig. 4.17A. The characteristic curve crosses above $(-1, 0)$ and the system is assessed to be stable. However, the bandwidth should be considered in the stability analysis, and 2×2 extreme combinations of the maximum and minimum bands of the amplitude and phase characteristics are considered as the worst cases conservatively. They are shown as four red dashed curves and one of them crosses below $(-1, 0)$. In total, the system is unstable for $L_g = 3.2$ mH. The characteristics curves for $L_g = 3.0$ mH are shown in Fig. 4.17B. All of them cross above $(-1, 0)$ with some margins, and the system is stable.

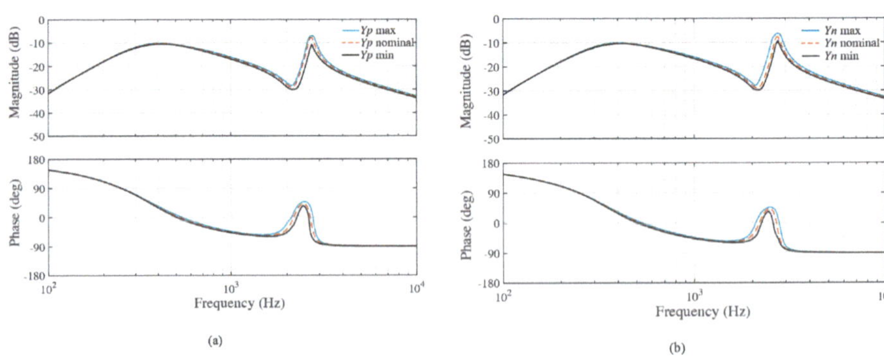

Figure 4.16 Computed results for positive and negative sequence-related admittances with tolerance bands; (A) positive sequence and (B) negative sequence.

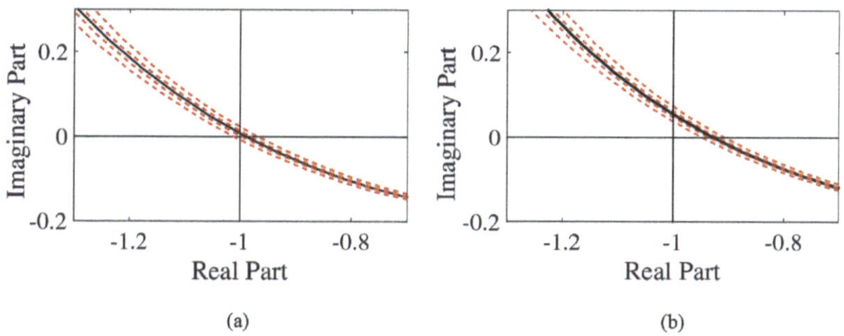

Figure 4.17 Stability analysis considering tolerance bands; (A) unstable case for $L_g = 3.2$ mH, and (B) stable case for $L_g = 3.0$ mH.

The same discussion can be done for the negative related component. Utilizing the characteristics that consider parameter tolerances, circuit and control parameters should be designed to be stable with enough stability margins.

4.3.3 Extension of method for nonlinear function blocks

This Section addresses an extension of the above discussed method based on a small-signal model for nonlinear function blocks, such as multiplier and sinusoidal function blocks [37]. These functions can be linearized at an operating point for frequency analysis, and the extended method is quite general for numerical processing as far as the objective converter is expressed with combinations of the linear or linearized models. The method only requires node-connection and element-model information of the circuit and control parts, and consists of two steps. At the first step, transient analysis is applied to an objective system to find an operating point. The system is linearized at the point and a small-signal system model is derived. This step can be omitted for a linear system. At the second step, frequency analysis is applied to the derived small-signal model to calculate each model for a desired response characteristic at each selected frequency.

4.3.3.1 Stamps of nonlinear function blocks

The general analysis method is investigated to be extended to a circuit with nonlinear control function blocks. Nonlinear function blocks cannot be analyzed directly. However, they can be analyzed when they are expressed with small-signal models after linearization at an operating point. Stamps of nonlinear function blocks are derived from small-signal models as explained below.

For example, the product operator can be linearized as follows and its stamp is shown in Fig. 4.18A.

$$x_k = \left.\frac{\partial x_k}{\partial x_i}\right|_{x_j=X_j} x_i + \left.\frac{\partial x_k}{\partial x_j}\right|_{x_i=X_i} x_j = X_j x_i + X_i x_j \tag{4.24}$$

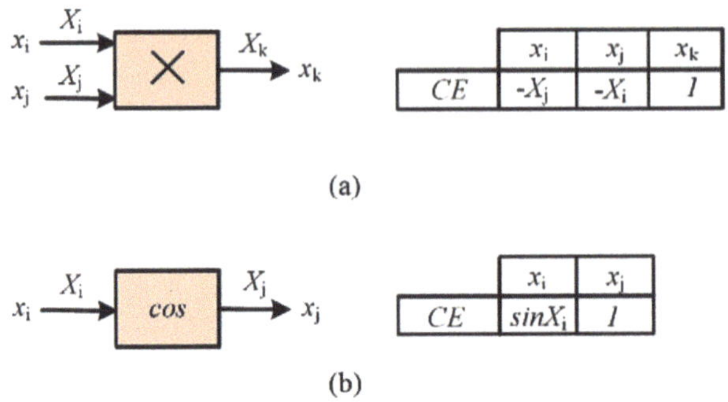

Figure 4.18 Stamp for nonlinear function block; (A) product operator and (B) cosine function.

where x_i, x_j are two inputs and x_k is the output, and X_i, X_j are the input values at the operating point. The stamp of cosine function between input x_i and output x_j is derived as shown in Fig. 4.18B.

$$x_j = \left.\frac{\partial x_j}{\partial x_i}\right|_{x_i=X_i} x_i = -\sin X_i x_i \tag{4.25}$$

The stamps for small-signal models of other nonlinear functions can be derived similarly.

4.3.3.2 Application example

The extended analysis method is applied to a GCC with a passive control system, where stability of the GCC is assured because the real part of the output admittance is positive and the Nyquist criterion is satisfied. The control consists of an *FB* and a feedforeward (*FF*) control loops as shown in Fig. 4.19A [37]. The *FB* control loop is designed to track the output current to a desired sinusoidal waveform. This may not be enough to ensure stability because the resulted admittance due to the *FB* loop at the output of GCC (Y_{fb}) may not be passive. Then the *FF* loop is added to make the total output admittance passive by adding the admittance term of Y_{ff}, as depicted in Fig. 4.10A. The *FF* loop $H[z]S_m(s)V_g(s)$ in Fig. 4.19A is designed to be resistive enough, or the z-transfer function $H[z]$ is designed to make a passive GCC.

The design steps of the passive control are described. Similar to Eq. (4.3), the state equations of the system are described below.

$$s\boldsymbol{X}(s) = \boldsymbol{A}\boldsymbol{X}(s) + \boldsymbol{b}U(s) + \boldsymbol{h}V_g(s) \tag{4.26}$$

$$U(s) = -\ \boldsymbol{k}_{fb}S_m\boldsymbol{X}(s) + G\left(i_r^* - \boldsymbol{c}S_m\boldsymbol{X}(s)\right) + HS_mV_g(s) \tag{4.27}$$

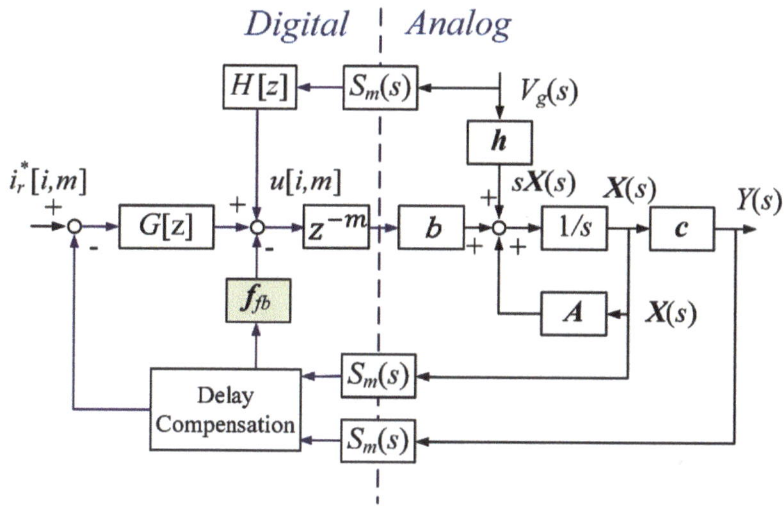

(a)

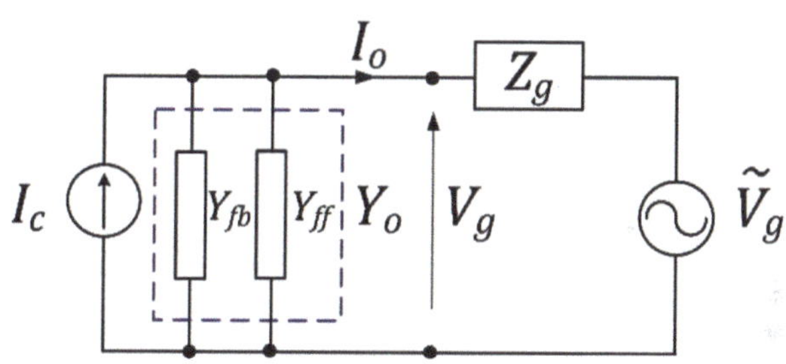

(b)

Figure 4.19 A GCC with digital *FB* and *FF* control loops; (A) control block diagram and (B) equivalent circuit.

From the above two equations, the following equation is derived.

$$\left(s\boldsymbol{I} - \boldsymbol{A} + \boldsymbol{b}\boldsymbol{k}_{fb}S_m + \boldsymbol{b}G\boldsymbol{c}S_m\right)^{-1}\boldsymbol{X}\,(s) = \boldsymbol{h}V_g(s) + \boldsymbol{b}HS_mV_g(s) + \boldsymbol{b}Gi_r^* \tag{4.28}$$

Thus, the converter output admittance Y_c becomes as follows.

$$\begin{aligned} Y_c = -\frac{I_{L_2}}{V_g} &= -\boldsymbol{c}\,\left(s\boldsymbol{I} - \boldsymbol{A} + \boldsymbol{b}\boldsymbol{k}_{fb}S_m + \boldsymbol{b}G\boldsymbol{c}S_m\right)^{-1}\boldsymbol{h} \\ &-\boldsymbol{c}\left(s\boldsymbol{I} - \boldsymbol{A} + \boldsymbol{b}\boldsymbol{k}_{fb}\,S_m + \boldsymbol{b}G\boldsymbol{c}S_m\right)^{-1}\boldsymbol{b}HS_m \end{aligned} \tag{4.29}$$

The output admittance Y_c is divided into the *FB* component Y_{fb} and the *FF* component Y_{ff} as follows.

$$\begin{aligned} Y_{fb} &= -\boldsymbol{c}\left(s\boldsymbol{I} - \boldsymbol{A} + \boldsymbol{b}\boldsymbol{k}_{fb}S_m + \boldsymbol{b}\boldsymbol{G}\boldsymbol{c}\, S_m\right)^{-1}\boldsymbol{h} \\ Y_{ff} &= -\boldsymbol{c}\left(s\boldsymbol{I} - \boldsymbol{A} + \boldsymbol{b}\boldsymbol{k}_{fb}S_m + \boldsymbol{b}\boldsymbol{G}\boldsymbol{c}\, S_m\right)^{-1}\boldsymbol{b}HS_m \equiv \overline{Y}_{ff}\, H \end{aligned} \tag{4.30}$$

A conductive characteristic Y_{ff} is added around the nonpassive frequency band by the *FF* control loop to make the total characteristic passive. The conductive characteristic Y_{ff} is designed to be R_d at the resonant frequency and to become smaller as it separates the frequency [37].

$$\widehat{Y}_{ff} \simeq \frac{1}{R_d + sL_d + \dfrac{1}{sC_d}} \tag{4.31}$$

The following transfer frequency characteristic $H[z]$ consists of the main characteristic and the adjustment filter characteristics, $\widehat{H}[z]$ and $F[z]$, respectively [37].

$$H[z] = \widehat{H}[z]F[z] \tag{4.32}$$

where

$$\widehat{H}[z] = \frac{a_0 + a_1z^{-1} + a_2z^{-2}}{1 + b_1z^{-1} + b_2z^{-2}};\ F[z] = \frac{c_0 + c_1z^{-1}}{1 + d_1z^{-1}} \tag{4.33}$$

The coefficients of $\widehat{H}[z]$, a_0, a_1, a_2, b_1, and b_2, are optimally fitted so that Y_{ff} is stable and close to $\widehat{Y}_{ff}$. It is not possible to fit Y_{ff} over the wide frequency range and Y_{ff} is optimally fitted to $\widehat{Y}_{ff}$ for a limited band over $f_1 \leq f \leq f_2$. The coefficients of $F[z]$, c_0, c_1, and d_1 are determined to reduce the *FF* side effects outside this fitting band.

The proposed frequency analysis method is applied to the described single-phase GCC (Fig. 4.11A) equipeted with passive control system to investigate its stability. The GCC and its control system parameters are given in Ref. [37]. Frequency characteristics of the analyzed results are shown in Fig. 4.20. The admittance is passive because the phase stays within ±90 degrees. The solid curves represent the analytical results which are close to those with the proposed approach (dashed curves).

An application example for using the discussed method to frequency analysis of a GCC with nonlinear control elements for power and frequency controls is given in Ref. [38].

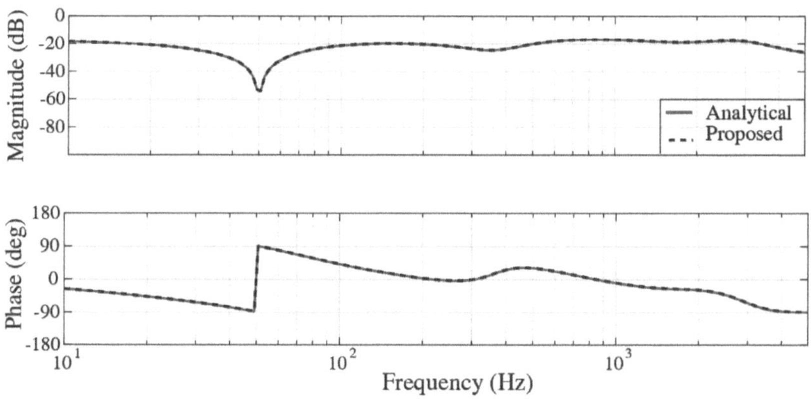

Figure 4.20 Frequency characteristics of the closed-loop system.

4.4 Poincaré map-based stability assessment

In this section, a time domain-based method using the Poincaré map is investigated to stability analysis of a GCC with nonlinear characteristics [25]. The basic principle and physical meanings are described in Section 4.1.4. The analysis methodology which is based on eigenvalue analysis of a sensitivity matrix between the periodic boundary values is explained. The obtaining of sensitivity matrix is an important step, since it must cover the sensitivities of the variations of operating points-based switching instances in a nonlinear GCC system. The analysis method checks whether the maximum absolute value of the eigenvalues exceeds unit to investigate amplification of state variations.

In this method, the periodic steady-state analysis is finally reduced to a solution of a nonlinear boundary value equation and it is solved often by the shooting method which directly solves the equation by a numerical algorithm such as the Newton–Raphson (NR) algorithm. In this process, the Jacobian or the sensitivity matrix is dependent on a nonlinear operating condition which is sensitive, especially, to switching instance variations when the GCC is under control. The periodic steady-state analysis process must consider the mentioned nonlinear effects, and precise required computation techniques for the sensitivity matrix [25,39,40].

4.4.1 Stability analysis methodology

4.4.1.1 Averaged analysis of a linear GCC with state feedback controller

Consider a simple example of an averaged GCC system controlled by a linear state-feedback for explaining the basic principle of stability analysis using Poincaré map:

$$\dot{x} = Ax + bu + gv_g \tag{4.34}$$

where u is an input, $\boldsymbol{A}$ is a system matrix, $\boldsymbol{b}$ is an input coefficient vector, and v_g is the grid voltage. The control law is as follows, where $\boldsymbol{k}_{fb}$ is a gain vector of state feedback and u_r is a reference input.

$$u = -\boldsymbol{k}_{fb}\boldsymbol{x} + u_r \tag{4.35}$$

The closed-loop system state equation is

$$\dot{\boldsymbol{x}} = \boldsymbol{A}_c x + \boldsymbol{b}u_r + \boldsymbol{g}v_g, \tag{4.36}$$

where $\boldsymbol{A}_c = \boldsymbol{A} - \boldsymbol{b}\boldsymbol{k}_{fb}$. The final boundary value $\boldsymbol{x}_T$ is as follows, where $\boldsymbol{h}$ is a vector which does not dependent on $\boldsymbol{x}_0$.

$$\boldsymbol{x}_T = e^{\boldsymbol{A}_c T}\boldsymbol{x}_0 + \boldsymbol{h} \tag{4.37}$$

The sensitivity matrix between the two boundary values can be obtained as

$$\frac{\partial \boldsymbol{x}_T}{\partial \boldsymbol{x}_0} = e^{\boldsymbol{A}_c T}. \tag{4.38}$$

In this case, the sensitivity matrix is identical to the system transition matrix and it is constant and does not dependent on $\boldsymbol{x}_0$, because the system is linear.

4.4.1.2 Principle of the shooting method

Periodic steady-state analysis is finally reduced to find an initial value vector $\boldsymbol{x}_0$ which satisfies the boundary value condition between $t = 0$ and $t = T$. For example, the *shooting method* can solve this problem by shooting the solution $\boldsymbol{x}_0$.

$$\boldsymbol{x}_0 = \boldsymbol{x}(T, \boldsymbol{x}_0) = \boldsymbol{P}(\boldsymbol{x}_0) = \boldsymbol{x}(T) \tag{4.39}$$

$$\widehat{\boldsymbol{P}}(\boldsymbol{x}_0) \equiv \boldsymbol{x}_0 - \boldsymbol{P}(\boldsymbol{x}_0) = 0 \tag{4.40}$$

The nonlinear equation is solved by the *N–R* algorithm and it is iteratively linearized along the operating condition. The superscript $^{(j)}$ stands for *j*-th iteration and the process is iterated until the norm of $\Delta\boldsymbol{x}_0^{(j+1)}$ becomes small enough.

$$\left[\boldsymbol{I} - \frac{\partial \boldsymbol{P}\left(\boldsymbol{x}_0^{(j)}\right)}{\partial \boldsymbol{x}_0^{(j)}}\right]\Delta\boldsymbol{x}_0^{(j+1)} = -\left[\boldsymbol{x}_0^{(j)} - P\left(\boldsymbol{x}_0^{(j)}\right)\right] \tag{4.41}$$

$$\boldsymbol{x}_0^{(j+1)} = \boldsymbol{x}_0^{(j)} + \Delta\boldsymbol{x}_0^{(j+1)} \tag{4.42}$$

The term $\frac{\partial P\left(x_0^{(j)}\right)}{\partial x_0^{(j)}}$ in the Jacobian is called a sensitivity matrix which is reevaluated at each iterated linearization process. Each column of the matrix is computed by one-period transient analysis of each sensitivity circuit which corresponds to one state variable with unit excitation. When the number of the state-variables is n, n sensitivity circuits have to be analyzed to get n columns of the matrix.

4.4.1.3 Computation of sensitivity matrix

Computation of the sensitivity matrix is an essential step of the shooting method. This step is explained using an example of a buck converter in Fig. 4.21A, which is

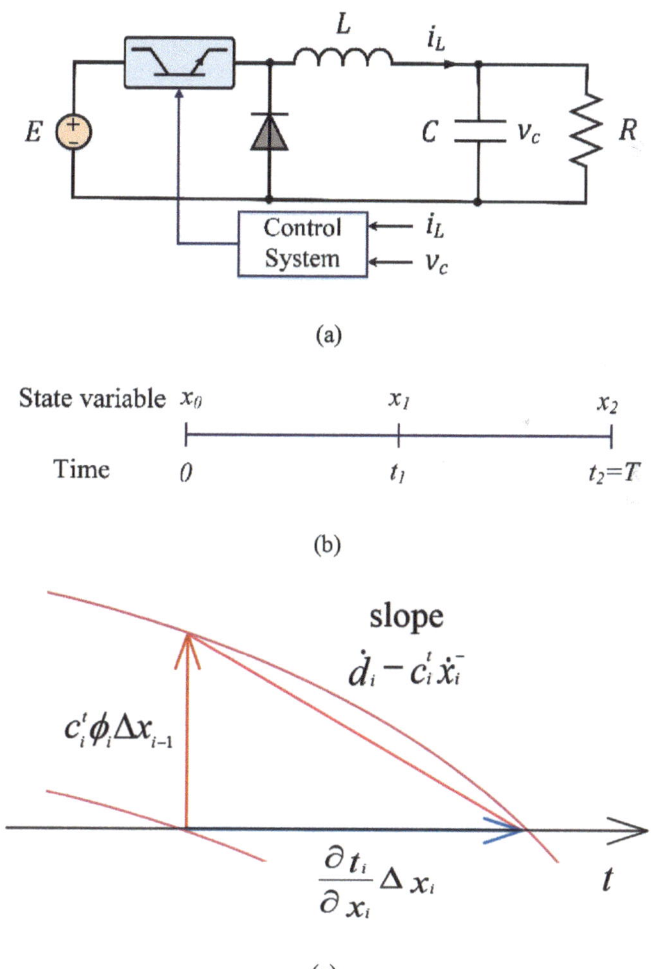

Figure 4.21 Switch modes and state variable variations of an internally controlled buck converter; (A) closed-loop system, (B) switch modes and state variables, and (C) change of the state variable due to the change of the initial value.

internally controlled by a feedback rule, to show how the sensitivity matrix is computed. It is assumed that the converter is operated in a continuous mode, and it has two switch topological modes for one period T. Switching instances and state values for the two modes are denoted as $t_0, t_1, t_2 = T$ and $\boldsymbol{x}_0$, $\boldsymbol{x}_1$,$\boldsymbol{x}_2$ as shown in Fig. 4.21B. The two transition matrices are denoted as ϕ_1 and ϕ_2 and an excitation vector for mode 1 is $\boldsymbol{\psi}_1$ where the matrices and the vector are dependent on x_0.

$$\boldsymbol{x}_1 = \boldsymbol{\phi}_1\boldsymbol{x}_0 + \boldsymbol{\psi}_1 \ ; \ \boldsymbol{x}_2 = \boldsymbol{\phi}_2\boldsymbol{x}_1 \tag{4.43}$$

When time lengths of two modes are constant and do not dependent on $\boldsymbol{x}_0$, the sensitivity matrix is the direct multiplication of the two transition matrices. However, when the converter is internally controlled and the time lengths of the two modes are dependent on the initial value $\boldsymbol{x}_0$, the sensitivity of $\boldsymbol{x}_T$ with respect to $\boldsymbol{x}_0$ is not simple multiplication of the two transition matrices. The two mode lengths are dependent on $\boldsymbol{x}_0$ and this nonlinear effects must be considered.

At the end of the time interval of mode 1, its length is flexible to satisfy the switching condition and the time t_1 is extended by $\frac{\partial t_1}{\partial \boldsymbol{x}_0}\Delta\boldsymbol{x}_0$. This effect of the switching instance variation to $\Delta\boldsymbol{x}_1$ is expressed as $\left(\dot{\boldsymbol{x}}_1^- - \dot{\boldsymbol{x}}_1^+\right)\frac{\partial t_1}{\partial \boldsymbol{x}_0}\Delta\boldsymbol{x}_0$ in Fig. 4.21C. Therefore the sensitivity for one period is as follows:

$$\frac{\partial \boldsymbol{x}_2}{\partial \boldsymbol{x}_0} = \boldsymbol{\phi}_2\left(\left(\dot{\boldsymbol{x}}_1^- - \dot{\boldsymbol{x}}_1^+\right)\frac{\partial t_1}{\partial x_1} + \boldsymbol{I}\right)\boldsymbol{\phi}_1 \neq \boldsymbol{\phi}_2\boldsymbol{\phi}_1 \tag{4.44}$$

where $\dot{\boldsymbol{x}}_1^-$ and $\dot{\boldsymbol{x}}_1^+$ are left- and right-hand derivatives at $t = t_1$, respectively.

4.4.1.4 Generalized sensitivity matrix computation for analog control

The derived sensitivity expression is generalized for a case where the switching instances are determined by an analog control method including the triangular modulation method. When one period consists of N_T switch cycles and the i-th switch cycle consists of m_i circuit modes, the sensitivity is expressed as follows.

$$\frac{\partial \boldsymbol{x}_T}{\partial \boldsymbol{x}_0} = \prod_{i=1}^{N_T}\left(\prod_{k=1}^{m_i}\left(\dot{\boldsymbol{x}}_{ki}\frac{\partial t_{ki}}{\partial \boldsymbol{x}_{ki}} + I\right)\boldsymbol{\phi}_{ki}\right) \tag{4.45}$$

where

$$\boldsymbol{x}_{ki} = \begin{cases} \dot{\boldsymbol{x}}_{ki}^- - \dot{\boldsymbol{x}}_{ki}^+ & k \neq m_i \\ \dot{\boldsymbol{x}}_{ki}^- & k = m_i \end{cases} \tag{4.46}$$

The switching instance sensitivity $\frac{\partial t_{ki}}{\partial \boldsymbol{x}_{ki}}$ is derived from the linearized equation of the switch condition as described above.

$$\boldsymbol{c}_{ki}^{T}\boldsymbol{x}_{ki} = d_{ki}(k = 1, \cdots, m_i) \tag{4.47}$$

where $\boldsymbol{c}_{ki}$ is a coefficient vector, T is transpose of a matrix, and d_{ki} is constant or time-function. The both sides are partially differentiated with respect to $\boldsymbol{x}_{ki}$, and the switching instance sensitivity is derived as follows:

$$\boldsymbol{c}_{ki}^{T}\left(\boldsymbol{I} + \dot{\boldsymbol{x}}_{\overline{ki}}\frac{\partial t_{ki}}{\partial \boldsymbol{x}_{ki}}\right) = \dot{d}_{ki}\frac{\partial t_{ki}}{\partial \boldsymbol{x}_{ki}} \tag{4.48}$$

The partial derivative value becomes as follows:

$$\frac{\partial t_{ki}}{\partial \boldsymbol{x}_{ki}} = \begin{cases} \dfrac{\boldsymbol{c}_{ki}^{T}}{\dot{d}_{ki} - \boldsymbol{c}_{ki}^{t}\dot{\boldsymbol{x}}_{\overline{ki}}} & \text{(variable)} \\ 0 & \text{(fixed)} \end{cases} \tag{4.49}$$

The coefficients, $\boldsymbol{c}_{ki}^{T}$ and $\dot{d}_{ki}$, can be usually given to data input of the circuit information. Transition matrices ϕ_{ki} for $i = 1, \cdots, N_T$ are computed by one-cycle transient analysis of the system.

4.4.1.5 Generalized sensitivity matrix computation for digital control

A digital control case is different from the above analog control case because time lengths of the switch modes are determined at each digital clock time for one switch cycle. For example, when one period consists of N_T switch cycles and all switch cycles consist of three circuit modes, the sensitivity matrix is expressed as follows.

$$\frac{\partial \boldsymbol{x}_T}{\partial \boldsymbol{x}_0} = \prod_{i=1}^{N_T} \boldsymbol{\phi}_{3i}\left(\boldsymbol{\phi}_{2i}\left(\boldsymbol{\phi}_{1i} + \dot{\boldsymbol{x}}_{1i}\frac{\partial t_{1i}}{\partial \boldsymbol{x}_{0i}}\right) + \dot{\boldsymbol{x}}_{2i}\frac{\partial t_{2i}}{\partial \boldsymbol{x}_{0i}}\right) \tag{4.50}$$

When each switch cycle consists of m_i modes, the matrix is as follows.

$$\frac{\partial \boldsymbol{x}_T}{\partial \boldsymbol{x}_0} = \prod_{i=1}^{N_T} \boldsymbol{\phi}_{m_i i}\left(\boldsymbol{\phi}_{m_i-1\,i}\left(\cdots\left(\boldsymbol{\phi}_{1i} + \dot{\boldsymbol{x}}_{1i}\frac{\partial t_{1i}}{\partial \boldsymbol{x}_{0i}}\right)\cdots\right) + \dot{\boldsymbol{x}}_{m_i-1i}\frac{\partial t_{m_i-1\,i}}{\partial \boldsymbol{x}_{0i}}\right) \tag{4.51}$$

4.4.2 Application example

As an application example, consider a single-phase GCC with an LCL filter which is shown in Fig. 4.1. It is assumed that the control system shown in Fig. 4.22 is used to control this GCC.

4.4.2.1 State-space equations of the GCC system

A set of state equations of a GCC example in Fig. 4.22A is derived to explain control principle of the system. The symbols E, L_1, L_2, C, v_g stand for DC source voltage, inductors, capacitor, and line voltage values, and R_1, R_2 stand for inductor resistances. The state-space model is given in Eq. (4.52), where $\boldsymbol{x}$ is the state variable vector of the inductor currents i_{L_1}, i_{L_2} and the capacitor voltage v_C.

$$\begin{aligned}\dot{\boldsymbol{x}}(t) &= \boldsymbol{A}\boldsymbol{x}(t) + \boldsymbol{b}u(t) + \boldsymbol{h}v_g(t)\\ y(t) &= \boldsymbol{c}x(t)\end{aligned} \tag{4.52}$$

where $\boldsymbol{A}$, $\boldsymbol{b}$, $\boldsymbol{c}, \boldsymbol{h}$ are

$$\boldsymbol{A} = \begin{bmatrix} 0 & -1/L_1 & 0 \\ 1/C & 0 & -1/C \\ 0 & 1/L_1 & 0 \end{bmatrix}; \boldsymbol{b} = \begin{bmatrix} E/L_1 \\ 0 \\ 0 \end{bmatrix}; \boldsymbol{c} = [0 \quad 0 \quad 1]; \boldsymbol{h} = \begin{bmatrix} 0 \\ 0 \\ -1/L_2 \end{bmatrix}$$

It is discretized as in Eq. (4.53) at a sample rate and a switch cycle T.

$$\begin{aligned}\boldsymbol{x}[i+1] &= \boldsymbol{A}_d\boldsymbol{x}[i] + \boldsymbol{b}_d u[i] + \boldsymbol{h}_d v_g[i]\\ y[i] &= \boldsymbol{c}_d\boldsymbol{x}[i]\end{aligned} \tag{4.53}$$

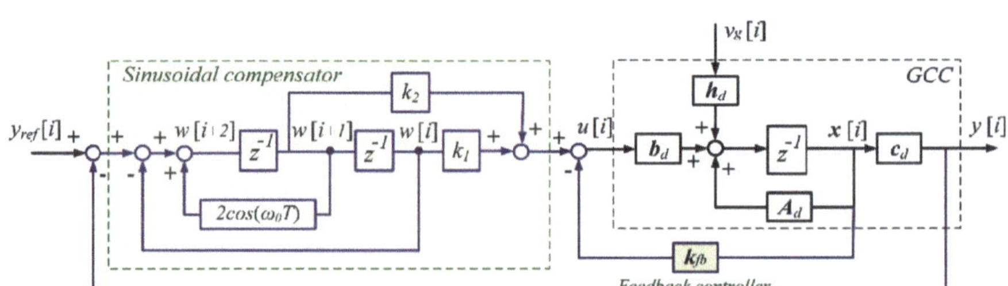

Figure 4.22 Control system of the single-phase GCC.

The coefficient matrix and vectors, $\boldsymbol{A}_d$, $\boldsymbol{b}_d$, $\boldsymbol{h}_d, \boldsymbol{c}_d$ are given as

$$\boldsymbol{A}_d = e^{A\,T_s};\ \boldsymbol{b}_d = \int_0^{T_s} e^{A\,\tau} d\tau \boldsymbol{b};\ y(t) = \boldsymbol{c}\boldsymbol{x}(t);\ \boldsymbol{h}_d = \int_0^{T_s} e^{A\,\tau} d\tau \boldsymbol{h};\ \boldsymbol{c}_d = \boldsymbol{c}. \quad (4.54)$$

4.4.2.2 Control principle of the GCC with sinusoidal compensator

The GCC is controlled according to the block diagram in Fig. 4.22. The system equipped with the sinusoidal compensator to eliminate its steady-state errors for a sinusoidal output of the angular frequency ω [41]. It is controlled with two *FB* gains for the second-order compensator and three *FB* gains $\boldsymbol{k}_{fb}$ for the three state variables, and the control law becomes

$$\begin{aligned} \boldsymbol{u}[i] &= -\boldsymbol{k}_{fb}\boldsymbol{x}\ [i] + k_1 w[i] + k_2 w[i+1] \\ w[i+2] &= -w[i] + 2\cos\omega T w[i+1] + \left(y_{\text{ref}}[i] - y[i]\right) \end{aligned} \quad (4.55)$$

The transfer function of the compensator is as follows.

$$G[z] = \frac{k_2 z + k_1}{z^2 - 2\cos(\omega T)z + 1} \quad (4.56)$$

The order of the control system is fifth and the five gains must be designed using an appropriate control strategy.

4.4.2.3 Dedicated simulator in the MATLAB environment

Using the above-mentioned principles, a simulator is developed for the given example. It solves the above state equations and the switching conditions using MATLAB functions and finds a steady-state solution. Then the sensitivity matrix between the boundary state values based on the Poincaré map is computed. Finally, the matrix is analyzed for eigenvalues whose maximum absolute value is less than unit or not.

4.4.2.4 Stability analysis results

Stability analysis results of the GCC system by the proposed method are investigated and compared for both linear (averaged) and nonlinear systems. The circuit parameters of the main circuit are $E = 200$ V, $L_1 = L_2 = 0.76$ mH, $C = 9.3\ \mu$F and the *FB* gains are designed to be $\boldsymbol{k}_{fb} = [0.0402, -0.0004, 0.0248], k_1 = -0.0185, k_2 = 0.0216$, considering their resistive components.

The line inductor value is gradually increased and the system is analyzed for its stability by the proposed method. The critical limit for the line inductor in linear system

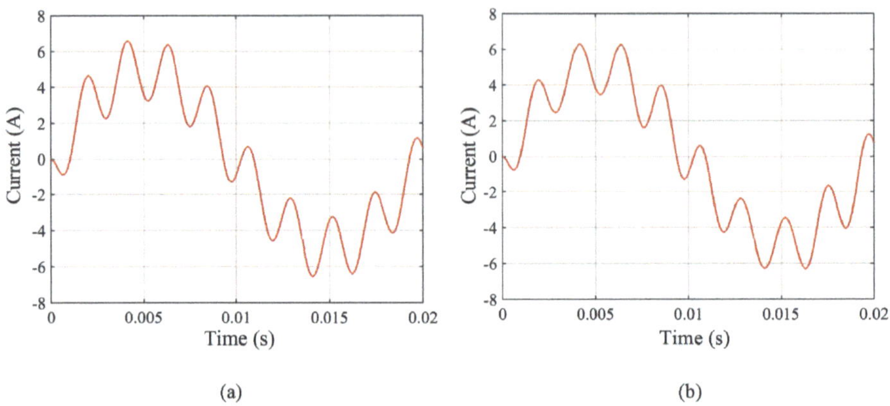

Figure 4.23 Simulated transient current i_{L_2} waveforms; (A) linearized and (B) precise.

case is $L_g = 3.684$ mH. The initial transient output current i_{L_2} is shown in Fig. 4.23A where an oscillation component around 450 Hz or the ninth harmonic order is heavily superposed. The oscillation component gradually attenuates, and finally the output becomes purely sinusoidal and coincides with the reference waveform as in Fig. 4.24A. Eigenvalues of the sensitivity matrix of the two boundary values by the Poincaré map are shown in Table 4.2a and their maximum absolute value is just one which shows a critical state.

Using this line inductor value instabilizes the nonlinear system case. The inductor value $L_g = 3.402$ mH is critical for this case. The initial transient of the output current is shown in Fig. 4.23B and it also has the large oscillation component of the ninth harmonic. Its steady-state waveform is shown in Fig. 4.24B where the ninth component remains due to inevitable harmonic components because the averaging technique is not applied. The reason why $L_g = 3.402$ mH is critical is that the harmonic components affect the GCC output saturation as shown in Fig. 4.25 where duties of the output voltage pulses are shown for one-cycle transient. Due to the harmonics of i_{L_2}, the

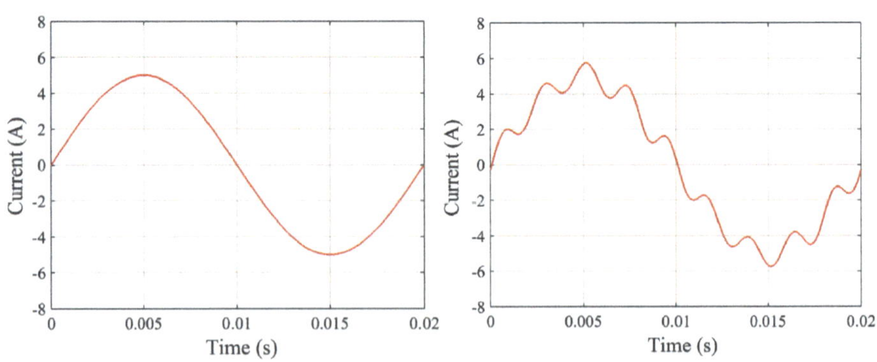

Figure 4.24 Simulated steady-state current i_{L_2} waveforms; (A) linearized and (B) precise.

Table 4.2 Eigenvalues of the sensitivity matrix.

(a) Averaged for L_g = 3.684 mH		(b) Precise for L_g = 3.402 mH		(c) Averaged for L_g = 3.402 mH	
Eigenvalues	**Abs. val.**	**Eigenvalues**	**Abs. val.**	**Eigenvalues**	**Abs. val.**
0.9407 + j0.3392	1.0000	−0.0912 + j0.3468	0.3586	0.0450 + j0.3379	0.3409
0.9407 − j0.3392	1.0000	−0.0912 − j0.3468	0.3586	0.0450 − j0.3379	0.3409
0.0000 + j0.0000	0.0000	0.0000 + j0.0000	0.0000	0.0000 + j0.0000	0.0000
0.0000 + j0.0000	0.0000	0.0000 + j0.0000	0.0000	0.0000 + j0.0000	0.0000
0.0000 + j0.0000	0.0000	0.0000 + j0.0000	0.0000	0.0000 + j0.0000	0.0000

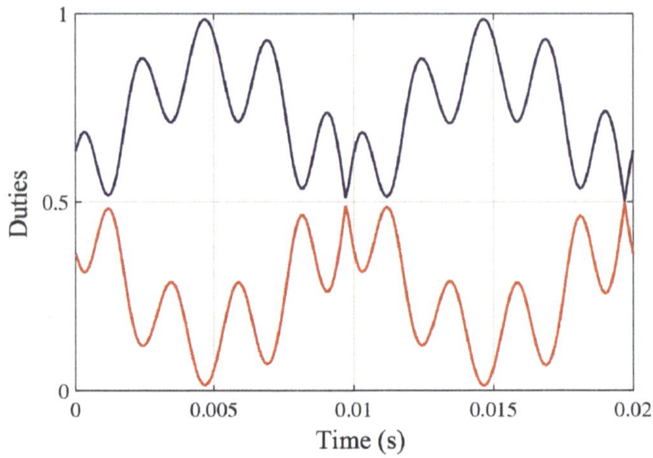

Figure 4.25 Critical condition for transient duties of the output voltage pulses which are close to the saturation limit $\left(L_g = 3.402\text{ mH}\right)$.

control input $u[i]$ becomes to be unit which is the limit of the output voltage. This means that the critical condition is determined by the harmonic components and the inverter output saturation. Eigenvalues of the sensitivity matrix of the two boundary values are shown in Table 4.2b, and their maximum absolute value is 0.3586 and less than one. The eigenvalues for linear system case with the same inductor value (shown in Table 4.2c), and the maximum absolute value is 0.3409 which almost coincides with the case (b).

The stability analysis is practically possible using the averaged circuit. However, the stability analysis without the averaging technique by the addressed method is useful to derive a precise critical condition by simulation, considering harmonic and converter saturation effects.

4.5 Sensitivity analysis

As discussed before, sensitivity analysis is an important part of the stability analysis in a GCC system. In Section 4.3, this issue is considered in the study of system parameters tolerances impacts on the GCC stability. The sensitivity of GCC is computed by considering the 5% variation in the GCC output admittance parameters using the Monte Carlo method. Sensitivity analysis should be taken into account for a safer stability margin during the synthesis process.

In Section 4.4, the importance of sensitivity analysis in the stability investigation of a GCC is emphasized. In the addressed method, the eigenvalues of a sensitivity matrix between the periodic boundary values are analyzed. When all absolute eigenvalues of the matrix at a periodic steady-state condition are less than one, the system is stable because variations of the initial values are not magnified. The sensitivity matrix is

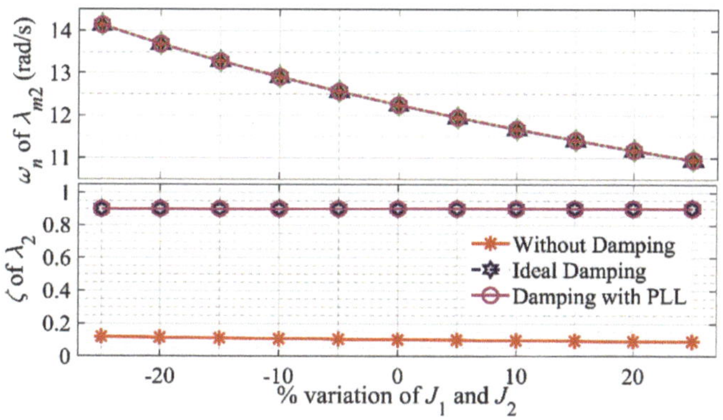

Figure 4.26 Sensitivity of dominant poles respect to variation of J_1 and J_2.

efficiently computed by a periodic steady-state analysis method such as the shooting method. It is seen that in case of having a nonlinear block, the Jacobian or the sensitivity matrix is dependent on a nonlinear operating condition which makes it very sensitive to the GCC switching instance variations and the control system parameters. Section 4.4 introduced a precise computation technique for the sensitivity matrix of GCCs.

Here, to clarify the usefulness and importance of sensitivity analysis, let to recall the addressed $P-\ \omega$ model in Chapter 3, which the relevant stability analysis is discussed in Section 4.2. Here, a brief sensitivity analysis is given to illustrate how the closed-loop poles are affected by intentional or nonintentional change of some main parameters.

In order to investigate how the poles are affected by intentional or nonintentional parameter variation, two sensitivity studies are performed. As shown in Ref. [28], λ_1 and λ_2 describe almost all important dynamics; therefore, the following discussions are focused on these poles for a microgrid with two GCC-based DGs, which its structure and parameters are given in Ref. [28].

In Fig. 4.26, we consider an intentional variation of the GCCs inertia parameters, J_1 and J_2. In this case, the variation is applied to GCC one and GCC two equally, and all control methods are redesigned according to Eq. (4.13) to keep ζ of λ_1 constant. Therefore, as shown in Fig. 4.26, ζ of λ_1 is kept unchanged, except that the case of no damping shows a decreasing ζ because the absence of dedicated damping. This variation coincides with the expression shown in Ref. [27]. On the other hand, ω_n of λ_1 increases while J_i increases. The observed variation also well coincides with Eq. (4.13).

4.6 Summary

In this chapter, the GCC stability assessment and sensitivity analysis are discussed. After presenting an overview on the most popular stability analysis issues, several stability analysis methods in both frequency and time domains are introduced.

The discussion is started from the basic stability analysis method using the eigenvalues/poles graph and then continued to address the more effective approaches based on the GCC frequency characteristics and the Poincare' map. Finally, the significance of sensitivity analysis in the stability study is emphasized.

This chapter only focused on those GCC stability issues that are caused by its dynamics and improper operating of the relevant control loops, without considering the dynamic of power grid. The dynamic impact of GCCs on power grid stability and performance is discussed in Chapter 5.

References

[1] Y.A. Familiant, J. Huang, K.A. Corzine, M. Belkhayat, New techniques for measuring impedance characteristics of three-phase ac power systems, IEEE Trans. Power Electron. 24 (7) (2009) 1802–1810.

[2] J. Sun, Impedance-based stability criterion for grid connected inverters, IEEE Trans. Power Electron. 26 (11) (2011) 3075–3078.

[3] M. Cespedes, J. Sun, Impedance modeling and analysis of grid connected voltage-source converters, IEEE Trans. Power Electron. 29 (3) (2014) 1254–1261.

[4] T. Kato, K. Inoue, Y. Akiyama, K. Ohashi, Stability analysis for grid connected three-phase inverter with LCL filters, in: IEEE 16th Workshop on Control and Modeling for Power Electronics (COMPEL), 2015, pp. 1–7.

[5] A. Rygg, M. Molinas, C. Zhang, X. Cai, A modified sequence-domain impedance definition and its equivalence to the dq-domain impedance definition for the stability analysis of ac power electronic systems, IEEE J. Emerg. & Selec. Topic. Power Electron. 4 (4) (2016) 1383–1396.

[6] R. Ortega, A. Loría, P.J. Nicklasson, H. Sira-Ramírez, Passivity-based Control of Euler-Lagrange Systems, Springer Verlag, 2010.

[7] S.R. Sanders, G.C. Verghese, Lyapunov-based control for switched power converters, IEEE Trans. Power Electron. 7 (1) (1992) 17–24.

[8] T. Kato, K. Inoue, M. Ishida, Investigation of stabilities of Lyapunov-based digital control for grid connected inverter, IEEE Energy Conver. Congr. & Expos. (2015) 2394–2399.

[9] L. Harnefors, A.G. Yepes, A. Vidal, J. Douval-Gandoy, Passivity-based stabilization of resonant current controllers with consideration of Time Delay, IEEE Trans. Power Electron. 29 (12) (2014) 6260–6263.

[10] L. Harnefors, X. Wang, A.G. Yepes, F. Blaabjerg, Passivity-based stability assessment of grid connected VSCs - an overview -, IEEE J. Emerg. Sel. Topics Power Electron. 4 (1) (2016) 116–125.

[11] X. Wang, F. Blaabjerg, P.C. Loh, Passivity-based stability analysis and damping injection for multi-paralleled voltage-source converters with LCL filters, IEEE Trans. Power Electron. 32 (11) (2017) 8922–8935.

[12] T. Kato, K. Inoue, Y. Nakajima, Stabilization of grid connected inverter system with feed-forward control, IEEE Energy Conv. Congr. & Expos. (ECCE) (2017) 3375–3382.

[13] R. Tymerski, Frequency analysis of time-interval-modulated switched networks, IEEE Trans. Power Electron. 6 (2) (1991) 287–295.

[14] R. Tymerski, Application of the time-varying transfer function for exact small-signal analysis, IEEE Trans. Power Electron. 9 (2) (1994) 196–205.

[15] B.K.H. Wong, H.S.H. Chung, A systematic graphing technique for small-signal low-frequency characterization of PWM DC/DC converters, IEEE Trans. Ind. Electron. 47 (1) (2000) 45–54.
[16] C.H. Ho, A.E. Ruehli, P.A. Brennan, The Modified nodal approach to network analysis, IEEE Trans. Circ. & Syst. (1975) 504–509. CAS-22.
[17] L.O. Chua, P.M. Lin, Computer Aided Analysis of Electronic Circuits: Algorithms and Computational Techniques, Prentice-Hall, 1975.
[18] J. Vlach, K. Singhal, Computer Method for Circuit Analysis and Design, Van Nostrand Reinhold, 1983.
[19] T. Kato, K. Inoue, H. Kawabata, Stability analysis of grid connected inverter system, Control & Model. Power Electr. (COMPEL) (2012) 1–5.
[20] Y. Takami, T. Kato, K. Inoue, General-purpose computation method of a power converter for frequency characteristics: application to stability analysis of a grid inverter, IEEE J. Emerg. & Selec. Topic. Power Electron. 5 (4) (2017) 1466–1473.
[21] G.D. Hachtel, R. Brayton, F.G. Gustavson, The sparse tableau approach to network analysis and design, IEEE Trans. Circ. Theor. CT-18 (1971) 101–113.
[22] O. Dranga, B. Buti, I. Nagy, Stability analysis of a feedback-controlled resonant DC-DC converter, IEEE Trans. Ind. Electron. 50 (1) (2003) 141–152.
[23] Y. Kuroe, Computer methods to analyze stability and bifurcation phenomena, in: Sec. 4.5 of Nonlinear Phenomena in Power Electronics by, IEEE Press, 2001.
[24] S. Eren, M. Pahlevaninezhad, A. Bakhshai, P.K. Jain, Composite nonlinear feedback control and stability analysis of a grid connected voltage source inverter with LCL Filter, IEEE Trans. Ind. Electron. 60 (11) (2013) 5059–5074.
[25] T. Kato, K. Inoue, Y. Takami, Stability analysis using Poincaré map in the time-domain for grid connected inverter, in: 2017 IEEE 18th Workshop on Control and Modeling for Power Electronics (COMPEL), 2017, pp. 1–7.
[26] T. Kato, K. Inoue, T. Sakiyama, Stability analysis methods of a grid connected inverter in time and frequency domains, IEEE Internat. Power Electron. Conf. (IPEC-NIIGATA2018-ECCE Asia) (2018) 2186–2192.
[27] J. Liu, Y. Miura, T. Ise, Fixed-parameter damping methods of virtual synchronous generator control using state feedback, IEEE Access 7 (2019) 99177–99190.
[28] J. Liu, Y. Miura, H. Bevrani, T. Ise, A unified modeling method of virtual synchronous generator for multi-operation-mode analyses, IEEE J. Emerg. Sel. Topics Power Electron. (n.d.). https://doi.org/10.1109/JESTPE.2020.2970025.
[29] J. Liu, H. Bevrani, T. Ise, A design-oriented QV response modeling approach for grid-forming distributed generators considering different operation modes, IEEE J. Emerg. Sel. Topics Power Electron. (n.d.). https://doi.org/10.1109/JESTPE.2021.3057517.
[30] H. Bevrani, B. Francois, T. Ise, Microgrid Dynamics and Control, Wiley, NJ, USA, July 2017.
[31] D. Pan, X. Ruan, C. Bao, W. Li, X. Wang, Capacitor-current-feedback active damping with reduced computation delay for improving robustness of LCL-type grid connected inverter, IEEE Trans. Power Electron. 29 (7) (July 2014) 3414–3427.
[32] C. Zou, B. Liu, S. Duan, R. Li, Influence of delay on system stability and delay optimization of grid connected inverters with LCL filter, IEEE Trans. Ind. Inf. 10 (3) (August 2014) 1775–1784.
[33] J. Wang, J.D. Yan, L. Jiang, J. Zou, Delay-dependent stability of single-loop controlled grid connected Inverters with LCL filters, IEEE Trans. Power Electron. 31 (1) (January 2016) 743–757.

[34] R. Spence, R.S. Sion, Tolerance Design of Electronic Circuits, Imperial College Press, 1997.

[35] N. Femia, G. Spagnuolo, True worst-case circuit tolerance analysis using genetic algorithms and affine arithmetic, IEEE Transac. Circ. & Syst. I: Fundament. Theor. & Appl. 47 (No. 9) (2000) 1285−1296.

[36] T. Fukuyama, T. Kato, Tolerance computation of a power electronic circuit by higher order sensitivity analysis method, Trans. IEEJ 121-D (8) (2001) 835−840.

[37] A. Kaneko, K. Oue, T. Kato, K. Inoue, Computation of frequency characteristics of grid inverters by general small-signal analysis method, in: IEEE 20th Workshop on Control and Modeling for Power Electronics (COMPEL), Paper No. 32, 2019.

[38] K. Oue, S. Sano, T. Kato, K. Inoue, Stability analysis of grid-forming inverter in dq frequency domain, in: IEEE Power and Energy Society General Meeting, IEEE 20th Workshop on Control and Modeling for Power Electronics (COMPEL), Paper No. 38, 2019.

[39] D.G. Bedrosian, J. Vlach, An accelerated steady-state method for networks with internally controlled switches, IEEE Trans. Circ. Syst. I 39 (7) (1992) 520−530.

[40] T. Kato, W. Tachibana, Periodic steady-state analysis of an autonomous power electronic system by a modified shooting method, IEEE Trans. Power Electron. 13 (3) (1998) 522−527.

[41] T. Kato, K. Inoue, Y. Donomoto, Fast current-tracking control for grid connected inverter with an LCL filter by sinusoidal compensation, IEEE Energy Conver. Congr. & Expos. (2011) 2543−2548.

Dynamic impacts modeling and evaluation of grid connected converters

5

5.1 Dynamic timescales and stability classification

5.1.1 Dynamic timescales and stability in renewable integrated power grids

Fig. 5.1 shows the timescales for various dynamic control and phenomena in power grids. It can be seen that the timescale related to the controls of grid connected converters (GCCs) ranges from a few microseconds to several milliseconds, thus encompassing wave and electromagnetic phenomena. The GCCs, due to their faster dynamics, will gain more prominence when analyzing renewable integrated power grid dynamics compared to a conventional power grid with the timescale of several milliseconds to minutes [1].

Compared to the large synchronous machines (synchronous generators [SGs]), the GCCs with primary sources are able to change their output much faster arresting system's frequency changes before any frequency-based relay and load shedding is triggered. Usually to analyze dynamics behavior of a conventional power grid in the timescales of electromechanical and thermodynamic, the fast dynamics associated with the GCCs are not considered, and this issue strongly simplifies the relevant dynamic models. As shown in Fig. 5.1, to consider the GCC dynamics timescale, it is required to extend the bandwidth to cover faster dynamics within electromagnetic timescales.

Power system stability is defined as [2] "the ability of an electric power system, for a given initial operating condition, to regain a state of operating equilibrium after being subjected to a physical disturbance, with most system variables bounded so that practically the entire system remains intact."

Many reasons may cause instability in a renewable integrated power grid, such as (1) undamped oscillations between GCCs and the grid in the subsynchronous/above-synchronous frequencies, (2) loss of synchronism between the GCCs and the grid, (3) loss of fault ride through capability of the GCCs and trip during grid faults, (4) undesirable slow/fast dynamic interactions between the GCCs and the grid, and (5) erroneous contribution to grid support functions (voltage/frequency control and synthetic inertia).

Power system stability phenomena can be divided into four categories: *rotor angle stability*, *voltage stability*, *frequency stability*, and *resonance stability*. Fig. 5.2 shows a classification of the various types of power grid stability, considering the original classifications presented in Refs. [1,2]. The angle stability, frequency stability, voltage stability, and resonance stability are known as four main stability classes. In Ref. [1],

Grid Connected Converters. https://doi.org/10.1016/B978-0-323-99902-1.00004-5

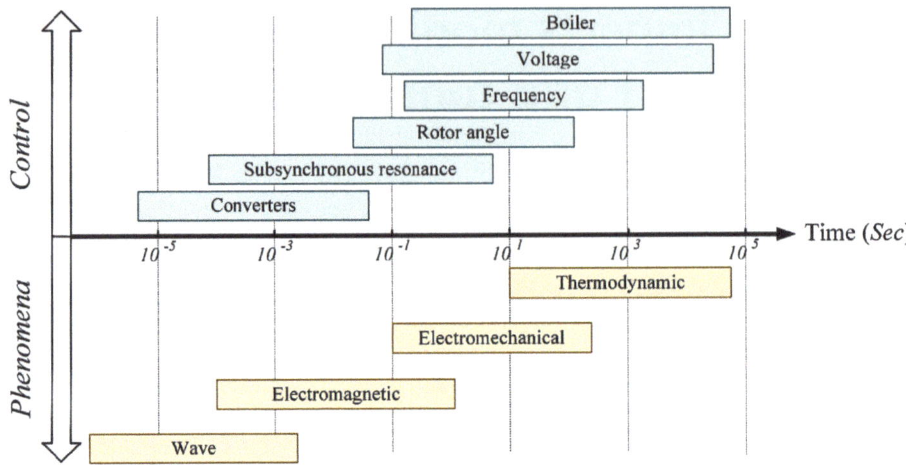

Figure 5.1 Power grid timescales.

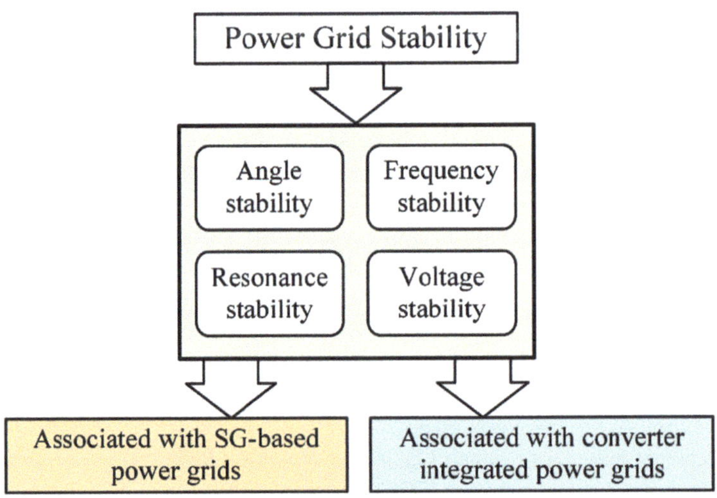

Figure 5.2 Power grid stability classification.

"Converter-driven stability" is introduced as the fifth class; however, since the dynamic impact of penetrated GCC-based distributed generators (DGs) in a power grid can directly affects all above four stability classes, a separate class may not be required for this issue. Here, the four stability classes are briefly explained.

Rotor angle stability is the ability of power grid to maintain synchronization after being subjected to a disturbance. In case of transient/large-disturbance angle stability, a severe disturbance does allow a generator to deliver its output electricity power into the grid. Small signal/steady-state angle stability is the ability of the power grid to maintain synchronization under small disturbances.

Voltage stability is the ability of a power grid to maintain steady acceptance voltages at all grid buses after being subjected to a disturbance from an assumed initial equilibrium point. A grid enters a state of voltage instability when a disturbance changes the grid condition to make a progressive fall or rise of voltages of some buses. Loss of load in an area, tripping of transmission lines, and other protected equipment are possible results of voltage instability [3].

Frequency stability is the ability of a power grid to maintain frequency within the specified operating limits. Generally, frequency instability is a result of a significant imbalance between load and generation, and it is associated with poor coordination of control and protection equipment, insufficient generation reserves, and inadequacies in equipment responses [3].

Resonance stability involves subsynchronous resonance associated with an electromechanical or electrical resonance. The subsynchronous resonance can be appeared because of a resonance between series compensated electrical network and the modes of torsional oscillations on the turbine-generator shaft, or due to a resonance between series compensation and the electrical characteristics of the generating unit. The second one is a purely electrical resonance and termed as *induction generator effect*, which is of interest in a renewable integrated power grid [1].

5.1.2 Impact of GCCs on power grid stability

As mentioned and shown in Fig. 5.1, high penetration of GCCs extends the power grid timescales to include some faster dynamics which are directly related to the GCCs and other power electronics switching components. The presence of GCCs affects all mentioned four types of stability in different ranges. For example, the power networks of Ireland [4], Texas [5], and South Australia [6] are already facing stability or performance problems in the presence of high penetration of GCC-based DGs in certain periods of the day.

The complexity of controlling thousands of GCCs and the potential of high overcurrent in GCCs and techniques for explicitly protecting against overloading are known as two major issues in renewable integrated power grids. The first one can be addressed through local measurement-based autonomous control techniques (to eliminate the need for fast control systems) which significantly reduce the mentioned complexity. The second issue is made by the fact that GCC-based DGs have no direct control of current, which this concern can be also addressed by appropriate feedback control methods for relieving GCC overload [7].

Displacement of SGs by GCCs reduces power grid inertia and has an impact on rotor angle stability. In addition to inertia reduction issue, the angle stability also can be influenced by changing the tie-line flows (which affects the damping of inter-area modes and transient stability), and changing the electromechanical modes of the grid.

The inertia reduction in a renewable integrated power grid also affects the frequency response of system, particularly the rate of change of frequency (RoCoF). On the other hand, the GCCs could play a supportive role in the grid frequency control by increasing the margin of reserve. Providing virtual inertia, faster primary frequency response, and smaller droop settings (compared with the SGs) are other facts for suffering the frequency stability [8].

On affecting the voltage stability, it is enough to refer the capability of GCCs in shaping the injected reactive power and performing supportive and corrective dynamics such those so far provided by power electronic systems such as flexible AC transmission systems (FACTS).

As mentioned, resonance stability is originally introduced for traditional subsynchronous resonance class; however, it is now highlighted due to the increased use of GCC-based DGs, particularly for the part of electrical resonance. That is why, significant presence of GCCs in renewable integrated power grids has extended down the dynamics timescale of interest to the electromagnetic phenomena range.

It was investigated that variable speed induction generators used in the doubly fed induction generator (DFIG) wind-turbines would be highly susceptible to induction generator effect-based electrical resonance [9]. The reason of this resonance is in the existing of a negative resistance of the induction generator rotor (seen on the stator side) as well as the action of the DFIG controls governing the connected converter between the stator and rotor circuits [1].

5.1.3 GCC stability

The dynamic behavior of GCCs is completely different from the SGs. A typical GCC relies on control loops and algorithms with fast response times, such as the phase-locked loop (PLL) and the inner-current control loops. In this regard, the wide timescale related to the controls of GCCs can result in cross couplings with both the electromechanical dynamics of machines and the electromagnetic transients of the network, over a wide frequency range from few Hz to hundreds of Hz. In Ref. [1], the GCC-driven stability from a bulk power grid perspective is divided into two forms of fast and slow interaction converter-driven stabilities.

For instance, the interactions of the fast inner-current loops of GCCs with passive system components may cause high-frequency oscillations, may lead to a *fast interaction converter-driven instability*, which is also known as *harmonic instability*. Triggering the GCC output LCL power filter resonance and connected feeder capacitors by the high-frequency switching of GCCs, or the local control dynamics may cause this instability. This type of instabilities can be prevented and/or mitigated by active damping strategies.

On the other hand, the *slow-interaction converter-driven instability* is initiated by slow dynamic interactions of the GCC control systems with slow-response components of the power grid. For instance, the dynamic interaction between the GCC outer power/voltage control loops and the PLL may lead to unstable low-frequency oscillations [1]. Instability issues related to the power transfer limits between the GCCs and a weak grid are also included in this category. This type of instabilities can be prevented and/or mitigated by tuning the PLLs, adding required equipment to enhance network strength, and using supplementary controls.

5.2 A dynamic model for the GCCs integration evaluation

In previous GCC studies, the grid is usually modeled as an infinite bus or a single SG [3]. In other studies, the given multiarea small-signal modeling methods with all types of grid-interfaced devices are too complex and difficult to use for understanding how a certain device affects the system. Although some reduced-order model techniques have been proposed to simplify the mentioned dynamic models by applying the singular perturbation theory to remove nondominant modes, the derivation of the final model is still a tedious work. However, these studies overlook the specific interaction between GCCs and multiple SGs.

This section addresses a simple modeling method for evaluation the GCCs integration impact focusing on active power and frequency dynamics. In this method, the model of a given renewable integrated power grid is simplified to state-space small-signal model of a radical *n* generators single-load. Modal analysis gives insight into how overall system dynamics are synthesized by the individual generators dynamics, and how different designs of GCCs affect the system oscillation modes.

A simple modeling framework is introduced to obtain a reduced-order small-signal state-space model of a radical *n* generators single-load model. Despite different state variables in different generating unit models, a simple universal formulation is given, which makes it as a useful dynamic model to evaluate a power grid composed of both GCCs and SGs. Finally, the modeling approach is verified by time-domain and modal analysis. Interested readers can find an extended discussion with more detail and examples in Ref. [10].

5.2.1 Multi-generator power grid description

Consider a radical *n* generators single-load system as shown in Fig. 5.3. Here, X_{adji} is the equivalent reactance from the transient *emf* to the adjacent bus of the *i*th generator. For an SG, it can be approximated by the *d*-axis transient reactance X'_d; for a GCC, it is the sum of the filter reactance and virtual reactance [11]. X_{gi} is the equivalent grid reactance, which will be discussed later. The sum of X_{adji} and X_i, denoted as X_i, is an important parameter to calculate the synchronizing power coefficient K_i. As the present work is focused on dynamics of active power and frequency, for simplification, line resistance and capacitance are ignored, and all voltage nodes are assumed to be set at 1 pu. Besides, to avoid identifying the equilibrium point δ_{0i}, it is assumed that $\cos \delta_{0i} \approx 1$. It is reasonable because $\cos \delta_{0i} > 0.916$ for a typical $X_i \leq 0.4$ pu, and generally $X_{adji} \leq 0.3$ pu no matter an SG or a GCC. Based on these assumptions, the synchronizing power coefficient can be considered as $K_i = 1/X_i$ in per-unit system.

The small-signal state-space model of the power system shown in Fig. 5.3 can be derived from respective grid connected generating unit (single generator infinite bus) models. Hence, it is necessary to deduce the small-signal state-space models of all generators, as shown in Fig. 5.4.

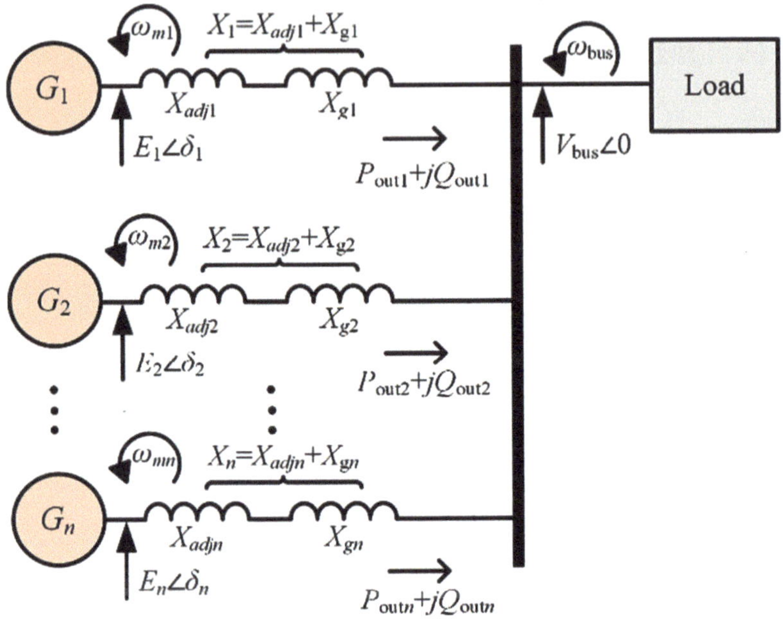

Figure 5.3 A multi-generator power system (radical n generators single-load model).

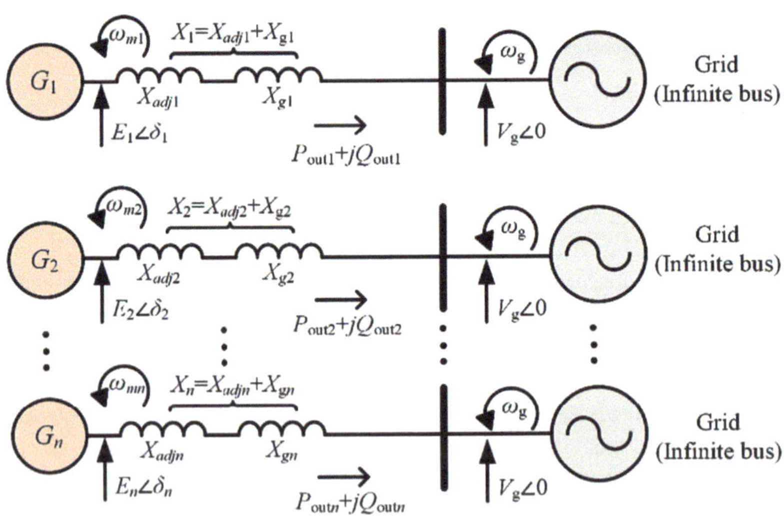

Figure 5.4 Grid connected generators (single generator infinite bus systems).

These models are in the following forms [11]:

$$\begin{cases} \dot{\boldsymbol{x}}_i = \boldsymbol{A}_i\boldsymbol{x}_i + \boldsymbol{B}_i u_i + \boldsymbol{E}_i w_i \\ \boldsymbol{y}_i = \boldsymbol{C}_i\boldsymbol{x}_i \end{cases} \tag{5.1}$$

where

$$u_i = \Delta P_{0i} \tag{5.2}$$

$$w_i = \Delta\omega_{\text{bus}} \tag{5.3}$$

$$\boldsymbol{y}_i = [\Delta\omega_{mi} \quad \Delta P_{\text{out}i}]^T \tag{5.4}$$

$$\boldsymbol{x}_i = \left[\Delta\omega_{mi} \quad \Delta P_{\text{out}i} \quad \boldsymbol{x}_{ai}^T\right]^T \tag{5.5}$$

To obtain the overall power system model, the following matrices are defined based on the grid connected generators models: $\boldsymbol{x}_i'$, $\boldsymbol{B}_i'$, and $\boldsymbol{E}_i'$ are obtained by eliminating the row related to $\Delta P_{\text{out}i}$ from $\boldsymbol{x}_i$, $\boldsymbol{B}_i$, and $\boldsymbol{E}_i$, respectively; $\boldsymbol{A}_i'$ is obtained by eliminating both the row and column related to $\Delta P_{\text{out}i}$ from $\boldsymbol{A}_i$; $\boldsymbol{A}_i''$ is obtained by eliminating the row related to $\Delta P_{\text{out}i}$ from the column vector of $\boldsymbol{A}_i$ related to $\Delta P_{\text{out}i}$.

5.2.2 Unified GCC-based DGs integrated power grid model

In a multigenerator system composed of n GCC-based DG, $\Delta\omega_{\text{bus}}$ should be eliminated from the disturbance input because it becomes dependent on system states thus not a disturbance anymore; instead, we need to consider the load power as a disturbance [11]. This is done by considering the following relationship.

$$\Delta\omega_g = \Delta\omega_{mi} - \frac{1}{K_i}\cdot\frac{\mathrm{d}\Delta P_{\text{out}i}}{\mathrm{d}t}(\forall i \in [1, n] \subset \mathbb{N}) \tag{5.6}$$

Besides,

$$\Delta P_{\text{load}} = \sum_{i=1}^{n} \Delta P_{\text{out}i} \tag{5.7}$$

Eqs. (5.6)–(5.7) yield

$$w_i = \Delta\omega_g = -\frac{1}{K_\Sigma}\cdot\frac{\mathrm{d}\Delta P_{\text{load}}}{\mathrm{d}t} + \sum_{i=1}^{n}\frac{K_i}{K_\Sigma}\Delta\omega_{mi}. \tag{5.8}$$

where $K_\Sigma = \sum_{i=1}^{n} K_i$.

By eliminating $\Delta\omega_g$ in the grid connected model of each GCC and adding Eq. (5.6) as extra state equations, the overall model for an arbitrary number of generators can be obtained as

$$\begin{cases} \dot{\boldsymbol{x}_m} = \boldsymbol{A}_m\boldsymbol{x}_m + \boldsymbol{B}_m\boldsymbol{u}_m + \boldsymbol{E}_m w_m \\ \boldsymbol{y}_m = \boldsymbol{C}_m\boldsymbol{x}_m + \boldsymbol{F}_m w_m \end{cases} \tag{5.9}$$

where

$$\boldsymbol{u}_m = [\,\Delta P_{01} \qquad \Delta P_{0n}\,]^T \tag{5.10}$$

$$w_m = \Delta P_{\text{load}} \tag{5.11}$$

$$\mathbf{y}_m = [\,\Delta\omega_{m1} \quad \ldots \quad \Delta\omega_{mn} \;\vdots\; \Delta P_{\text{out1}} \quad \ldots \quad \Delta P_{\text{out}n}\,]^T \tag{5.12}$$

$$\boldsymbol{x}_m = \left[\begin{array}{c} \boldsymbol{x}'_1 + (1/K_\Sigma)\boldsymbol{E}'_1\Delta P_{\text{load}} \\ \vdots \\ \boldsymbol{x}'_n + (1/K_\Sigma)\boldsymbol{E}'_n\Delta P_{\text{load}} \\ \hline \Delta P_{\text{out1}} - (K_1/K_\Sigma)\Delta P_{\text{load}} \\ \vdots \\ \Delta P_{\text{out}(n\text{-}1)} - (K_{n\text{-}1}/K_\Sigma)\Delta P_{\text{load}} \end{array}\right] \tag{5.13}$$

$$\boldsymbol{A}_m = \left[\begin{array}{ccc|cccc} & & & \boldsymbol{A}''_1 & \boldsymbol{0}_{(d_1+1)\times 1} & \cdots & \boldsymbol{0}_{(d_1+1)\times 1} \\ \boldsymbol{A}_{m11} & \cdots & \boldsymbol{A}_{m1n} & \boldsymbol{0}_{(d_2+1)\times 1} & \boldsymbol{A}''_2 & \cdots & \boldsymbol{0}_{(d_2+1)\times 1} \\ \vdots & & \vdots & \vdots & \vdots & \ddots & \vdots \\ \boldsymbol{A}_{mn1} & \cdots & \boldsymbol{A}_{mnn} & \boldsymbol{0}_{(d_{n-1}+1)\times 1} & \boldsymbol{0}_{(d_{n-1}+1)\times 1} & \cdots & \boldsymbol{A}''_{n-1} \\ & & & -\boldsymbol{A}''_n & -\boldsymbol{A}''_n & \cdots & -\boldsymbol{A}''_n \\ \hline \boldsymbol{A}_{m(n+1)1} & \cdots & \boldsymbol{A}_{m(n+1)n} & & & & \\ \vdots & & \vdots & & \boldsymbol{0}_{(n-1)\times(n-1)} & & \\ \boldsymbol{A}_{m(2n-1)1} & \cdots & \boldsymbol{A}_{m(2n-1)n} & & & & \end{array}\right] \tag{5.14}$$

$$
\boldsymbol{B}_m = \left[\begin{array}{cccc}
\boldsymbol{B}'_1 & \mathbf{0}_{(d_1+1)\times 1} & \cdots & \mathbf{0}_{(d_1+1)\times 1} \\
\mathbf{0}_{(d_2+1)\times 1} & \boldsymbol{B}'_2 & \cdots & \mathbf{0}_{(d_2+1)\times 1} \\
\vdots & \vdots & \ddots & \vdots \\
\mathbf{0}_{(d_n+1)\times 1} & \mathbf{0}_{(d_n+1)\times 1} & \cdots & \boldsymbol{B}'_n \\
\hline
\multicolumn{4}{c}{\mathbf{0}_{(n-1)\times n}}
\end{array}\right] \tag{5.15}
$$

$$
\boldsymbol{E}_m = \left[\begin{array}{c}
(1/K_\Sigma)\left(K_1\boldsymbol{A}''_1 - \sum_{j=1}^{n}\boldsymbol{A}_{m1j}\boldsymbol{E}'_j\right) \\
\vdots \\
(1/K_\Sigma)\left(K_n\boldsymbol{A}''_n - \sum_{j=1}^{n}\boldsymbol{A}_{mnj}\boldsymbol{E}'_j\right) \\
\hline
-(1/K_\Sigma)\sum_{j=1}^{n}\boldsymbol{A}_{m(n+1)j}\boldsymbol{E}'_j \\
\vdots \\
-(1/K_\Sigma)\sum_{j=1}^{n}\boldsymbol{A}_{m(2n-1)j}\boldsymbol{E}'_j
\end{array}\right] \tag{5.16}
$$

$$
\boldsymbol{C}_m = \left[\begin{array}{cccc|c}
\begin{bmatrix}1 & \mathbf{0}_{1\times d_1}\end{bmatrix} & \mathbf{0}_{1\times(d_2+1)} & \cdots & \mathbf{0}_{1\times(d_n+1)} & \\
\mathbf{0}_{1\times(d_1+1)} & \begin{bmatrix}1 & \mathbf{0}_{1\times d_2}\end{bmatrix} & \cdots & \mathbf{0}_{1\times(d_n+1)} & \mathbf{0}_{n\times(n-1)} \\
\vdots & \vdots & \ddots & \vdots & \\
\mathbf{0}_{1\times(d_1+1)} & \mathbf{0}_{1\times(d_2+1)} & \cdots & \begin{bmatrix}1 & \mathbf{0}_{1\times d_n}\end{bmatrix} & \\
\hline
\multicolumn{4}{c|}{\mathbf{0}_{n\times(\sum_{i=1}^{n} d_i+n)}} & \begin{array}{c}\boldsymbol{I}_{n-1} \\ -1 \;\cdots\; -1\end{array}
\end{array}\right] \tag{5.17}
$$

$$F_m = \begin{bmatrix} -E'_{1(1)}/K_\Sigma \\ \vdots \\ -E'_{n(1)}/K_\Sigma \\ \hline K_1/K_\Sigma \\ \vdots \\ K_n/K_\Sigma \end{bmatrix} \tag{5.18}$$

where $\boldsymbol{I}_{n-1}$ is the $(n-1)\times(n-1)$ identity matrix, $E'_{i(1)}$ is the first element of $\boldsymbol{E}'_i$, and the submatrices in $\boldsymbol{A}_m$ and $\boldsymbol{E}_m$ are

$$\boldsymbol{A}_{mij} = \begin{cases} \boldsymbol{A}'_i + \left[(K_j/K_\Sigma)\boldsymbol{E}'_i \quad \boldsymbol{0}_{(d_i+1)\times d_j} \right] \text{ if } i=j, \\ \left[(K_j/K_\Sigma)\boldsymbol{E}'_i \quad \boldsymbol{0}_{(d_i+1)\times d_j} \right] \text{ otherwise,} \end{cases} \tag{5.19}$$

$$\text{for } i,\ j \in [1, n].$$

$$\boldsymbol{A}_{m(n+i)j} = \begin{cases} \left[K_i(K_\Sigma - K_j)/K_\Sigma \quad \boldsymbol{0}_{1\times d_j} \right] \text{ if } i=j, \\ \left[-K_iK_j/K_\Sigma \quad \boldsymbol{0}_{1\times d_j} \right] \text{ otherwise,} \end{cases} \tag{5.20}$$

$$\text{for } i \in [1, n-1] \text{ and } j \in [1, n].$$

It can be confirmed that Eq. (5.9) conforms to the studies in Ref. [11], as the cases $n = 1$ and 2 correctly match the single- and multigenerator grid models reported in Ref. [11]. Moreover, it can be observed from Eq. (5.13) that the multigenerator system is a $\left(\sum_{i=1}^{n} d_i + 2n - 1 \right)$ *th* order system, in which the first $\sum_{i=1}^{n} d_i + n$ state variables, that is, $\boldsymbol{x}'_i + (1/K_\Sigma)\,\boldsymbol{E}'_i \Delta P_{\text{load}}$ $(i \in [1, n])$, come from the individual generating units, and the remained $n - 1$ ones, that is, $\Delta P_{\text{out}i} - (K_i/K_\Sigma)\Delta P_{\text{load}}$ $(i \in [1, n-1])$, are generated by the synchronizing effect between the n generators as shown in Eq. (5.6).

5.2.3 Model simplification

Here, in order to apply the proposed state-space model of a radical n generators single-load system to a power grid with a more complicated structure, a disturbance-oriented simplification method is used. To describe the simplification approach, consider the simple IEEE 9-bus test system shown in Fig. 5.5 [12,13] as an example. If we intend to obtain the time-domain response for a disturbance located at Bus 9, the equivalent reactances from the generator adjacent buses to Bus 9 are calculated and applied to X_i shown in Fig. 5.3. The equivalent reactances are calculated in a similar way to short circuit ration calculation, for instance

$$X_1 = X_{T1} + \frac{(X_{4\_6} + X_{6\_9})(X_{4\_5} + X_{5\_7} + X_{7\_8} + X_{8\_9})}{X_{4\_6} + X_{6\_9} + X_{4\_5} + X_{5\_7} + X_{7\_8} + X_{8\_9}} \tag{5.21}$$

The other load buses in the system, considered as constant PQ nodes, can be neglected. It is noteworthy that if a load disturbance at a different bus is investigated, a different state-space model will be obtained due to different X_i. As the system modes should be unique, the location of disturbance cannot be important; this implies that the simplified system does not perfectly reproduce the modes of the original system.

The main reason for the modeling errors lies in the fact that the electrical distance between two remote generators is extended after simplification, causing underestimated magnitude and frequency of the swinging between these two generators. For instance, in Fig. 5.5, the equivalent reactance between G_1 and G_2 is not too large; whereas, for the disturbance at Bus 9, it becomes $X_1 + X_2$, which is a large value. However, for the time-domain results, the simplified model can still give satisfactory

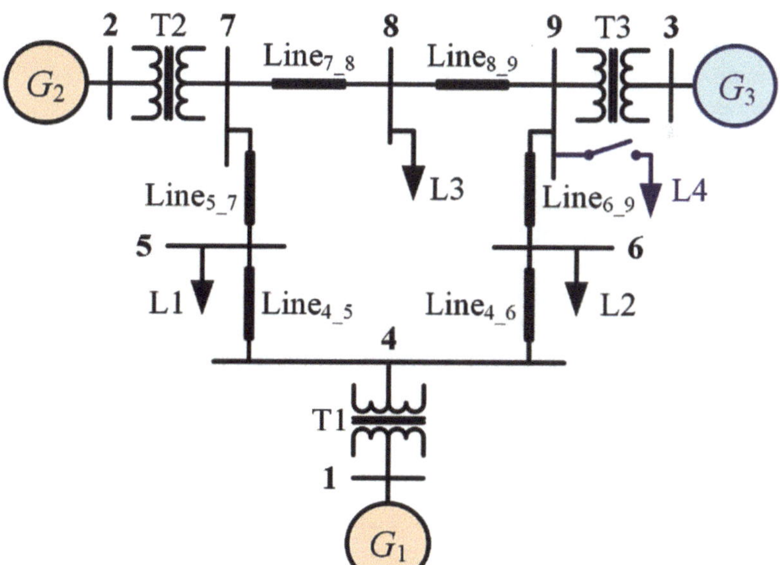

Figure 5.5 IEEE 9-bus system.

accuracy, because generally, the oscillation modes due to remote generators are only slightly excited. It is also necessary to mention that the simplified model correctly reproduces the other modes that less dependent on the electrical distance. This will be later verified using the IEEE 9-bus system.

For a large-scale power system, it is important to aggregate the coherent generators before topology simplification [3,14], to mitigate the electrical distance modifications. Here it should be emphasized that by employing the center-of-gravity (COG) concept [15] for obtaining X_i parameters in the grid shown in Fig. 5.3, one can see that if the COG is chosen as the disturbance bus, the simplified system will have the smallest modeling errors. Thus, the mentioned modal analysis-based model is incorporated with the COG-based representation of the system to obtain the best results. Indeed, COG representation reinterprets original system, including n area, as n sets of single generator infinite bus model of Fig. 5.3, which in turn makes the proposed model independent of topology. This further justifies applicability of the proposed method for large-scale systems.

5.2.4 Simulation study

Exploratory studies to assess the effectiveness of the proposed modeling framework, as well as the grid integration performance of GCCs, are conducted on the IEEE 9-bus test system. To initially demonstrate the accuracy of the proposed modeling method and evaluate the grid integration performance of various types of GCC, three case studies based on the IEEE 9-bus system are discussed.

Here, to investigate the impacts of the integration of GCCs with grid-following and grid-forming control schemes, the IEEE 9-bus system (Fig. 5.5) is used as a test system. The rated power of G_1, G_2, and G_3 are 512, 270, and 125 MVA, respectively. Detailed parameters of the test system are given in Ref. [16].

To evaluate the grid integration performance of GCCs, three case studies are discussed. In case 1, it is assumed that all three generators in Fig. 5.5 are SGs. Case 2 is the same as case 1; except G_3 is replaced by a cluster of coherent grid-following-based GCCs with the same overall rated power. Finally, case 3 is similar to case 2; except the grid-forming control is used to control the replaced GCCs.

The results obtained from the reduced-order models are well matched to the simulation results in the PSCAD (Power Systems Computer Aided Design) environment, except for slight errors in fast oscillations (1−2 Hz) observed in output powers. Both the magnitude and frequency of these oscillations are slightly underestimated. This verifies the satisfactory accuracy of the proposed model and the given discussion on its errors.

Fig. 5.6 shows the G_1 and G_2 frequency and output active power responses, when a 10 MW step load increase is applied to Bus 9. Comparing Case 1 and Case 2, it is clear

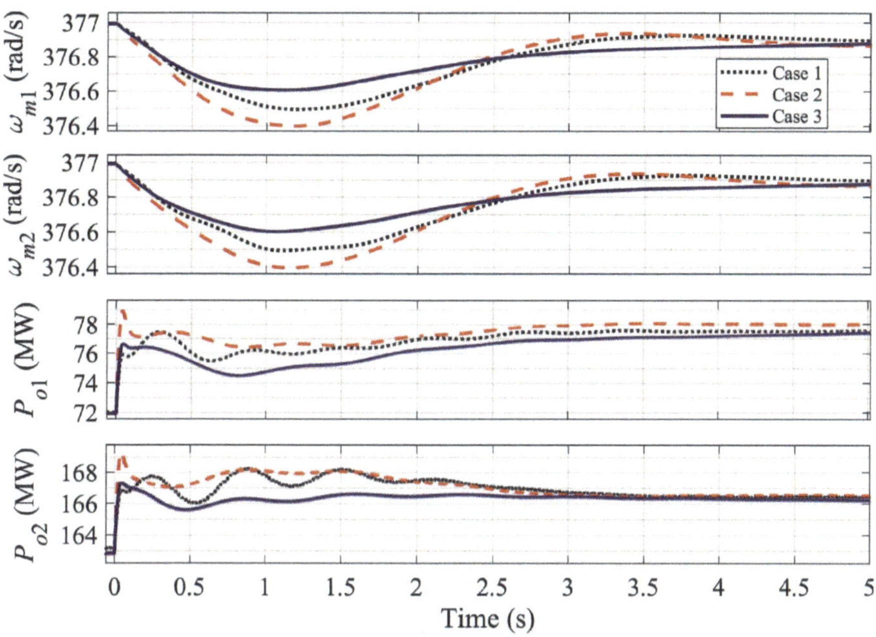

Figure 5.6 System response to 10 MW load increase at Bus 9.

that increased penetration of GCCs in grid-following form leads to having a more severe frequency nadir and higher RoCoF following the disturbance. The results of Case 3 show that these issues can be addressed by the grid-forming type of GCCs with much mitigated RoCoF and frequency nadir. Moreover, Cases 3 has even better performance than Case 1, regarding the improved frequency nadir.

The grid integration performance of GCCs can be also evaluated based on Fig. 5.6. By comparing Case 2 to Case 1, it is clear that increased penetration of grid-following GCCs leads to more severe frequency nadir and higher RoCoF right after the disturbance. It is noteworthy that ω_{m3} is missing in Case 2 because such a quantity does not exist in grid-following GCCs. The results of Case 3 show that these issues can be addressed by the GCC control with much mitigated RoCoF and frequency nadir. Moreover, Case 3 has even better performance than Case 1, regarding the improved frequency nadir. The response of P_{out3} implies that this improvement in frequency regulation is owing to a higher transient output power from the GCC, because a GCC has a faster governor than an SG. Further comparison between Cases 2 and 3 shows that a fast oscillation mode is better damped by the GCC with damping property, owing to the improved damping ratio.

The system performance is also investigated following a 20 MW step decrease in power control input of G_3. The system responses are shown in Fig. 5.7. Comparisons between the studied cases further verify the conclusion obtained with the mentioned scenario. The GCC improves frequency nadir and reduced transient RoCoF.

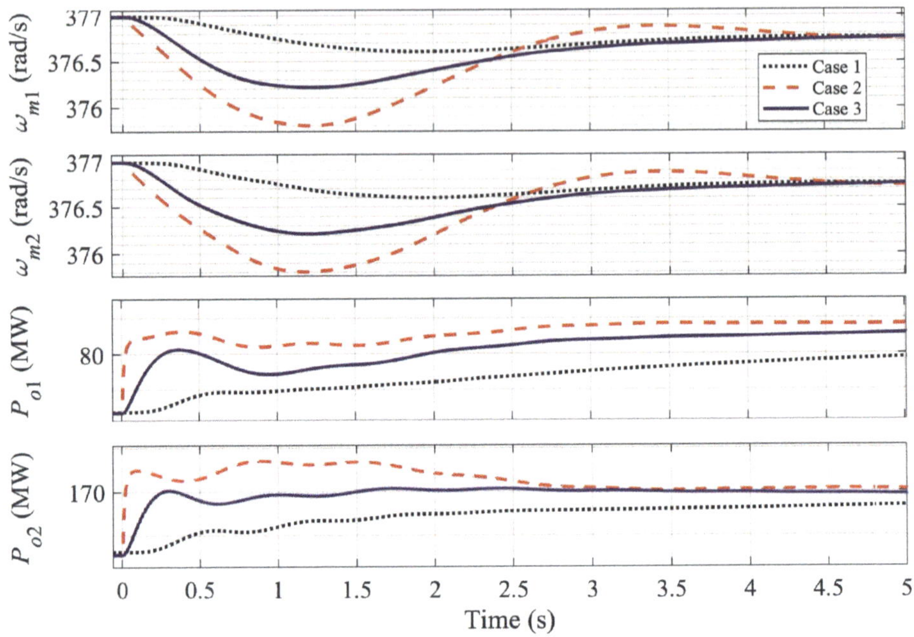

Figure 5.7 System response to 20 MW step decrease in G_3 power command.

The proposed modeling can also be used to investigate the dynamic response to a control input P_{0i} change. Hence, step responses to a 20 MW decrease of P_{03} are shown in Fig. 5.7. As this scenario is not related to a specific bus, the proposed method is applied to the COG-based simplification to obtain the best results. It is noteworthy that for Case 2, as G_2 is not included in the model, such a transient is modeled as a load increase at the location of COG.

Fig. 5.7 shows an even better accuracy of the proposed model owing to the COG-based simplification. Comparisons between the studied cases further verify the conclusion obtained with Scenario 1. Both GCCs outperform the grid-following GCC, with improved frequency nadir and reduced transient RoCoF, whereas GCC with damping has an even better performance owing to the improved damping on the fast oscillation mode related to G_3. On the other hand, Case 1 shows completely different dynamics, because changing P_0 of an SG is delayed by its slow governor and turbine responses.

Detailed parameters of generators, governors, steam turbines, and exciters of the SGs in the test system are shown in Fig. 5.8, and discussed in Refs. [10,17]. The relevant parameters are defined in Ref. [16]. The model of an SG including a governor, a turbine, and a modified second-order generator model considering the disturbance of

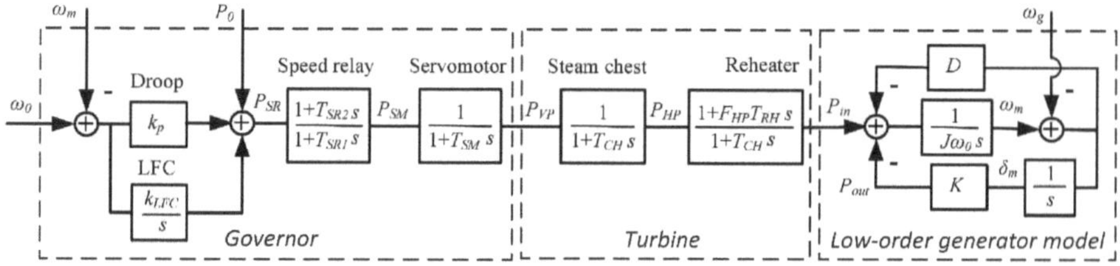

Figure 5.8 SG model including a steam governor and turbine system.

grid (bus) frequency ω_g is presented in Fig. 5.8, whereas those with and without damping GCCs are available in Refs. [11,18], respectively. For simplicity, the symbol Δ denoting a small perturbation is omitted in Fig. 5.8. It is noteworthy that the effect of the exciter, which produces negative damping, is neglected in this work. If the effects of the exciter and/or a power system stabilizer need to be considered, a Heffron-Phillips model should be used instead of the second-order generator model [14].

The state-space model of a grid connected generator model according to Fig. 5.8 can be expressed as follows:

$$\begin{cases} \dot{\boldsymbol{x}} = \boldsymbol{A}\boldsymbol{x} + \boldsymbol{B}u + \boldsymbol{E}w \\ y = \boldsymbol{C}\boldsymbol{x} \end{cases} \tag{5.22}$$

where

$$\boldsymbol{x} = \begin{bmatrix} \Delta\omega_m \\ \Delta P_{\text{out}} \\ \Delta P_{\text{in}} - F_{HP}\Delta P_{HP} \\ \Delta P_{HP} \\ \Delta P_{VP} \\ \Delta P_{SM} - \left(T_{SR2}/T_{SR1}\right)\Delta P_{SR} \\ \Delta P_{SR} + k_p\Delta\omega_m - \Delta P_0 \end{bmatrix}^{\mathrm{T}} \tag{5.23}$$

$$A = \begin{bmatrix} -\dfrac{D}{J\omega_0} & -\frac{1}{J\omega_0} & \frac{1}{J\omega_0} & \frac{F_{HP}}{J\omega_0} & 0 & 0 & 0 \\ K & 0 & 0 & 0 & 0 & 0 & 0 \\ 0 & 0 & -\frac{1}{T_{RH}} & \frac{1-F_{HP}}{T_{RH}} & 0 & 0 & 0 \\ 0 & 0 & 0 & -\frac{1}{T_{CH}} & \frac{1}{T_{CH}} & 0 & 0 \\ -\dfrac{k_p T_{SR2}}{T_{SM} T_{SR1}} & 0 & 0 & 0 & -\frac{1}{T_{SM}} & \frac{1}{T_{SM}} & \frac{T_{SR2}}{T_{SM} T_{SR1}} \\ \dfrac{k_p}{T_{SR1}}\left(\dfrac{T_{SR2}}{T_{SR1}} - 1\right) & 0 & 0 & 0 & 0 & -\frac{1}{T_{SR1}} & \frac{1}{T_{SR1}}\left(1 - \frac{T_{SR2}}{T_{SR1}}\right) \\ -k_{LFC} & 0 & 0 & 0 & 0 & 0 & 0 \end{bmatrix}^T \tag{5.24}$$

$$B = \begin{bmatrix} 0 & 0 & 0 & 0 & \frac{T_{SR2}}{T_{SM} T_{SR1}} & \frac{1}{T_{SR1}}\left(1 - \frac{T_{SR2}}{T_{SR1}}\right) & 0 \end{bmatrix}^T \tag{5.25}$$

$$E = \begin{bmatrix} \dfrac{D}{J\omega_0} & -K & 0 & 0 & 0 & 0 & 0 \end{bmatrix}^{\mathrm{T}}; \; \boldsymbol{C} = [\boldsymbol{I}_2 \;\; 0_{2\times 5}]^{\mathrm{T}}. \tag{5.26}$$

It is noteworthy that this model includes the load-frequency control (LFC) [8]. Without LFC, the seventh-order system shown above is reduced to a sixth-order system, by deleting the columns and rows related to $x_7 = \Delta P_{SR} + k_p \Delta\omega_m - \Delta P_0$.

The SG parameters for the IEEE 9-bus system shown in Fig. 5.5 follow [16]; except an LFC system with $k_{LFC} = 2.5$ pu is applied to G_1, and the damping factor D is calculated based on impedance parameters [14] instead of a fixed value of 2 pu. The subtransient open-circuit time constant T''_{qo} of G_1 and G_2 are changed to 0.05 s. The droop coefficient k_p of G_1 is changed to 30 pu. The original values will result in unstable individual generator poles.

As a grid-following GCC outputs a constant power, that is, does not participate in frequency regulation, Case 2 can be modeled as a multi-generator system consisting of only G_1 and G_2.

5.3 An updated frequency response model for a GCC-based DGs integrated power system

Sustained off-normal frequency variations for a long time may negatively affect power grid operation, stability, security, and performance. Depending on the size and time of frequency variation, different types of frequency controllers are needed to stabilize and restore the power grid frequency. The "size" of frequency deviation refers to the amplitude of Δf; which accordingly shows the size of disturbance and generation-load imbalance. Under normal operation, small frequency changes can be compensated by the primary control loop through the governor system of the existing SGs. Following a disturbance, the mentioned regulation loop of the SGs responds for a few seconds. However, due to the proportional characteristic of SGs droops, the grid frequency will settle down to a value different from the nominal frequency [19].

Fig. 5.9 depicts a simplified frequency response model (FRM) including primary control loop [8]. The R_G, D_G, and M_G are the aggregated droop characteristic, damping factor, and inertia of power system, respectively. The T_{Gg} and T_{Gt} are the aggregated time constants of governors and turbines of the power system SGs, respectively. The ΔP_G, ΔP_{G-ref}, and $\Delta\omega_G$ are the total power generation change, power reference change, and power system frequency change, respectively.

The high penetration of GCC-based DGs in power grids introduces technical challenges due to their high uncertainty, intermittency, and nonsynchronous grid connection. This type of sources increases the necessity of flexibility in operation and regulation power requirements. Furthermore, the replacement of SGs by power electronic-based DGs/RESs reduces system rotational inertia. To have a secure and reliable power grid with high penetration of GCCs, it is vital to use the significant potential of these converters in providing regulation power and frequency control supports.

The GCCs can receive the frequency and power set-points and other required operation/control references from the corresponding electric utility or system operator to produce the required frequency regulation support. These references are distributed between the GCCs to determine the amount of contribution for each participant power source in the grid frequency regulation. The required amount of GCC-based regulation power is mainly determined by considering the amount of reserve from those SGs that can be relocated [19].

Fig. 5.10A depicts the frequency response dynamics of a GCC with a first-order power measurement filter [20]. The unrelated parameters and dynamics to the frequency and active power as well as the fast dynamics are ignored. The equivalent FRM

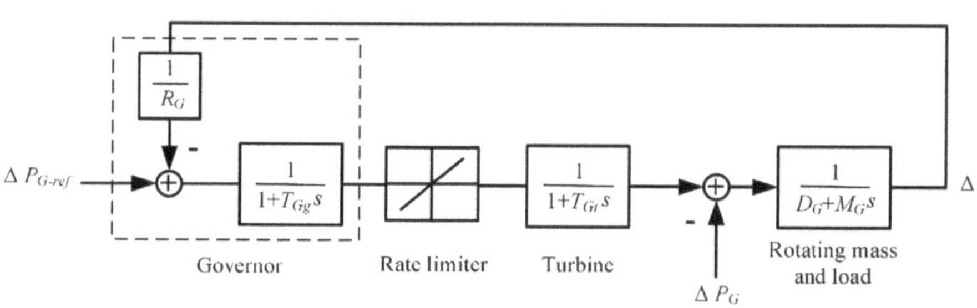

Figure 5.9 Equivalent FRM for a power system with SGs.

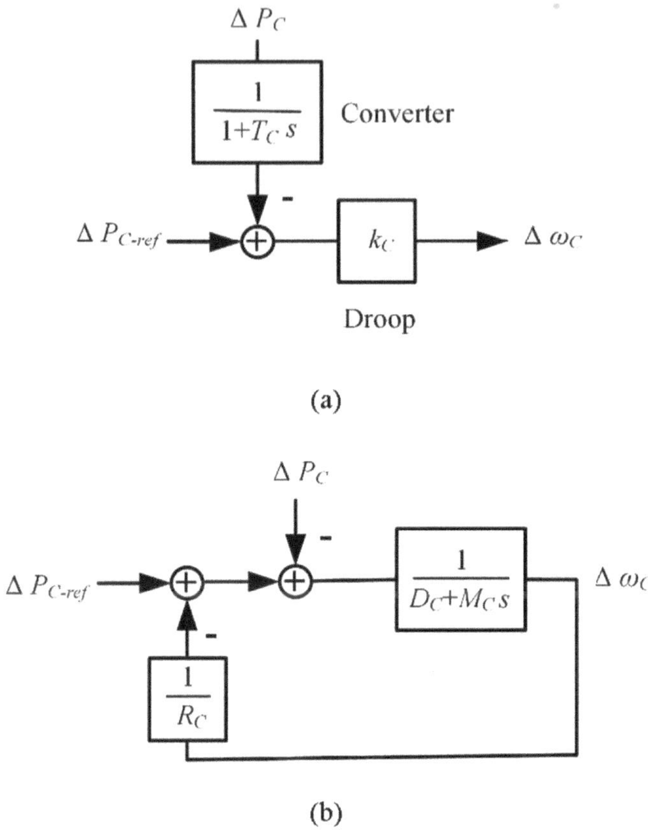

Figure 5.10 FRM for (A) a GCC, and (B) a GCC equivalent dynamics.

concerning the equivalent inertia and damping factors is shown in Fig. 5.10B. In this model, the inertia, droop, and damping parameters can be calculated as follows [21]:

$$M_C \cong \frac{T_C}{k_C},\ R_C \cong k_C,\ D_C \cong \frac{1}{k_C} \tag{5.27}$$

As mentioned, in modeling of GCC, the fast dynamics of inner control loops, frequency measurement (PLL), etc. are not considered since they are relatively fast and do not affect the timescale of a power grid frequency response dynamics.

An updated FRM that includes both conventional SGs and GCC-based DGs in a power grid by considering their different FRMs is presented in Fig. 5.11. Similar to the idea given in Ref. [7], the rate of GCCs penetration (r) can be included in the updated FRM of the renewable integrated power grid. Therefore, the SG power rating reduces to $(1-r)$ pu and the GCC rating to "r" pu to meet a constant 1pu system load. This scales the inertia, droop, and damping values of SG by $(1-r)$ and the GCC inertia, droop, and damping terms by the factor of r.

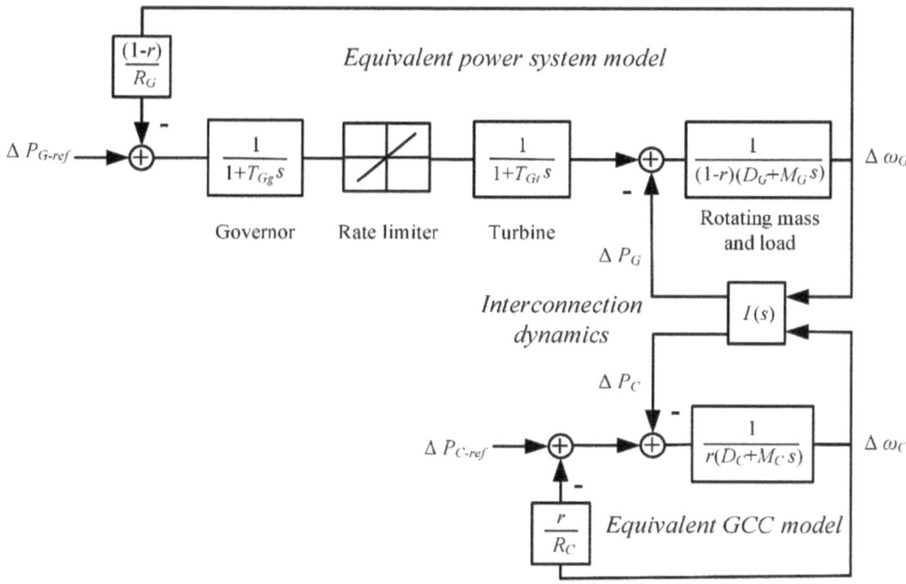

Figure 5.11 Equivalent FRM for a GCCs integrated power system.

The interconnection dynamics between two different FRMs can be represented as

$$\Delta P(s) = I(s)\Delta\omega(s) \tag{5.28}$$

where

$$\Delta P = \begin{bmatrix} \Delta P_G \\ \Delta P_C \end{bmatrix},\ \Delta\omega = \begin{bmatrix} \Delta\omega_G \\ \Delta\omega_C \end{bmatrix},\ I = \begin{bmatrix} i_{11} & i_{12} \\ i_{21} & i_{22} \end{bmatrix}$$

The updated FRM shows that the GCC provides damping feedback on the grid frequency in the system. In addition to damping, the GCC provides a source of inertia for the grid which addresses the concern of low inertia in the renewable integrated power system. The emulated inertia and damping characteristics improve the system frequency dynamic performance. In Ref. [7], it is showed that the dominant characteristic of the GCCs in a power system is damping, and virtually providing delays in the form of inertia is unnecessary and may not be useful.

In the introduced updated FRM (Fig. 5.11), both conventional generating units and GCCs are participating in the frequency regulation. In a renewable integrated power system, the GCCs are needed to actively participate in frequency control issue and maintaining system reliability along with conventional generation.

The contribution of GCCs in active power and frequency regulation refers to the ability of these units to regulate their power output, either by disconnecting a part of generation or by an appropriate control action. More effective practical algorithms and control methodologies are needed to do these issues, properly.

5.4 Summary

High penetration of GCC-based DGs extends the power grid timescales to include some faster dynamics which are directly related to the GCCs and other power electronics switching components. The presence of GCCs affects all types of stability in different ranges. Displacement of SGs by the GCC-based DGs reduces power grid inertia and particularly affects rotor angle and frequency stabilities. In addition to inertia reduction issue, these stabilities can be also influenced by changing the tie-line flows, and changing the electromechanical modes of the grid. On the other hand, the GCCs can support the grid frequency control by emulating inertia and increasing the reserve margin. In this chapter, first dynamic timescales and stability in renewable integrated power grids are discussed. Then simple modeling methods are presented for evaluation the GCCs integration impact on active power and frequency dynamics of the power grid.

References

[1] IEEE Task Force, Stability Definitions and Characterization of Dynamic Behavior in Systems with High Penetration of Power Electronic Interfaced Technologies, PES-TR77, April 2020.
[2] P. Kundur, et al., Definition and classification of power system stability, IEEE Trans. Power Syst. 19 (3) (May 2004) 1387−1401.
[3] H. Bevrani, M. Watanabe, Y. Mitani, Power System Monitoring and Control, Wiley-IEEE Press, Hoboken, NJ, USA, 2014.
[4] EirGrid and SONI, DS3: system services review TSO recommendations, EirGrid Tech. Rep. (May 2013) 1−77.
[5] J. Matevosyan, P. Du, Wind integration in ERCOT, in: Integration of Large-Scale Renewable Energy into Bulk Power Systems, Springer, 2017, pp. 1−25.
[6] Australian Energy Market Operator, Black system South Australia 28 September 2016 - final report, Tech. Rep. (March 2017) 1−271.
[7] R. Lasseter, Z. Chen, D. Pattabiraman, Grid-forming inverters: a critical asset for the power grid, IEEE J. Emerg. & Select. Top. Power Electr. 8 (2) (2019) 925−935.
[8] H. Bevrani, Robust Power System Frequency Control, second ed., Springer, Gewerbestrasse, Switzerland, 2014.
[9] A.E. Leon, J.A. Solsona, Sub-synchronous interaction damping control for DFIG wind turbines, IEEE Trans. Power Syst. 30 (1) (January 2015) 419−428.
[10] J. Liu, H. Golpira, H. Bevrani, T. Ise, Grid integration evaluation of virtual synchronous generators using a disturbance-oriented unified modeling approach, IEEE Trans. Power Syst. 36 (5) (September 2021) 4660−4671.

[11] J. Liu, Y. Miura, H. Bevrani, T. Ise, A unified modeling method of virtual synchronous generator for multi-operation-mode analyses, IEEE J. Emerg. Sel. Topics Power Electron. to be published. https://doi.org/10.1109/JESTPE.2020.2970025.
[12] B. Dembart, et al., Power System Dynamic Analysis – Phase I, Electric Power Research Institute, Palo Alto, CA, USA, July 1977. Rep. EL-484.
[13] H. Bevrani, M. Imanaka, T. Kato, Grid dynamics shaping using flexible controlled power converters, Energy Rep. 6 (Suppl. 9) (2020) 1490–1495.
[14] J. Machowski, J. Bialek, J.R. Bumby, Power System Dynamics and Stability, second ed., John Wiley & Sons, Chichester, UK, 2008, pp. 169–206, 491–534, 557–592.
[15] H. Golpira, A.R. Messina, H. Bevrani, Renewable Integrated Power System: Stability and Control, IEEE-Wiley, 2021.
[16] P. Demetriou, M. Asprou, J. Quiros-Tortos, E. Kyriakides, IEEE 9-bus Modified Test System, 2013 [Online] Available: https://www2.kios.ucy.ac.cy/testsystems/index.php/ieee-9-bus-modified-test-system/.
[17] S. Dong, Y.C. Chen, Adjusting synchronverter dynamic response speed via damping correction loop, IEEE Trans. Energy Convers. 32 (2) (June 2017) 608–619.
[18] J. Liu, Y. Miura, T. Ise, Fixed-parameter damping methods of virtual synchronous generator control using state feedback, IEEE Access 7 (2019) 99177–99190.
[19] H. Bevrani, et al., Power system frequency control: an updated review of current solutions and new challenges, Elec. Power Syst. Res. 194 (2021).
[20] P.J. Hart, R.H. Lasseter, T.M. Jahns, Reduced-order harmonic modeling and analysis of droop-controlled distributed generation networks, in: 2016 IEEE 7th International Symposium on Power Electronics for Distributed Generation Systems (PEDG), 2016, pp. 1–9.
[21] S. D'Arco, J.A. Suul, Equivalence of virtual synchronous machines and frequency-droops for converter-based MicroGrids, IEEE Trans. Smart Grid 5 (1) (January 2014) 394–395.

Part II

Control synthesis for stabilizing and performance enhancement

Control structure of grid connected converters

6

6.1 Overall control structure

6.1.1 Grid-following-based GCC

As mentioned before, the implemented functions by the control system of grid connected converters (GCCs) must be distinguished according to the operating mode of the distributed generator (DG). For the grid-following mode, the AC voltage and frequency are supplied by the grid. So the GCC can be controlled in a "P/Q mode" to inject the assigned powers [1].

A grid-following GCC acts as a current source which injects active and reactive power to the grid according to defined power, or current set-points. The overall control structure of a grid-following GCC is shown in Fig. 6.1. The control structure usually consists of an outer power control loop that generates reference values for the inner current control loop. The current loop therefore needs to be faster than the power control loop.

Like the synchronous generators (SGs), during the start-up, the GCCs should be synchronized with the grid at the coupling point, and in order to form an appropriate output sinusoidal voltage, the phase angle of the grid voltage should be estimated. Furthermore, a frame transform $P(\theta)$ can be used to simplify the realization of the control system by changing the reference system. In order to use the *dq*-transformation in the control of a GCC, information regarding the grid frequency is needed in order to apply the *abc* to *dq*-transformation.

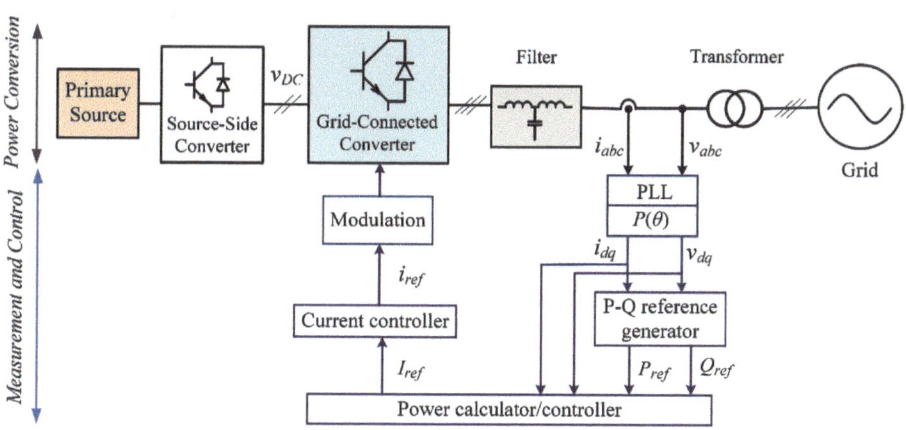

Figure 6.1 A typical grid-following-based GCC with some control blocks.

Grid Connected Converters. https://doi.org/10.1016/B978-0-323-99902-1.00003-3

The phase-locked loop (PLL) serves the purpose of providing the correct phase angle and frequency to the *dq*-transformation making it possible to synchronize the GCC to the grid specifically in grid-following control. Other alternatives can be used instead of the PLL to detect the grid angle. In any case, an estimate of the grid angle is needed either to generate the three-phase sinusoidal references or the angle to be used in the park transform [2].

According to Fig. 6.1, the voltages V_d and V_q can be written as

$$V_d(t) = V\cos(\omega t + \theta_0 - \phi);\ V_q(t) = V\sin(\omega t + \theta_0 - \phi) \tag{6.1}$$

where ϕ is the rotation of the *dq*-frame. If synchronized, $\phi = \omega t + \theta_0$ and it can be observed from Eq. (6.1) that $V_q(t) = 0$. Therefore, regulating the V_q voltage to zero can be used to provide the correct rotation to the *dq*-transformation. This can be formulated in a feedback (FB) law as $\omega = K(s)V_q(t)$.

where $K(s)$ is a linear transfer function and can be implemented as a proportional-integral (PI) regulator. When ω is obtained, since $d\phi/dt = \omega$, for calculating ϕ an integrator is needed. This integrator is implemented as a voltage-controlled oscillator (VCO) and can be seen as a wrapped state integrator that limits its output between 0 and 2π and resets the output to zero when 2π is reached. Fig. 6.2A shows an implementation of the PLL and how it provides the correct rotation to the *abc-dq* transformation [3].

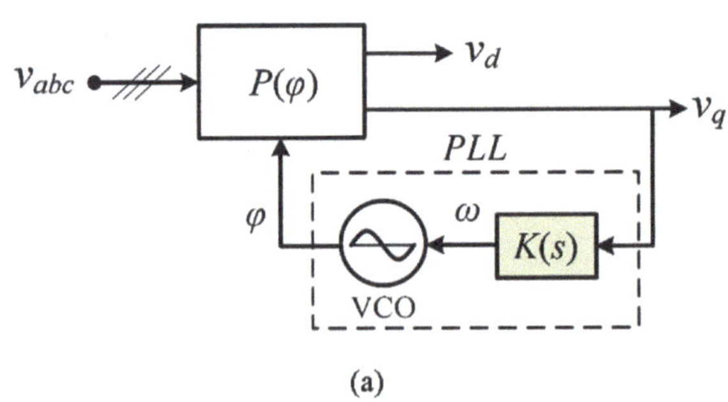

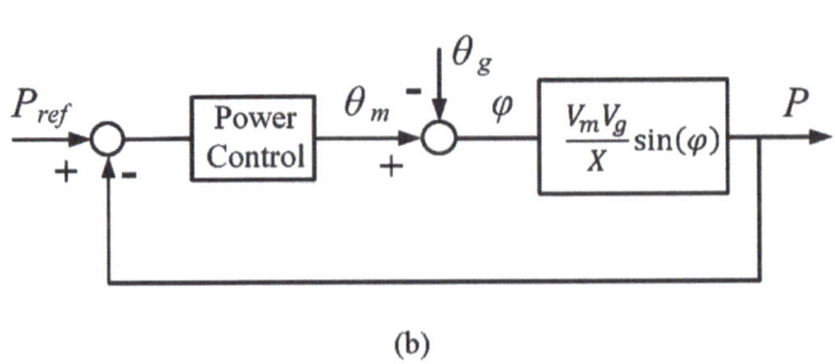

Figure 6.2 The schematic diagram of (A) a PLL and *abc-dq* transformation, (B) power control loop.

Although the PLL function may affect the interconnected system dynamics during the synchronization process, it is frozen after synchronization and will not affect the control system. The grid-following-based GCCs requires synchronization signal to stay synchronized with the grid during operation [4].

Indeed, the aim of the PLL in control systems of grid-following GCC and grid-forming GCC is not the same. The grid-following control uses an estimate of the grid angle since the injected current has to be referenced to the grid voltage. In the grid-forming control, there is an internal angle reference. The information about the grid angle or frequency is helpful only to reject the grid disturbance and to achieve a better active power control. This kind of information is of highest importance on the functionalities which are achieved by a GCC in the power system [2].

Let to use the second formulation of the active power to explain the basics of the second type of control. The active power is linked with the modulated angle δ_m, and as shown in Fig. 6.2B, the grid angle δ_g acts as a disturbance on the power control [2].

The magnitude reference for V_m is independent of the active power control, it can be modified with respect to a voltage or reactive power loop. Hence, contrary to the previous control, a variation on the grid voltage has no direct effect on the modulated voltage since the active power only generates an angle reference and not the magnitude. To modify the difference of angle between the modulated voltage angle δ_m and the grid voltage angle δ_g, the control has to modify slightly the modulated voltage frequency. An internal frequency has to be defined as ω_m, which is equal to the grid frequency in steady state.

For the grid-forming mode which is the interested operation mode in the present book, the GCC has to control its output voltage and frequency using the local produced references like a voltage source converter (VSC).

6.1.2 Grid-forming-based GCC

The most important control blocks of a typical grid-forming based GCC are shown in Fig. 6.3. This control structure is briefly discussed in Chapter 2. The current/voltage/power controls are known as *inner control loops*, while DC-link voltage control, power control (maximum power point tracking-MPPT), reference generator unit and dynamic characteristic emulator (to support the power grid) are known as *outer control loops*. Other control function such as active/reactive power control to support the grid, harmonic compensator, and synchronizing block are not shown in Fig. 6.3.

In grid-forming type, there are several well-known control schemes for the GCC control system. This control system may use the VSG concept to emulate the swing equation of SGs using the dynamic characteristic emulator block. In this case, for normal operation, the PLL is often not necessary. Power synchronization control is another control scheme, which provides inertia through the power synchronization loop, and PLL may be used for current control purpose during the grid faults. *Direct power control* is a simple control scheme to provide the voltage and angle of the converter [5].

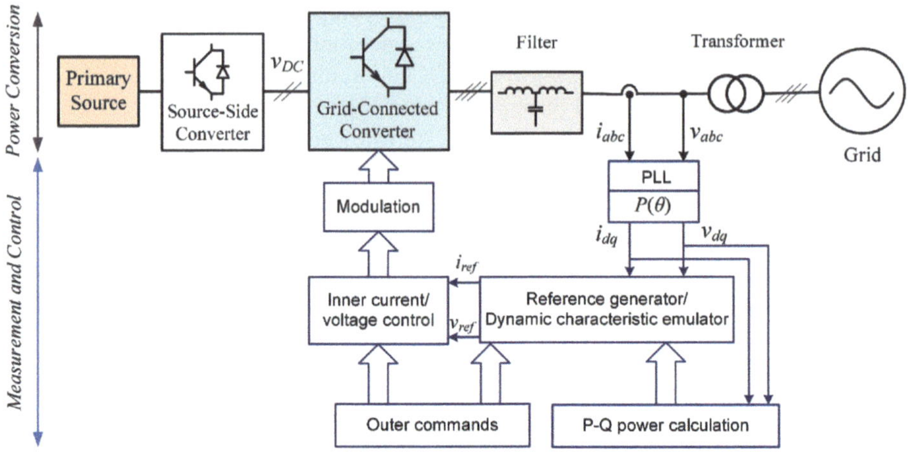

Figure 6.3 A typical grid-forming-based GCC with some control blocks.

Most of GCCs are working as VSCs. The primary source includes a variety of energy sources such as wind-turbine generators, photovoltaic (PV) units, and storage systems. Hence, unlike the SGs, the energy source is not necessarily a mechanical turbine-generator that is converting rotational kinetic energy to electrical energy.

The overall performance and dynamic characteristics of GCCs is dominated by the relevant control systems. The control systems may enable GCCs to fully and independently controlling the exchanged active and reactive powers with the grid [6]. Fig. 6.3 depicts a typical GCC with both energy conversion, and measurement-control systems. The main parts of the energy conversion system are DC bus and grid side converter that converts the DC current to AC for injection into the grid. The PLL locks onto the grid fundamental frequency and thus keeps the converter synchronized with the network.

The "Inner current/voltage control" unit regulates the active and reactive current to be injected into the network based on the commands issued by the "Reference generator/Dynamic characteristic emulator" and "Outer commands" blocks. These two blocks which are usually slower than the inner current/voltage controls include all the required control functions such as frequency-active power control, voltage/reactive power control, ride-through functionality, and dynamics emulator schemes.

A simple dynamic characteristic emulator unit in the GCC is a droop control unit which exhibits a linear trade-off between frequency and voltage versus real and reactive power, much like a typical SG does in steady state. The droop control gives rise to system-wide synchronization (to reach the same frequency) and power sharing (to provide power by each DG in proportion to its capacity). The VSG (for emulating the dynamic characteristics of an SG [7]), and virtual oscillator controller (for emulating a nonlinear oscillator [8]) provide more complex versions of the dynamic characteristics emulator unit. Despite some differences, the output terminal behavior of a GCC with one of mentioned control schemes resembles a voltage source with an amplitude and frequency that varies with reactive power generation and the system load, respectively [9].

6.2 Main control loops and objectives

A comprehensive description of the GCC control loops can be found in several references such as [1,10–13]. In this section, the most important control loops of a grid-forming-based GCC and the relevant objectives are briefly described. A GCC with overall control structure is shown in Fig. 6.4.

As can be seen, the main control loops can be summarized into two blocks, "*P-ω* droop and relevant control loops" block and "*Q-V* droop and relevant control loops" block. The most important control loops inside each block are explained in Chapter 2, which are more discussed in the coming subsections.

6.2.1 P-ω droop and relevant control loops

The *P-ω* droop and relevant control loops cover those control functions that affect the active power and frequency response of the system and are discussed in Section 3.2 of Chapter 3. These functions include *P-ω* droop, damping characteristic, and inertia; that are copied from the dynamic behaviors of an SG and can be emulated by the following differential equation [13].

$$P_0 - P_g = J\omega_0 \frac{d\omega_m}{dt} + k_p(\omega_m - \omega_0) + D(\omega_m - \omega_g) \tag{6.2}$$

where ω_g and P_g are the measured grid frequency and active power. ω_m is the modulated frequency; k_p and D are the droop gain and damping factor. Finally, ω_0

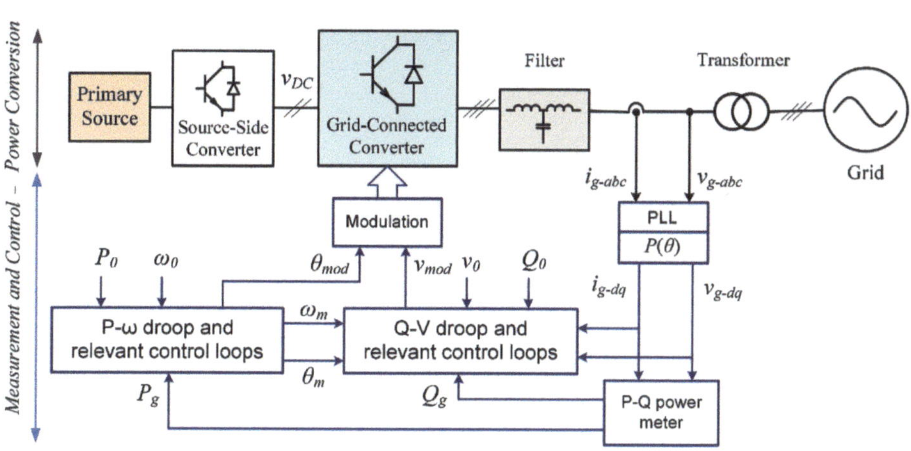

Figure 6.4 A GCC with overall control structure.

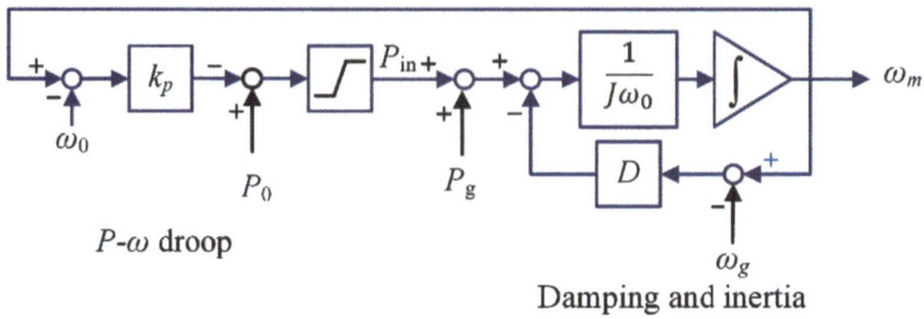

Figure 6.5 A realization scheme for P-ω droop, damping, and inertia.

and P_0 are the nominal grid frequency and active power, respectively. Above dynamic equation can be virtually realized by the simple control loops shown in Fig. 6.5.

6.2.2 Q-V droop and relevant control loops

The "Q-V droop and relevant control loops" block in Fig. 6.4, as discussed in Section 3.3 of Chapter 3, can be mainly realized using two different schemes of inner loops in dq frame, that is, VI-based and Q-based methods; which are replotted in Fig. 6.6A and B, respectively.

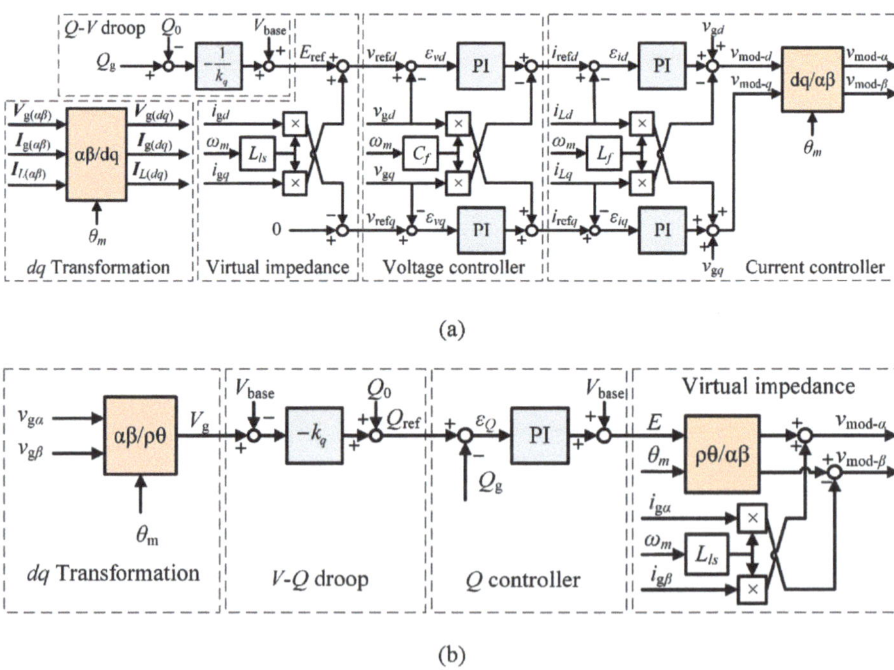

Figure 6.6 The "Q-V droop and relevant control loops" block with (A) inner VI-based control, and (B) Q-based control.

6.2.2.1 VI-based control and Q-based control

The *VI*-based control (Fig. 6.6A) needs FBs of $\boldsymbol{V}_g, \boldsymbol{I}_g$, and $\boldsymbol{I}_L$, which are the vectors of grid voltage, grid current, and load current at the point of common coupling, respectively, as shown in Fig. 6.6A. It begins with a *Q*-*V* droop control as follows [12].

$$E_{\text{ref}} = V_{\text{base}} - (1/k_q)(Q_g - Q_0) \tag{6.3}$$

where Q_g is the output reactive power, Q_0 is the reactive power command, E_{ref} is the reference of virtual electromotive force (*emf*), V_{base} is the rated voltage, and k_q is the *V*-*Q* droop coefficient. With the obtained E_{ref}, the voltage loop references $v_{\text{ref}d}$ and $v_{\text{ref}q}$ are obtained through the *dq* frame-based virtual impedance control as

$$v_{\text{ref}d} = E_{\text{ref}} + \omega_m L_{ls} i_{gq} \tag{6.4}$$

$$v_{\text{ref}q} = -\omega_m L_{ls} i_{gd} \tag{6.5}$$

where L_{ls} is the virtual inductance.

The voltage controller applies *PI* controls for both v_{gd} and v_{gq}, which can be expressed as

$$\varepsilon_{vd} = v_{\text{ref}d} - v_{gd} \tag{6.6}$$

$$i_{\text{ref}d} = K_{pv}[1 + 1/(T_{iv}s)]\varepsilon_{vd} - \omega_m C_f v_{gq} \tag{6.7}$$

$$\varepsilon_{vq} = v_{\text{ref}q} - v_{gq} \tag{6.8}$$

$$i_{\text{ref}q} = K_{pv}[1 + 1/(T_{iv}s)]\varepsilon_{vq} + \omega_m C_f v_{gd} \tag{6.9}$$

where ε_{vd} and ε_{vq} are the control errors of v_{gd} and v_{gq}, respectively, K_{pv} and T_{iv} are the proportional gain and integral time constant of voltage *PI* control, respectively.

Similarly, the current controller can be expressed as

$$\varepsilon_{id} = i_{\text{ref}d} - i_{Ld} \tag{6.10}$$

$$v_{pwmd} = K_{pi}[1 + 1/(T_{ii}s)]\varepsilon_{id} - \omega_m L_f i_{Lq} + v_{gd} \tag{6.11}$$

$$\varepsilon_{iq} = i_{\text{ref}q} - i_{Lq} \tag{6.12}$$

$$v_{pwmq} = K_{pi}[1 + 1/(T_{ii}s)]\varepsilon_{iq} + \omega_m L_f i_{Ld} + v_{gq} \tag{6.13}$$

where ε_{id} and ε_{iq} are the control errors of i_{Ld} and i_{Lq}, respectively; K_{pi} and T_{ii} are the proportional gain and integral time constant of current *PI* control, respectively.

Fig. 6.6B shows the control scheme of an inner *Q*-based control. It applies a *V-Q* droop control described in Eq. (6.14), which reverses the input and output of the *Q-V* droop control given in Eq. (6.3).

$$Q_{\text{ref}} = Q_0 - k_q(V_g - V_{\text{base}}) \tag{6.14}$$

Afterward, a *PI* controller for the reactive power is applied to generate the virtual *emf E* as

$$\varepsilon_Q = Q_{\text{ref}} - Q_g \tag{6.15}$$

$$E = V_{\text{base}} + K_{pq}[1 + 1/(T_{iq}s)]\varepsilon_Q \tag{6.16}$$

where ε_Q is the control error of Q_{outf}, K_{pq} and T_{iq} are the proportional gain and integral time constant of reactive power *PI* control, respectively.

With the obtained *E* and the phase angle θ_m from the *P-ω* control loop, an *αβ* frame-based virtual impedance control is applied to determine the final output voltage. It can be noticed that the *Q*-based control is much simpler than the *VI*-based control. Only one *PI* controller is applied, and *dq* transformation is not needed. Besides, the sensor for $\boldsymbol{I}_L$ can be omitted in the *Q*-based control.

6.2.2.2 Comparison of VI-based control and Q based control

From the given discussion in Section 3.3 and [12], the main differences between the two investigated *Q-V* control schemes can be concluded as follows:

(1) The *Q*-based control is simpler than the *VI*-based control.
(2) Inner inductor current sensors are required for the *VI*-based control, but not for the *Q*-based control.
(3) L_f contributes to the output impedance of the *Q*-based control but not to that of the *VI*-based control.
(4) The order of the *VI*-based control is larger than the order of *Q*-based control system.
(5) The *VI*-based control suffers from voltage deviation and inaccurate reactive power command following the voltage drop across *X*, but the *Q*-based control does not.
(6) The *VI*-based control is more difficult to tune due to higher pole sensitivity. Moreover, a small K_{pv} results in poor dynamics, and a large K_{pv} conflicts with the inner current loop. Contrarily, the tuning of K_{pq} of the *Q*-based control just adjusts the trade-off between response speed and damping.

Besides, previous studies [14,15] show that the *Q*-based control can share the reactive power between GCCs much better, also owing to its direct *Q-V* droop regulation based on V_g. In fact, the issue of poor reactive power sharing of the *VI*-based control also comes from the voltage drop across *X*.

It is also worthwhile to mention the different current limiting strategies for these two control schemes. To prevent overcurrent, the *VI*-based control can simply limit $\boldsymbol{i}_{\text{ref}(dq)}$ by limiting the output of the antiwindup PI controllers of the voltage loop, whereas a virtual impedance-based current limiting strategy proposed in Ref. [16] is effective for the *Q*-based control [17,18]. The mechanism of the virtual impedance-based current limiting

Table 6.1 Comparison between *VI* and *Q-V* control schemes.

Property	*VI*-based control	*Q*-based control
Controller complexity	Complex	Simple
Inductor current sensors	Necessary	Not necessary
Output impedance	L_f excluded	L_f included
Control system order	Third order	Second order
Reactive power command following	Inaccurate	Accurate
Bus voltage deviation	Large	Small
Parameter tuning	Sensitive	Robust
Reactive power sharing	Poor	Good
Harmonic control without dedicated control	Acceptable	Poor
Harmonic control with dedicated control	Good	Good
Current limiting strategy	PI saturation-based	Virtual impedance-based

is as follows: once the RMS value I_{out} exceeds a predefined threshold I_{th} (typically 1 pu), a temporary virtual impedance $\Delta Z_{ls} = \Delta R_{ls} + j\omega_m \Delta L_{ls}$ will be generated.

The differences between these two control schemes are summarized in Table 6.1. It implies that the *Q*-based control with a dedicated harmonic sequence control is a better candidate for a grid-forming GCC. Nevertheless, the *VI*-based control may still be a choice if a dedicated harmonic sequence control cannot be applied due to the specification limit of the adopted digital signal processor.

6.3 Feedforward and feedback control schemes

Using control loops to satisfy desired stability and performance characteristics of the GCC is a vital issue. In the previous sections and chapters of this book, the FB control loops are emphasized. However, sometimes there is a certain limit to meet all control objectives in employing only FB control loops, and adding a feedforward (FF) control loop may open a new design freedom to solve the limitation. The application of FF control loops is well known for many engineering systems; for example, to stabilize a DC system [19]. Over the recent years, the FF method has been also applied to the GCC systems [20–23]. For example, the system stability is improved by an FF control in Ref. [24], an FF impedance-shaping in Ref. [19], and a capacitor-voltage FF control in Ref. [25].

However, due to the complexity of the GCC dynamics, the FF stabilization method is in face of some practical difficulties. In order to tackle this problem, some solutions and ideas are presented in Refs. [20,21,26–28]. In Ref. [21], a combined FB and FF stabilization design method for a GCC system is presented, which the tracking performance characteristic is covered by an FB control loop and the overall stability is ensured by an FF control loop. It consists of three steps: in the first step, an optimal state-FB control gains for a single-phase control subsystem is designed, in the second step, a resistive characteristic is added by an FF control design based on a fitting method, and in the third step, combining two sets of the designed single-phase control subsystems, a three-phase control system is derived.

A two steps design method is also presented in Section 7.3 of Chapter 7. In the first step, a state FB control loop is designed to have a desirable sinusoidal tracking characteristic. This control loop satisfies the performance objective but may fail to stabilize the system in all operating frequency range. In the second step, a conductive characteristic function is added to the system by an FF control loop which makes the overall system passive and ensures the system stability.

6.3.1 A GCC example

In the present section, only an example to show the structure of a GCC system with FB and FF control loops is presented and synthesis methodologies are not discussed. For this purpose here, the digitally controlled GCC system given in Chapter 4 is recalled as an example. Consider a single-phase GCC with an LCL filter shown in Fig. 6.7 which is controlled with a digital control system.

The system structure for a GCC and its equivalent circuit are shown in Fig. 6.7. An LCL filter of (L_1, C, L_2) is used to sufficiently reduce the harmonic components, and to convert the voltage input $v_i = Eu$ to the current i_{L2} as the output of GCC. The linear differential equations for the GCC focusing on the LCL filter and neglecting the dynamics of switching elements and control system can be obtained as follows:

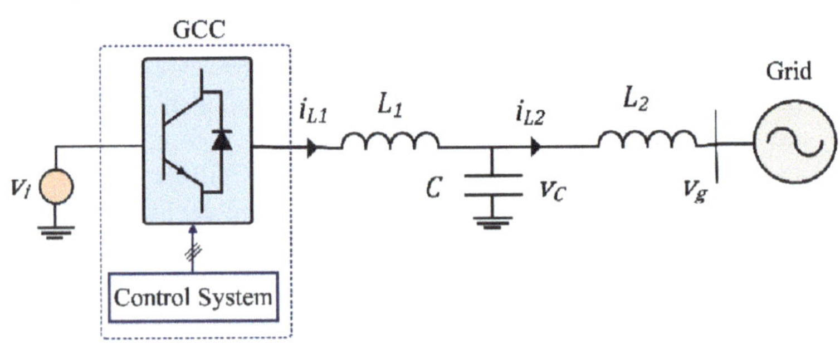

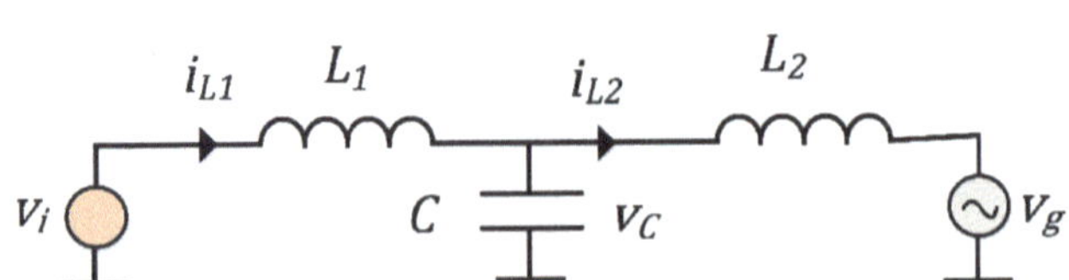

Figure 6.7 (A) A GCC with LCL filter, (B) Equivalent circuit.

$$L_2\frac{di_{L2}}{dt} = v_C - v_g$$

$$L_1\frac{di_{L1}}{dt} = v_i - v_C \qquad (6.17)$$

$$C\frac{dv_C}{dt} = i_{L1} - i_{L2}$$

Assume states vector, measured output, control input, and disturbance input as

$$\dot{\boldsymbol{x}}^T = [\, i_{L1} \quad v_C \quad i_{L2} \,];\; y = i_{L2} = x_3;\; u = v_i;\; w = v_g$$

where v_C is the capacitor voltage. The state-space equation model is obtained as follows, which its block diagram realization is shown in Fig. 6.8A.

$$\dot{\boldsymbol{x}}(t) = \boldsymbol{A}\boldsymbol{x}(t) + \boldsymbol{b}u(t) + \boldsymbol{h}v_g(t)$$

$$y(t) = \boldsymbol{c}x(t) \qquad (6.18)$$

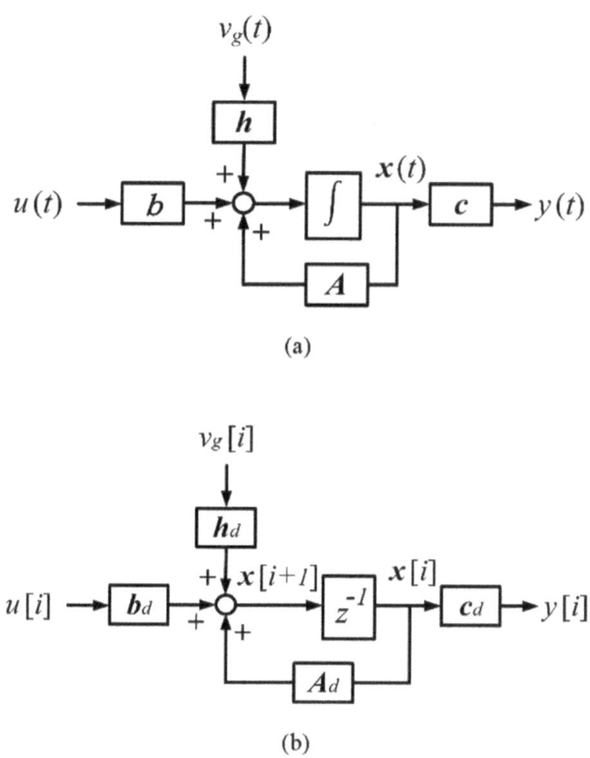

Figure 6.8 (A) The block diagram realization of GCC state-space model; (A) time-domain form (analog), (B) discretized form (digital).

where $\boldsymbol{A}$, $\boldsymbol{b}$, $\boldsymbol{c}$, $\boldsymbol{h}$ are

$$\boldsymbol{A} = \begin{bmatrix} 0 & -1/L_1 & 0 \\ 1/C & 0 & -1/C \\ 0 & 1/L_2 & 0 \end{bmatrix}; \boldsymbol{b} = \begin{bmatrix} E/L_1 \\ 0 \\ 0 \end{bmatrix}; \boldsymbol{c} = \begin{bmatrix} 0 & 0 & 1 \end{bmatrix}; \boldsymbol{h} = \begin{bmatrix} 0 \\ 0 \\ -1/L_2 \end{bmatrix}$$

The discretized form of state vector sampled at $t = iT$ is $\boldsymbol{x}[i]$, where T is a sampling time interval. Its input voltage $v_i[i] = Eu[i]$ is generated according to its input $u[i]$, where E is DC voltage of the GCC source. Discrete state equations for the output current of L_2, $y[i] = i_{L2}[i]$, are as follows.

$$\begin{aligned} \boldsymbol{x}[i+1] &= \boldsymbol{A}_d x[i] + \boldsymbol{b}_d u[i] + \boldsymbol{h}_d v_g[i] \\ y[i] &= \boldsymbol{c}_d x[i] \end{aligned} \tag{6.19}$$

The coefficient matrix and vectors, $\boldsymbol{A}_d$, $\boldsymbol{b}_d$, $\boldsymbol{h}_d$, $\boldsymbol{c}_d$ are given in Eq. (6.20), and its block diagram realization is shown in Fig. 6.8B.

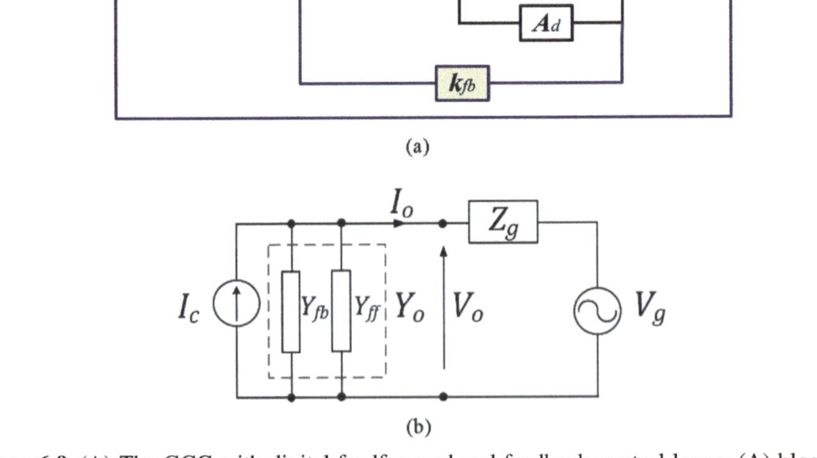

Figure 6.9 (A) The GCC with digital feedforward and feedback control loops; (A) block diagram representation, and (B) equivalent circuit.

$$A_d = e^{A\,T_s};\ b_d = \int_0^{T_s} e^{A\,\tau} d\tau b;\ y(t) = cx(t);\ h_d = \int_0^{T_s} e^{A\,\tau} d\tau h;\ c_d = c. \quad (6.20)$$

6.3.2 Feedforward and feedback control loops

A digital control scheme including both FF and FB control loops can be designed for the given GCC example (Fig. 6.7), as shown in Fig. 6.9. A methodology for designing these controllers is discussed in Chapter 7.

Each control loop may fallow a specific objective, and addition of the FF control loop to the system increases design freedom. To understand the dynamic impacts of the mentioned control loops, one may formulate the impacts as the output admittance (impedance) of the GCC, as shown in Fig. 6.9B. The output admittance shapes the dynamic characteristics of the GCC.

In the proposed synthesis method given in Chapter 7, first a state-FB control gains vector is designed to improve tracking performance. Since the designed characteristic of the output admittance Y_{fb} may not be passive over the interested frequency band, a conductive characteristic Y_{ff} is then added by the FF control loop to ensure stability.

6.4 Virtual synchronous generator

6.4.1 Rotational inertia reduction in modern power grids

Modern power grids face new technical challenges arising from the increasing penetration of GCC-based DGs and Microgrids. Paramount among this is the inertia requirement challenge, as inverter-connected renewable sources are increasingly replacing SGs. Unlike conventional SGs, GCC-based DGs do not provide physical (rotating) inertia support for the grid. Therefore, since SGs are gradually replaced by these DGs, the power grid is faced with the issue of lack of inertia, which intrinsically leads to a large rate of change of frequency (RoCoF) and even frequency instability in the grid [29].

Rotational inertia reduction in the grid may adversely affect frequency dynamics as well as small-signal stability and system control and leads to degrading performance of traditional control schemes. This, in turn, may result in large dynamic deviations and, probably load shedding, and even instability. Accordingly, the increasing penetration of GCC-based DGs motivates the need to develop additional ancillary control services to improve dynamic performance and stability.

To address this issue, the concept of the VSG, virtual synchronous machine, or synchronverter, has been proposed [1]. It is shown that by adding short-term energy storage to emulate the kinetic energy of a rotating mass and mimicking the swing equation of an SG in the control scheme, the GCCs can also provide inertia support for the grid to restrain its frequency fluctuation, in the same way as an SG.

Manipulation of inertia constant by the VSGs could enhance system dynamics and stability. This could be realized using inertia-based control schemes [7]. Indeed, inertia provision plays an important role in designing advanced control schemes in modern power grids. However, the successful implementation of advanced control schemes not only depends on the nature of the designed scheme but also the adequacy of the inertia provision sources.

As discussed in Chapter 9, the proper control of GCC is not only applicable to emulate inertia, and thus to improve the inertia-based power grid dynamics such as frequency and active power control; but it also applicable to other power grid dynamic issues such as improving voltage regulation, reactive power control, and power quality [15]. Therefore, it is true to say that, flexible smart GCCs capable to provide a significant capability to shape the various dynamics of a power grid in desirable directions [29].

6.4.2 Virtual synchronous generator concept and structure

The conceptual architecture of the VSG system is presented in Fig. 6.10. In this scheme, a primary power source (DG or storage system) is connected to the grid via a grid-forming type of GCC. As mentioned above, the VSG is expected to operate as a conventional SG by providing inertia and damping property virtually and by displaying the same reaction of SGs, when there is a sudden change of load or disturbance in the system. Therefore, the VSG control block regulates the output of the GCC based on the RoCoF and the difference between the reference frequency and grid frequency like the way conventional SGs are governed by the swing equation, which is appeared in Eq. (6.2) and can be presented as Eq. (6.21) [30].

$$\Delta P_{\mathrm{VSG}} = P_m - P_e = J\frac{d}{dt}\Delta\omega + D\Delta\omega = J\frac{d^2}{dt^2}\delta(t) + D\frac{d}{dt}\delta(t) \tag{6.21}$$

where J is the inertia coefficient, D is the damping coefficient, and δ is the rotor angular position difference from its reference position. The swing equation describes the relative motion of the rotor concerning the stator field as a function of time. The generator inertia reacts to the disturbance and plays a significant role in power system stability, which is the most important property of an SG.

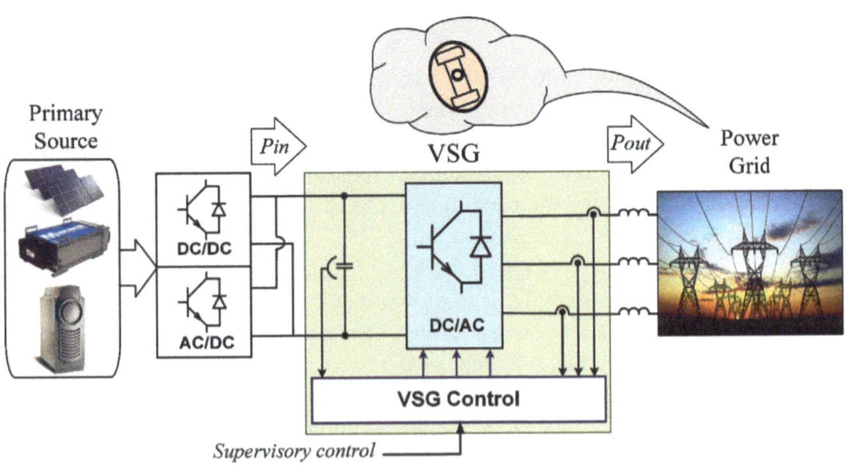

Figure 6.10 The conceptual and general structure of VSG.

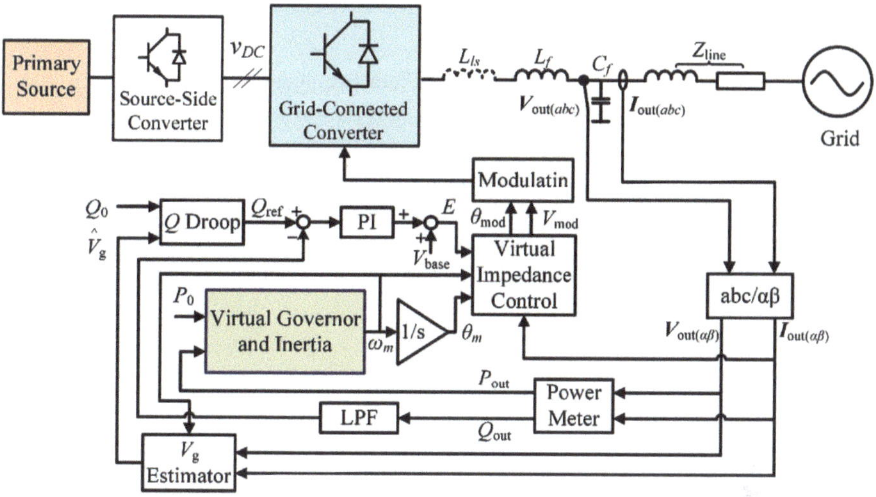

Figure 6.11 The basic control scheme of a VSG.

A typical control scheme of a VSG system is shown in Fig. 6.11 [29]. In the power generation part, the *Virtual Governor and Inertia* block is the core of VSG control. In the literature, various approaches using different damping technologies are proposed for this part. A basic approach, which is based on the emulation of SG swing function and the effect of SG damper windings, is expressed as

$$P_0 - P_{\text{out}} = J\omega_0 \frac{d\omega_m}{dt} + k_p(\omega_m - \omega_0) \tag{6.22}$$

In Fig. 6.11, a virtual voltage drop over virtual inductance L_{ls} is generated to adjust the equivalent output reactance X of the inverter as presented in Eq. (6.23).

$$X = \omega_0 \left(L_{LS} + L_f + L_{\text{line}} \right) \tag{6.23}$$

The block V_g *Estimator* estimates the bus voltage from the measurement of output voltage and current, to provide a common reference for the block *Q Droop*, in which the droop relation between voltage and reactive power is applied [1]. VSG control emulates Eq. (6.22) in the *Virtual Governor and Inertia* block.

Apart from the VSG control topology shown in Fig. 6.11, several different VSG controls for different applications exist in the literature [1,7]. Some applications are frequency regulation support of a remote AC grid via low-frequency AC transmission system [31], enhance the performance of a stand-alone gas-engine generator [32], HVDC-based low-frequency oscillation damping [33], providing vehicle-to-grid (V2G) services [34], and applications in PVs [35] and wind power generation [36] systems. Some applications are presented in Chapter 9.

6.5 Summary

In this chapter, as an introduction to the GCC control system, the control structure of GCCs is described and the main control loops and their role in the GCC dynamic characteristics in both grid-following and grid-forming types are introduced. The control objectives to ensure the system stability and desirable performance are discussed and an example for application of FB and FF control loops is explained. The importance of dynamic characteristics emulation capability in GCCs, the concept, and typical structure of VSG as a type of grid-forming GCC is emphasized.

References

[1] H. Bevrani, B. Francois, T. Ise, Microgrid Dynamics and Control, Wiley, NJ, USA, July 2017.
[2] T. Qoria, X. Guillaud, Control of power electronics driven power sources, in: Chapter 8 of Converter-Based Dynamics and Control of Modern Power Systems, Elsevier, 2021.
[3] A. Yazdani, R. Iravani, Voltage-Sourced Converters in Power Systems: Modeling, Control, and Applications, IEEE, 2010, ISBN 9780470551561, https://doi.org/10.1002/9780470551578.
[4] S. Peyghami, H. Mokhtari, F. Blaabjerg, Y. Yang, Grid synchronization for distributed generations, Encycl. Sustain. Technol. 4 (2017) 179−194.
[5] M. Ndreko, S. Ruberg, W. Winter, Grid forming control for stable power systems with up 100% inverter based generation: a paradigm scenario using IEEE 118-bus system, in: 17th Int. Workshop on Large-Scale Integration of Wind Power into Power Systems as Well as on Transmission Networks for Offshore Wind Power Plants, Sweden, October 2018.
[6] IEEE Task Force, Stability Definitions and Characterization of Dynamic Behavior in Systems with High Penetration of Power Electronic Interfaced Technologies, PES-TR77, April 2020.
[7] H. Bevrani, T. Ise, Y. Miura, Virtual synchronous generators: a survey and new perspectives, Int. J. Electr. Power Energy Syst. 54 (2014) (2014) 244−254.
[8] B.B. Johnson, et al., Synthesizing virtual oscillators to control islanded inverters, IEEE Trans. Power Electron. 31 (8) (2016) 6002−6015.
[9] Y. Lin, et al., Research Roadmap on Grid-Forming Inverters, NREL Technical Report, NREL/TP-5D00-73476, November 2020.
[10] A. Yazdani, R. Iravani, Voltage-sourced Converters in Power Systems, John Wiley & Sons, Hoboken, NJ, USA, 2010.
[11] R. Teodorescu, M. Liserre, P. Rodriguez, Grid Converters for Photovoltaic and Wind Power Systems, John Wiley & Sons, Hoboken, NJ, USA, 2011.
[12] J. Liu, H. Bevrani, T. Ise, A design-oriented Q-V response modeling approach for grid-forming distributed generators considering different operation modes, IEEE J. Emerg. Sel. Topics Power Electron. https://doi.org/10.1109/JESTPE.2021.3057517.
[13] J. Liu, Y. Miura, H. Bevrani, T. Ise, A unified modeling method of virtual synchronous generator for multi-operation-mode analyses, IEEE J. Emerg. Sel. Topics Power Electron. https://doi.org/10.1109/JESTPE.2020.2970025.
[14] Q.-C. Zhong, Robust droop controller for accurate proportional load sharing among inverters operated in parallel, IEEE Trans. Ind. Electron. 60 (4) (April 2013) 1281−1290.

[15] J. Liu, Y. Miura, H. Bevrani, T. Ise, Enhanced virtual synchronous generator control for parallel inverters in microgrids, IEEE Trans. Smart Grid 8 (5) (September 2017) 2268–2277.
[16] A.D. Paquette, D.M. Divan, Virtual impedance current limiting for inverters in microgrids with synchronous generators, IEEE Trans. Ind. Appl. 51 (2) (March–April 2015) 1630–1638.
[17] J. Liu, Y. Miura, T. Ise, Power quality improvement of microgrids by virtual synchronous generator control, in: Proc. Elect. Power Quality Supply Rel. Conf. (PQ), Tallinn, Estonia, 2016, pp. 119–124.
[18] J. Liu, Y. Miura, T. Ise, Parallel operation of a synchronous generator and a virtual synchronous generator under unbalanced loading condition in microgrids, in: Proc. 8th Int. Power Electron. Motion Control Conf. (IPEMC–ECCE Asia), Hefei, China, 2016, pp. 3741–3748.
[19] T. Kato, K. Inoue, M. Ishida, Investigation of stabilities of Lyapunov-based digital control for grid connected inverter, IEEE Energy Conv. Congr. & Expos. (2015) 2394–2399.
[20] T. Kato, K. Inoue, Y. Nakajima, Stabilization of grid connected inverter system with feedforward control, IEEE Energy Conv. Congr. & Expos. (ECCE) (2017) 3375–3382.
[21] T. Kato, K. Inoue, Y. Yamamoto, Optimal digital controller design for passive stabilization of a grid connected three-phase inverter with LCL filter, in: 2019 IEEE Energy Conversion Congress and Exposition (ECCE), September 2019, https://doi.org/10.1109/ECCE.2019.8912644. Baltimore, MD, USA.
[22] L. Harnefors, L. Zhang, M. Bongiorno, Frequency-domain passivity-based current controller design, IET Power Electron. 1 (4) (2008) 455–465.
[23] L. Harnefors, A.G. Yepes, A. Vidal, J. Doval-Gandoy, Passivity based controller design of grid connected VSCs for prevention of electrical resonance instability, IEEE Ind. Electron. 62 (2) (2015) 702–710.
[24] D. Yang, X. Ruan, H. Wu, Impedance shaping of the grid connected inverter with LCL filter to improve its adaptability to the weak grid condition, IEEE Trans. Power Electron. 29 (11) (2014) 5795–5805.
[25] T. Kato, K. Inoue, Y. Donomoto, Fast current-tracking control for grid connected inverter with an LCL filter by sinusoidal compensation, IEEE Energy Conv. Congr. & Expos. (2011) 2543–2548.
[26] X. Wang, F. Blaabjerg, P.C. Loh, Passivity-based stability analysis and damping injection for multi-paralleled voltage-source converters with LCL filters, IEEE Trans. Power Electron. 32 (11) (2017) 8922–8935.
[27] L. Harnefors, R. Finger, X. Wang, H. Bai, F. Blaabjerg, VSC input-admittance modeling and analysis above the nyquist frequency for passivity-based stability assessment, IEEE Trans. Ind. Electron. 34 (8) (2017) 6362–6370.
[28] H. Deng, R. Oruganti, D. Srinivasan, PWM methods to handle time delay in digital control of a UPS inverter, IEEE Trans. Power Electron. 3 (1) (March 2005) 1–6.
[29] H. Bevrani, M. Imanaka, T. Kato, Grid dynamics shaping using flexible controlled power converters, Energy Rep. 6 (Suppl. 9) (2020) 1490–1495.
[30] H. Bevrani, Robust Power System Frequency Control, second ed., Springer, Gewerbestrasse, Switzerland, 2014.
[31] M. Al-Tameemi, J. Liu, H. Bevrani, T. Ise, A dual VSG-based M3C control scheme for frequency regulation support of a remote AC grid via low-frequency AC transmission system, IEEE Access 8 (2020) 66085–66094.

[32] H.S. Hlaing, J. Liu, Y. Miura, H. Bevrani, T. Ise, Enhanced performance of a stand-alone gas-engine generator using virtual synchronous generator and energy storage system, IEEE Access 7 (2019) 176960−176970.

[33] Y. Cao, et al., A virtual synchronous generator control strategy for VSC-MTDC systems, IEEE Trans. Energy Convers. 33 (2) (2017) 750−761.

[34] J.A. Suul, S. D'Arco, G. Guidi, Virtual synchronous machine-based control of a single-phase bi-directional battery charger for providing vehicle-to-grid services, IEEE Trans. Ind. Appl. 52 (4) (2016) 3234−3244.

[35] S. Mishra, D. Pullaguram, S.A. Buragappu, D. Ramasubramanian, Single-phase synchronverter for a grid connected roof top photovoltaic system, IET Renew. Power Gener. 10 (8) (2016) 1187−1194.

[36] Y. Ma, W. Cao, L. Yang, F.F. Wang, L.M. Tolbert, Virtual synchronous generator control of full converter wind turbines with short-term energy storage, IEEE Trans. Ind. Electron. 64 (11) (2017) 8821−8831.

Stability and performance improvement of grid connected converters

7

7.1 Oscillation damping enhancement methods

A grid connected converter (GCC) may suffer from low-frequency oscillations due to interconnected dynamics [1] and resonant phenomena mostly because of output LCL filter [2–4]. Many active damping methods have been developed to suppress these oscillations/resonant effects [5–12]. This section addresses a simple but effective active damping method using a digital signal processor (DSP)-based control system [13].

7.1.1 Discrete state-space model of GCC

Consider a single-phase full-bridge GCC configuration which is connected to a utility source of voltage v_g (grid) as shown in Fig. 7.1. The GCC works as a voltage-source converter that uses a digital control scheme to provide the PWM switching signals. An LCL filter of L_1, C, L_2 is employed at the converter output to convert the voltage input $v_i = E$ to the current output i_{L2} and to reduce its harmonic components sufficiently.

The state variables associated to L_1, C, L_2 are denoted as i_{L_1}, v_C, i_{L_2} and their vector form sampled at $t = iT$ is as $x[i]$ where T is a sampling time interval. The input voltage $v_i[i] = E$ is a DC voltage source. As discussed in Section 6.3, the continuous and discrete state-space equations when the output is the current of L_2 ($y[i] = i_{L2}[i]$), can be obtained as Eqs. (7.1 and 7.2), respectively.

$$\begin{aligned} \dot{\boldsymbol{x}}(t) &= \boldsymbol{A}\boldsymbol{x}(t) + \boldsymbol{b}u(t) + \boldsymbol{h}v_g(t) \\ y(t) &= \boldsymbol{c}x(t) \end{aligned} \tag{7.1}$$

$$\begin{aligned} \boldsymbol{x}[i+1] &= \boldsymbol{A}_d\boldsymbol{x}[i] + \boldsymbol{b}_d u[i] + \boldsymbol{h}_d v_g[i] \\ y[i] &= \boldsymbol{c}_d x[i] \end{aligned} \tag{7.2}$$

Grid Connected Converters. https://doi.org/10.1016/B978-0-323-99902-1.00008-2

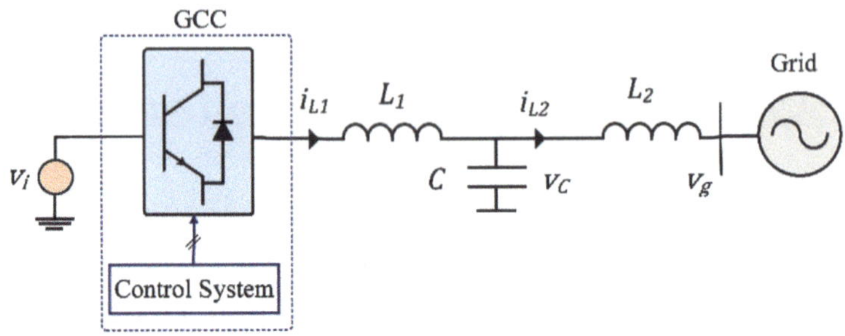

Figure 7.1 A single-phase full-bridge GCC with LCL filter.

where

$$\begin{aligned}
\boldsymbol{A}_d &= e^{AT_s};\\
\boldsymbol{b}_d &= \int_0^{T_s} e^{A\tau} d\tau \boldsymbol{b};\\
\boldsymbol{h}_d &= \int_0^{T_s} e^{A\tau} d\tau \boldsymbol{h};\\
\boldsymbol{c}_d &= \boldsymbol{c}.
\end{aligned} \tag{7.3}$$

and

$$\boldsymbol{A} = \begin{bmatrix} 0 & -1/L_1 & 0 \\ 1/C & 0 & -1/C \\ 0 & 1/L_2 & 0 \end{bmatrix};\ \boldsymbol{b} = \begin{bmatrix} E/L_1 \\ 0 \\ 0 \end{bmatrix};\ \boldsymbol{c} = [0 \quad 0 \quad 1];\ \boldsymbol{h} = \begin{bmatrix} 0 \\ 0 \\ -1/L_2 \end{bmatrix}$$

7.1.2 Digital control scheme

A digital control scheme as shown in Fig. 7.2A is used for the GCC which is described in previous subsection. The control scheme contains a sinusoidal compensator ($G[z]$) and static state feedback (FB) ($\boldsymbol{k}_{fb}$). The compensator is adopted such that the output y follows desired reference signal y_r [14]. s and z^{-1} are differential and one-sample delay operators, and $s = j2\pi f$ and $z = e^{-sT}$. All delays in blocks are unified into one delay block z^{-m} where the total delay time is $d = mT$ [μ sec] $(0 < m)$ from the i-th sampling to $u[i]$ output. The compensator is used to eliminate steady-state error for a reference sinusoidal waveform of an angular frequency ω_0. The transfer function

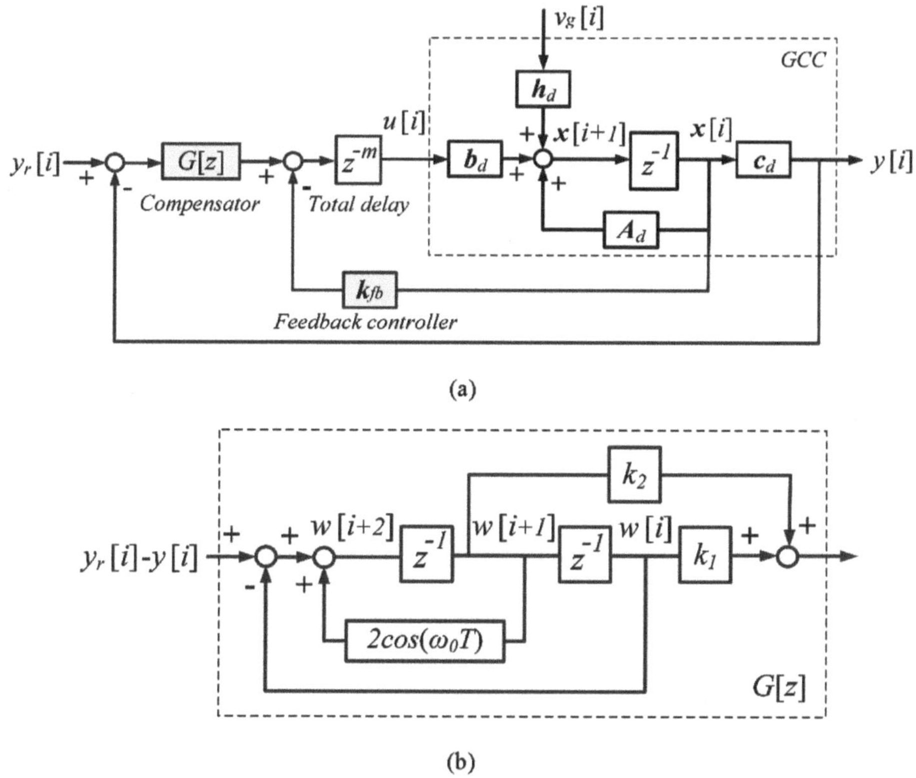

Figure 7.2 Block diagram of (A) GCC and digital control system, and (B) compensator $G[z]$.

of the compensator is as follows, and its denominator is identical to that of z-transform of a sinusoidal waveform [14].

$$G[z] = \frac{k_2 z + k_1}{z^2 - 2\cos(\omega_0 T)z + 1} \tag{7.4}$$

Two control gains k_1, k_2 in the numerator are the design parameters. The control input $u[i]$ is computed according to the following state FB rule, where $w[i]$ and $w[i+1]$ are two auxiliary variables.

$$\begin{aligned} u[i] &= -\boldsymbol{k}_{fb}x[i] + k_1 w[i] + k_2 w[i+1] \\ w[i+2] &= -w[i] + 2\cos\omega_0 T w[i+1] + (y_r[i] - y[i]) \end{aligned} \tag{7.5}$$

The state-space model of the closed-loop system can be obtained as follows:

$$
\left.
\begin{aligned}
\begin{bmatrix} \boldsymbol{x}[i+1] \\ w[i+1] \\ w[i+2] \end{bmatrix} &= \begin{bmatrix} \boldsymbol{A}_d & \boldsymbol{0} & \boldsymbol{0} \\ 0 & 0 & 1 \\ -\boldsymbol{c}_d & -1 & 2\cos\omega T \end{bmatrix} \begin{bmatrix} \boldsymbol{x}[i] \\ w[i] \\ w[i+1] \end{bmatrix} \\
&+ \begin{bmatrix} \boldsymbol{b}_d \\ 0 \\ 0 \end{bmatrix} \begin{bmatrix} -\boldsymbol{k}_{fb} & k_1 & k_2 \end{bmatrix} \begin{bmatrix} \boldsymbol{x}[i] \\ w[i] \\ w[i+1] \end{bmatrix} \\
&+ \begin{bmatrix} \boldsymbol{0} \\ 0 \\ 1 \end{bmatrix} y_r[i] + \begin{bmatrix} \boldsymbol{h}_d \\ 0 \\ 0 \end{bmatrix} v_g[i] \\
&= \begin{bmatrix} \boldsymbol{A}_d - \boldsymbol{b}_d\boldsymbol{k}_{fb} & \boldsymbol{b}_d k_1 & \boldsymbol{b}_d k_2 \\ 0 & 0 & 1 \\ -\boldsymbol{c}_d & -1 & 2\cos\omega T \end{bmatrix} \begin{bmatrix} \boldsymbol{x}[i] \\ w[i] \\ w[i+1] \end{bmatrix} + \begin{bmatrix} \boldsymbol{0} \\ 0 \\ 1 \end{bmatrix} y_r[i] + \begin{bmatrix} \boldsymbol{h}_d \\ 0 \\ 0 \end{bmatrix} v_g[i]
\end{aligned}
\right\} \tag{7.6}
$$

or

$$
\widehat{\boldsymbol{x}}[i+1] = \widehat{\boldsymbol{A}}\widehat{x}\,[i] + \widehat{\boldsymbol{g}}\, y_r[i] + \widehat{\boldsymbol{h}}\, v_g[i]. \tag{7.7}
$$

Based on the above augmented system matrix $\widehat{\boldsymbol{A}}$ and the input vector $\widehat{\boldsymbol{h}}$, three FB gains $\boldsymbol{k}_{fb}$ and two compensator gains k_1, k_2 must be determined by the designer. An optimal control method to calculate these control gains is addressed in Chapter 8.

7.1.3 Damping improvement using virtual impedance

To improve oscillations damping, in addition to the well-known FB controllers for the GCCs, as mentioned in Chapter 2, a solution is to use virtual impedance, such as a virtual resistor R_1 in series with the inductor L_1 as shown in the equivalent circuit of the understudy PWM-based GCC in Fig. 7.3A. The virtual resistor can be virtually created by a proper control mechanism, and unlike an actual resistor does not cause losses. As mentioned, by an appropriate control scheme, the damping resistor can be emulated, and the converter output voltage v_i can be updated to v_i^*:

$$v_i^* = v_i - R_1 i_{L1} \tag{7.8}$$

The resistor R_1 is effective to suppress the resonant peak created by the filter characteristic, and attenuates the oscillations. Desirable values of the state FB gains $\boldsymbol{k}_{fb} = [f_1\, f_2\, f_3]$, and compensator gains (k_1, k_2), can be calculated by the well-known control techniques, for example, optimal control methods. Following the addition of the virtual resistor, the updated FB gain $\breve{f}_1$ of i_{L_1} becomes $\breve{f}_1 = f_1 + R_1/E$.

This virtual damping method can be extended to a more general case by considering a series virtual resistor for L_2 and a parallel virtual conductor for G_C of the filter, as shown in Fig. 7.3B. Therefore, now there are three virtual elements (R_1, R_2, G_C). A simple idea to design the static state FB gain vector (for the GCC in Fig. 7.1) considering the virtual resistors (in Fig. 7.3B) is to set identical eigenvalues, for both static FB controllers with and without resistors cases, so that their damping effects behave similarly. The design process is described below.

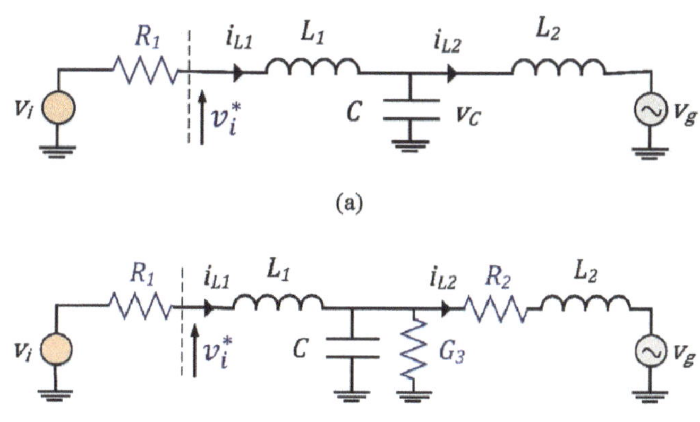

Figure 7.3 Equivalent circuit of GCC with (A) a virtual resistor (R_1), (B) three virtual resistors (R_1, R_2, G_c).

The system equation of the system with the impact of virtual resistors is expressed as follows.

$$\dot{\boldsymbol{x}}(t) = \breve{\boldsymbol{A}}\boldsymbol{x}(t) + \boldsymbol{b}u(t) + \boldsymbol{h}v_g(t) \tag{7.9}$$

where

$$\breve{\boldsymbol{A}} = \begin{bmatrix} -\dfrac{R_1}{L_1} & -\dfrac{1}{L_1} & 0 \\ \dfrac{1}{C} & -\dfrac{G_C}{C} & -\dfrac{1}{C} \\ 0 & \dfrac{1}{L_2} & -\dfrac{R_2}{L_2} \end{bmatrix} \tag{7.10}$$

The system matrix with the static state FB control of $\breve{\boldsymbol{k}}_{fb} = \left[\breve{f}_1 \breve{f}_2 \breve{f}_3\right]$ is

$$\breve{\boldsymbol{A}} - \boldsymbol{b}\breve{\boldsymbol{k}}_{fb} = \begin{bmatrix} -\dfrac{R_1 + \breve{f}_1 E}{L_1} & -\dfrac{1 + \breve{f}_2 E}{L_1} & -\dfrac{\breve{f}_3 E}{L_1} \\ \dfrac{1}{C} & -\dfrac{G_C}{C} & -\dfrac{1}{C} \\ 0 & \dfrac{1}{L_2} & -\dfrac{R_2}{L_2} \end{bmatrix} \tag{7.11}$$

and the system characteristic equation becomes as follows.

$$\breve{\phi}(s) = \left|s\boldsymbol{I} - \breve{\boldsymbol{A}} - \boldsymbol{b}\breve{\boldsymbol{k}}_{fb}\right| = \begin{vmatrix} s - \breve{K}_1 & -\breve{K}_2 & -\breve{K}_3 \\ -\alpha & s - \gamma & \alpha \\ 0 & -\beta & s - \delta \end{vmatrix} \tag{7.12}$$

or

$$\begin{aligned} \breve{\phi}(s) = s^3 &- \left(\gamma + \delta + \breve{K}_1\right)s^2 + \left(\breve{K}_1\gamma + \gamma\delta + \breve{K}_1\delta + \alpha\beta - \breve{K}_2\alpha\right)s \\ &- \breve{K}_1\gamma\delta - \breve{K}_3\alpha\beta - \breve{K}_1\alpha\beta + \breve{K}_2\alpha\delta \end{aligned} \tag{7.13}$$

where

$$\left.\begin{aligned}
&\alpha = \frac{1}{C}, && \beta = \frac{1}{L_2} \\
&\gamma = -\frac{G_C}{C}, && \delta = -\frac{R_2}{L_2} \\
&\breve{K}_1 = -\frac{R_1 + \breve{f}_1 E}{L_1}, \quad \breve{K}_2 = -\frac{1 + \breve{f}_2 E}{L_1}, && \breve{K}_3 = -\frac{\breve{f}_3 E}{L_1}
\end{aligned}\right\}$$

The FB gains $\breve{\boldsymbol{k}}_{fb} = [\breve{f}_1 \breve{f}_2 \breve{f}_3]$ can be easily determined by an optimal control method such as linear quadratic regulator as described in Chapter 8. On the other hand, the system matrix of the original system (Fig. 7.1) with the static state FB control of $\boldsymbol{k}_{fb} = [f_1\, f_2\, f_3]$ is:

$$\boldsymbol{A} - \boldsymbol{b}\boldsymbol{k}_{fb} = \begin{bmatrix} -\dfrac{f_1 E}{L_1} & -\dfrac{1 + f_2 E}{L_1} & -\dfrac{f_3 E}{L_1} \\ \dfrac{1}{C} & 0 & -\dfrac{1}{C} \\ 0 & \dfrac{1}{L_2} & 0 \end{bmatrix} \tag{7.14}$$

and, its characteristic equation becomes

$$\phi(s) = \left| s\boldsymbol{I} - \boldsymbol{A} - \boldsymbol{b}\boldsymbol{k}_{fb} \right| = \begin{vmatrix} s - K_1 & -K_2 & -K_3 \\ -\alpha & s & \alpha \\ 0 & -\beta & s \end{vmatrix} \tag{7.15}$$

or

$$\phi(s) = s^3 - K_1 s^2 + (\alpha\beta - K_2\alpha)s - K_3\alpha\beta - K_1\alpha\beta \tag{7.16}$$

where

$$K_1 = -\frac{f_1 E}{L_1}, \quad K_2 = -\frac{1 + f_2 E}{L_1}, \quad K_3 = -\frac{f_3 E}{L_1}$$

The coefficients of characteristic Eqs. (7.13) and (7.16) are compared so that the constants K_1, K_2, and K_3 are set to have the same eigenvalues.

$$\left.\begin{aligned} K_1 &= \gamma + \delta + \breve{K}_3 \\ \alpha\beta - K_2\alpha &= \breve{K}_1\gamma + \gamma\delta + \breve{K}_1\delta \\ &\quad +\alpha\beta - \breve{K}_2\alpha \\ -K_3\alpha\beta - K_1\alpha\beta &= -\breve{K}_1\gamma\delta - \breve{K}_3\alpha\beta \\ &\quad -\breve{K}_1\alpha\beta + \breve{K}_2\alpha\delta \end{aligned}\right\} \tag{7.17}$$

K_1, K_2, and K_3 are calculated from $\breve{K}_1$, $\breve{K}_2$, and $\breve{K}_3$ as follows:

$$\left.\begin{aligned} K_1 &= \gamma + \delta + \breve{K}_1 \\ \alpha K_2 &= -\{\breve{K}_1\gamma + \gamma\delta + \breve{K}_1\delta - \breve{K}_2\alpha\} \\ \alpha\beta K_3 &= -\{-\breve{K}_1\gamma\delta - \breve{K}_3\alpha\beta - \breve{K}_1\alpha\beta \\ &\quad +\breve{K}_2\alpha\delta + (\gamma + \delta + \breve{K}_1)\alpha\beta\} \end{aligned}\right\} \tag{7.18}$$

Finally, the state FB gains with the proposed virtual damping become:

$$\mathbf{k}_{fb} = \left[-\frac{L_1}{E}K_1 \quad -\frac{K_2L_1 + 1}{E} \quad -\frac{L_1}{E}K_3 \right]. \tag{7.19}$$

7.1.4 Simulation and experimental results

The addressed active damping control method in the previous subsections is applied to a single-phase GCC to investigate its effectiveness via studying the frequency response of the closed-loop system transfer function using a simulation environment and an experimental setup. The input DC voltage and the grid voltage are considered as 200 and 100 V(rms), respectively. The low-pass filter (LPF) parameters are $L_1 = L_2 = 0.76$ mH and $C = 9.3\mu$ F. The switching frequency is 10 kHz, and the grid frequency is 50 Hz. The compensator and FB control gains for the design cases of one virtual resistor and three virtual resistors are determined as Eqs. (7.20 and 7.21), respectively [13]:

$$\mathbf{k}_{fb} = [0.0576 \quad 0.0006 \quad 0.0380],\ k_1 = -0.0241,\ k_2 = 0.0289. \tag{7.20}$$

$$\mathbf{k}_{fb} = [0.0566 \quad 0.0022 \quad 0.0360],\ k_1 = -0.0241,\ k_2 = 0.0289. \tag{7.21}$$

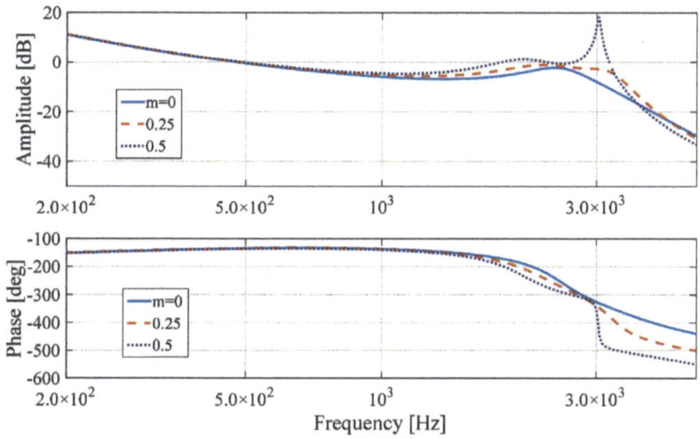

Figure 7.4 Simulated frequency response for one virtual resistor and different time delays.

As shown in the simulation results depicted in Fig. 7.4, the frequency response of the closed-loop system depends on the time delay (m). Here, one virtual resistor is used, and the results are shown for time delays of 0, 0.25, and 0.5 s, while $R_1 = 5\ \Omega$ and the converter internal resistance is 3Ω. Without considering time delay (an ideal scenario), the frequency response curve is almost flat. However, as m increases, an increasing peak and a wider varying phase are appeared.

The laboratory environment to evaluate the synthesis results in this section and other parts of the present chapter is shown in Fig. 7.5A. Experimental results for the case of with and without a virtual resistor in the presence of time delay are shown in Fig. 7.5B, which confirm the mentioned effects. In the performed experimental setup, the GCC with the same parameters as in simulation is controlled by a DSP-based digital control system with a time delay of 50 μs at a sampling time of 100 μs.

The dotted curve shows the system frequency response without the damping control ($R_1 = 0\ \Omega$), and it has a peak around the resonant frequency of 2.7 kHz. The dashed curve shows the frequency response with the designed damping control and still, it has a peak around the resonant frequency due to having a physical time delay. Compensation of time delay to solve this physical constraint and its undesirable effect is addressed in the next section.

7.2 Time delay compensation

The GCC is often controlled digitally, and as discussed in the previous section, there is a control time delay, which is inevitable due to analog to digital (A/D) conversion, computation times, etc. Effects of the delay have been extensively analyzed [15–19], and it is shown that time delay degrades the GCC performance and frequency response, and in worst cases, it may even lead to instability. In Section 7.1, it is

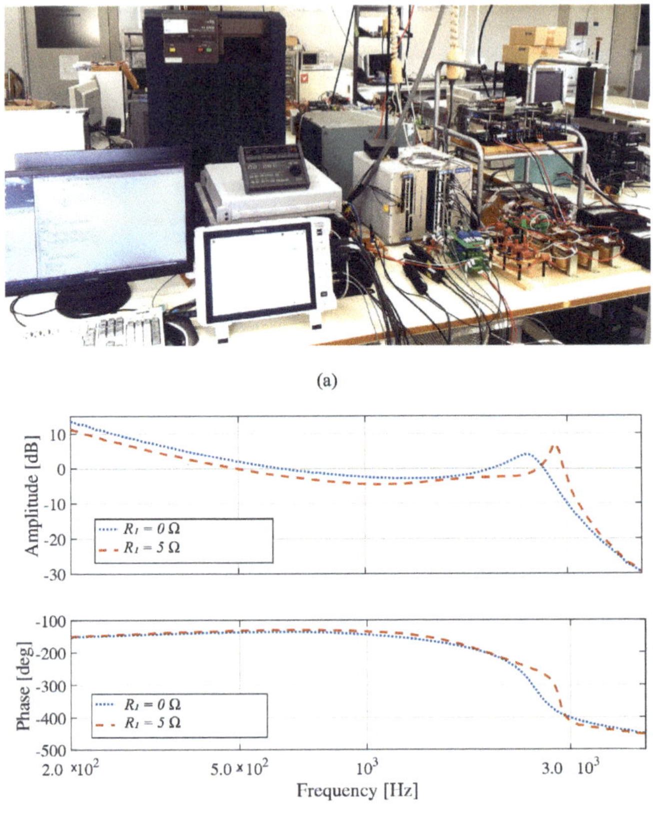

Figure 7.5 Experimental study; (A) laboratory environment, and (B) frequency response for the case of with/without a virtual resistor in the presence of delay.

investigated that the time delay especially affects the system frequency response around the resonant frequency of the converter output filter by making large peaks and phase shifts, which degrade the stability and damping characteristic.

Therefore, finding a solution to compensate the time delay in the closed-loop system is an important issue in the GCC digital control synthesis process. For this purpose in the relevant literature, several compensation techniques have been proposed using adjustment of the inverter output pulse width [20], modification of sampling methods [21,22], duty ratio compensation [23], phase compensation [24], and predictive control [25,26].

This section addresses an effective time delay compensation method based on a predictive estimation technique [13], which considers nonaveraged modes of transients of the state variables during the delay time. After description of the mentioned method, it is applied to the designed damping control system for a single-phase GCC with an LCL filter in the previous section.

7.2.1 Proposed method

The basic idea used here for developing a time delay compensation method is estimating future states value $\boldsymbol{x}[i,m]$ at $t=(i+m)T$ from the current states $\boldsymbol{x}[i]$ at $t=iT$. Then, the control input signal $\boldsymbol{u}$ can be computed as $\widehat{\boldsymbol{u}}[i,m]$ at $t=(i+m)T$, which is the advanced time by mT respect to $t=iT$. The future state value $\boldsymbol{x}[i,m]$ is estimated using the modified z-transform as follows, where $\boldsymbol{A}_m$, $\boldsymbol{b}_m$, and $\boldsymbol{h}_m$ are coefficients of discrete state equations from $t=iT$ to $t=(i+m)T$.

$$\begin{aligned}\widehat{\boldsymbol{x}}[i,m] &= \boldsymbol{A}_m\boldsymbol{x}[i]+\boldsymbol{b}_m\widehat{u}_d[i-1]+\boldsymbol{h}_m v_g[i]\\ \widehat{\boldsymbol{u}}[i,m] &= -\boldsymbol{k}_{fb}\widehat{\boldsymbol{x}}[i,m]+k_1w[i,m]+k_2w[i+1,m]\end{aligned} \tag{7.22}$$

where

$$\boldsymbol{A}_m=e^{A\,mT_s},\ \boldsymbol{b}_m=\int_0^{mT_s}e^{A\tau}d\tau\boldsymbol{b},\ \ \boldsymbol{h}_m=\int_0^{mT_s}e^{A\tau}d\tau\boldsymbol{h}. \tag{7.23}$$

This basic idea is illustrated in Fig. 7.6A. State values are sampled as $\boldsymbol{x}[i-1]$, $\boldsymbol{x}[i]$, $\boldsymbol{x}[i+1]$, $\cdots$. There is a time delay mT for generation of the GCC output pulse based on $\boldsymbol{u}[i]$. The input $\widehat{\boldsymbol{u}}[i,m]$ at $t=(i+m)T$ is computed from $\widehat{\boldsymbol{x}}[i,m]$ which is estimated by Eq. (7.22). A switching pulse is produced from $t=(i+m)T$ to $(i+1+m)T$ and can be synchronized with $\widehat{\boldsymbol{x}}[i,m]$.

It is noteworthy that the converter output duty length $\widehat{u}_d$ which is shaded in Fig. 7.6A must be precisely computed, considering timing relations. For example, when the time delay ratio is $m=0.5$, the dashed pulse in the figure v_u-v_v which is one of the dual output pulses from $t=\{i+0.25(1-|\widehat{u}|)\}T$ to $t=\{i+0.25(1+|\widehat{u}|)\}T$ is completely covered during the delay time. When m varies between $0\le m\le 0.5$, there are three cases for covering the pulse during the delay as shown in Fig. 7.6B, where, $\widehat{u}[i-1,m]$ is abbreviated simply as $\widehat{u}$. The averaged duty ratio $\widehat{u}_d[i-1]$ is expressed as follows:

$$|\widehat{u}_d[i-1]|=\begin{cases}0; & \text{if } m<0.25(1-|\widehat{u}|)\\ \dfrac{\{m+0.25(|\widehat{u}|-1)\}}{m}; & \text{if } 0.25(1-|\widehat{u}|)<m<0.25(1+|\widehat{u}|)\\ 0.5\dfrac{|\widehat{u}|}{m}; & \text{if } 0.25(1+|\widehat{u}|)<m<0.5\end{cases} \tag{7.24}$$

The term m in the denominator expresses averaging operation during mT. The sign of $\widehat{u}_d[i-1]$, or the polarity of the pulse, is directly dependent on that of $\widehat{u}$.

$$\widehat{u}_d[i-1]=\text{sign}(\widehat{u})|\widehat{u}_d[i-1]| \tag{7.25}$$

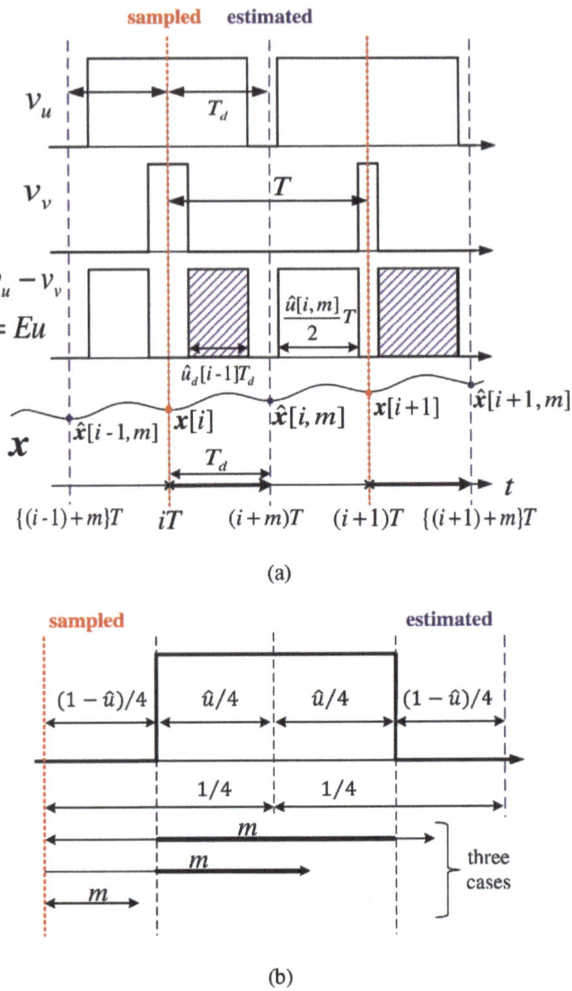

Figure 7.6 Proposed compensation principle: (A) graphical description of the basic idea, and (B) Three cases for computation of the duty $\hat{u}_d[i-1]$, where $\hat{u} = \hat{u}[i-1,m]$.

Generally, the expression of $\hat{u}_d[i-1]$ is dependent on the delay ratio m because the covered pulse length is dependent on m. A control block diagram with the proposed delay compensation is shown in Fig. 7.7. Based on the sampled values at $t = iT$, z^m-advanced state values at $t = (i+m)T$ are estimated, and they are used to compute $\hat{u}[i,m]$ which is the GCC control input at $t = (i+m)T$ produced by the digital controller. The input $\hat{u}[i,m]$ is ready at $t = iT$. However, there is a time delay of z^{-m}, and the $\hat{u}[i,m]$ is actually delivered to the converter at $t = (i+m)T$. Thus, a control delay z^{-m} is equivalently compensated by the *Delay Compensator* block according to the above description.

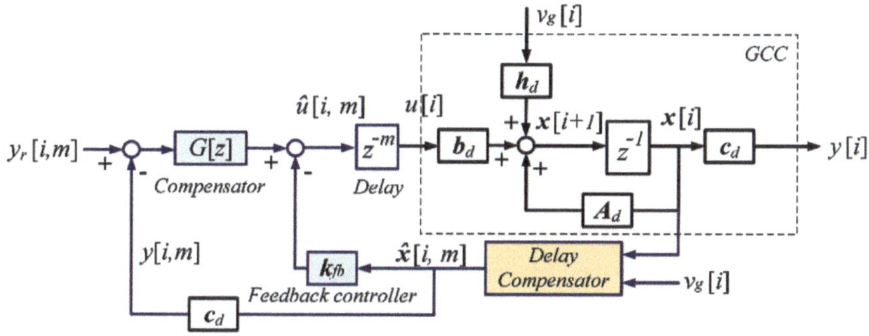

Figure 7.7 GCC digital control system with delay compensation.

7.2.2 State estimation enhancement

For obtaining the given results in the previous section, the averaged values of the state variables are used. Using the averaging method, the state variables may not be estimated precisely due to the mixed ripples and noises with the measured states. For example, for the case study at hand, the inductor current i_{L_1} is mixed with large ripples. To solve this problem, one can use other effective state estimation techniques such as observer design [27]. Assuming the system is observable, to design a full-order observer system, estimation of all state variables is needed. However, here for the mentioned GCC example, as discussed in Ref. [28], only the estimation of output current i_{L_1} is required to track the target objectives. Therefore, instead of a full-order observer system, the design of a minimum order one is enough.

As another solution for better state estimation, here a predictive estimation method is introduced. The state variables are estimated directly from the state equations without applying the averaging method. The developed control system has a time delay of 0.5 sample, or $T_d = 0.5T$. As shown in Fig. 7.8, the time interval T_d from $t = iT$ to $t = (i+m)T$ is divided into three transition periods (T_1, T_2, T_3), where T_2 is the shaded output pulse width $|\hat{u}[i-1, m]|T/2$ and the rest $mT - T_2$ is divided equally into T_1 and T_3. As can be seen in Fig. 7.8, the current first decreases, then increases, and finally decreases, assuming the converter output is a positive pulse. The situation is quite opposite when the converter output is negative.

The state variables can be precisely estimated from the following three equations which express state transients in the above three periods, respectively. The sampling time was assumed to be small enough and $v_g[i]$ can be assumed to be constant during each interval.

$$\begin{aligned}
\boldsymbol{x}_1[i] &= \boldsymbol{A}_1\boldsymbol{x}[i] + \boldsymbol{h}_1 v_g[i]; \quad \text{for } T_1 = 0.25(1 - |\hat{u}[i-1, m]|)T \\
\boldsymbol{x}_2[i] &= \boldsymbol{A}_2\boldsymbol{x}_1[i] + \boldsymbol{b}_2\, sign\,(u[i-1, m]) + \boldsymbol{h}_2 v_g[i]; \quad \text{for } T_2 = 0.5|\hat{u}[i-1, m]|T \\
\hat{\boldsymbol{x}}\,[i, m] &= \boldsymbol{A}_3\boldsymbol{x}_2[i] + \boldsymbol{h}_3 v_g[i]; \quad \text{for } T_3 = T_1
\end{aligned} \tag{7.26}$$

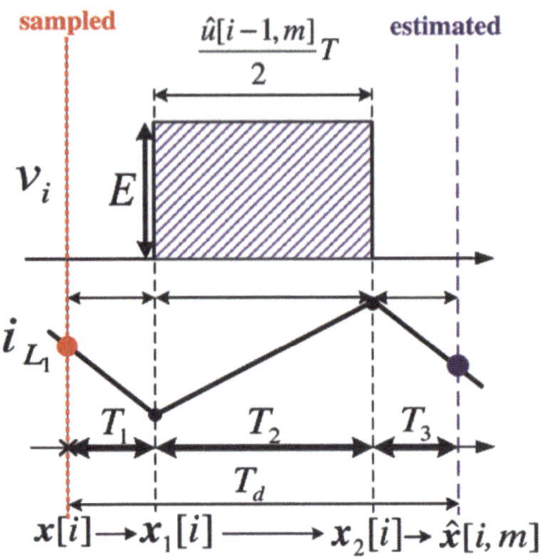

Figure 7.8 Precise estimation of the inductor current i_{L_1}, considering three modes.

where

$$A_j = e^{AT_j},\ \ b_j = \int_0^{T_j} e^{A\tau} d\tau b,\ \ h_j = \int_0^{T_j} e^{A\tau} d\tau h\ (j = 1,\ 2,\ 3). \tag{7.27}$$

These state modes are dependent on the time delay and the input duty. They have to be modified when they form a different combination of transition periods. For example, the proposed method can be applied for a three-phase GCC with a time delay of $m = 1$ sec. The basic idea is the same, that it is based on the proposed predictive estimation method which considers nonaveraged modes of transients of the state variables during the time delay. In the three-phase case, the computation process of the converter output voltage becomes more complex because it needs further consideration [13].

7.2.3 Experimental results for damping control with delay compensation

For the example at hand, the result of applying explained estimation method for the GCC output current i_{L1} in a half cycle is shown in Fig. 7.9A. The oscillatory waveform is the current $i_{L1}(t)$ with ripples, its sampled values $i_{L1}[i,m]$ are denoted as the upper curve inside the oscillatory waveform, and the estimated values $\widehat{i}_{L1}[i,m]$ are denoted as the lower curve. It can be seen that the $\widehat{i}_{L1}[i,m]$ is close enough to the actual value $i_{L1}[i,m]$. A closer view (zoomed) view with more information is shown in Fig. 7.9B.

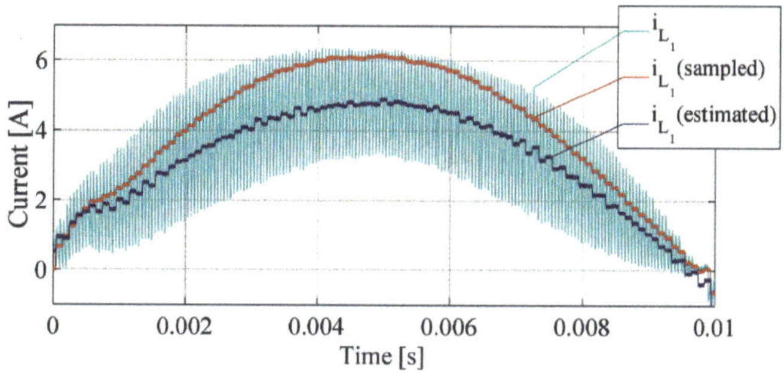

(a)

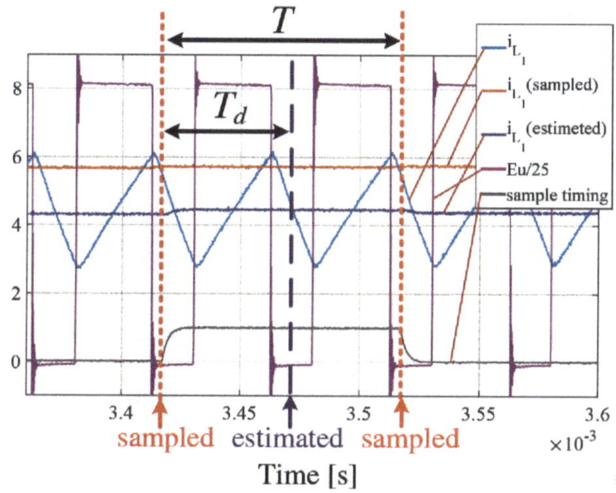

(b)

Figure 7.9 Sampled and estimated waveforms for current i_{L1}; (A) half cycle, and (B) a zoomed view.

The scaled GCC output voltage (0.04 *Eu*) can be seen as long amplitude pulses in Fig. 7.9B. The delay is half of the sampling time and the actual values of $i_{L1}[i,m]$ are located almost in the center of the ripples. The estimated $\widehat{i}_{L1}[i,m]$ crosses the actual $i_{L1}(t)$ waveform at the estimated time $t = (i + m)T$.

The effects of the proposed damping control with one virtual resistor are investigated, and the experimental closed-loop frequency responses are measured and plotted in Fig. 7.10. Without the delay compensation and damping control, the frequency response is shown as a dotted curve, which depicts a large resonant peak. After the compensation, the frequency response is shown as a solid curve and the peak is a little mitigated. After damping with one virtual resistor, the frequency response is shown as a dashed curve and the peak is significantly suppressed. However, the gain after using the damping control is a little reduced; and in practice, the designer may need to make a trade-off between damping effects and control responses.

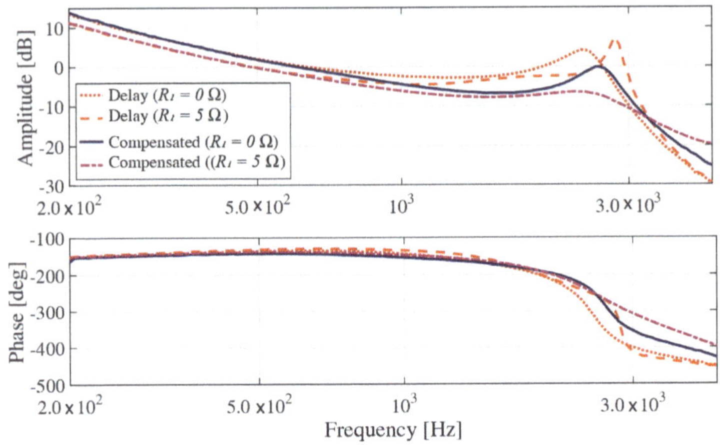

Figure 7.10 Frequency response of the closed-loop system with and without compensation.

In addition to studying the closed-loop frequency characteristics, the time—domain responses are also investigated. The current waveform in Fig. 7.11A before the compensation has oscillations near zero-cross timings where control performance is affected apparently due to dead-time effects. Such oscillations are effectively suppressed as shown in Fig. 7.11B after using the delay compensation, and in Fig. 7.11C after applying the damping with one-resistor. These observations validate the effectiveness of the proposed delay compensation.

Finally, the effectiveness of the proposed damping control with three virtual resistors, supported by the delay compensation method, is experimentally studied. The measured frequency response characteristics without considering time delay are shown and compared in Fig. 7.12A. Control gains are given in Eq. (7.21). The frequency responses are shown as a dashed curve with the one-resistor damping control and as a solid curve with the three-resistor damping control. As can be seen, using three virtual resistors provides a smaller peak at the resonant frequency. The GCC output current and the grid voltage waveforms with three virtual resistors-based damping control are shown in Fig. 7.12B.

7.2.4 Extension to a three-phase GCC

The discussed compensation method can be applied to the control system of a three-phase GCC which is shown in Fig. 7.13 [29]. In a three-phase GCC, one of the phase switching control signals is dependent on the other two signals and can be obtained as a linear combination of them. Thus, the GCC control can be described with only two phases of u, v, and w, (such as u, w). The control input of the third phase ($u_v[i]$) is calculated from the control input of the mentioned two phases ($u_u[i]$ and $u_w[i]$) to satisfy the following zero-sequence equation:

$$u_v[i] = -(u_u[i] + u_w[i]) \tag{7.28}$$

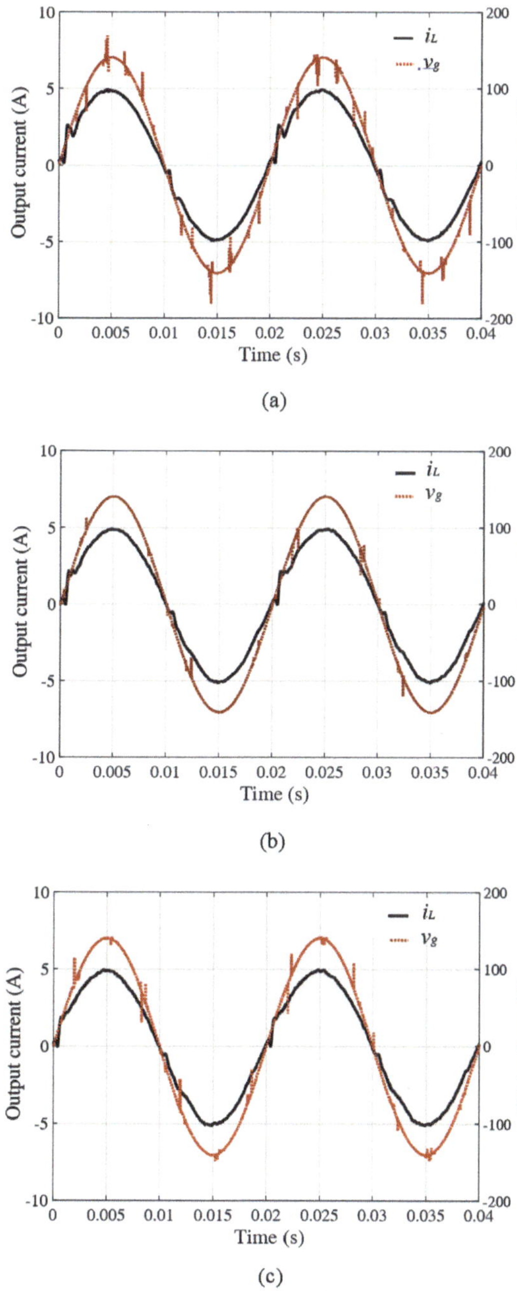

Figure 7.11 GCC output current and grid voltage waveforms; (A) no delay compensation and damping control, (B) using delay compensation, without damping control, and (C) using delay compensation and one virtual resistor damping control.

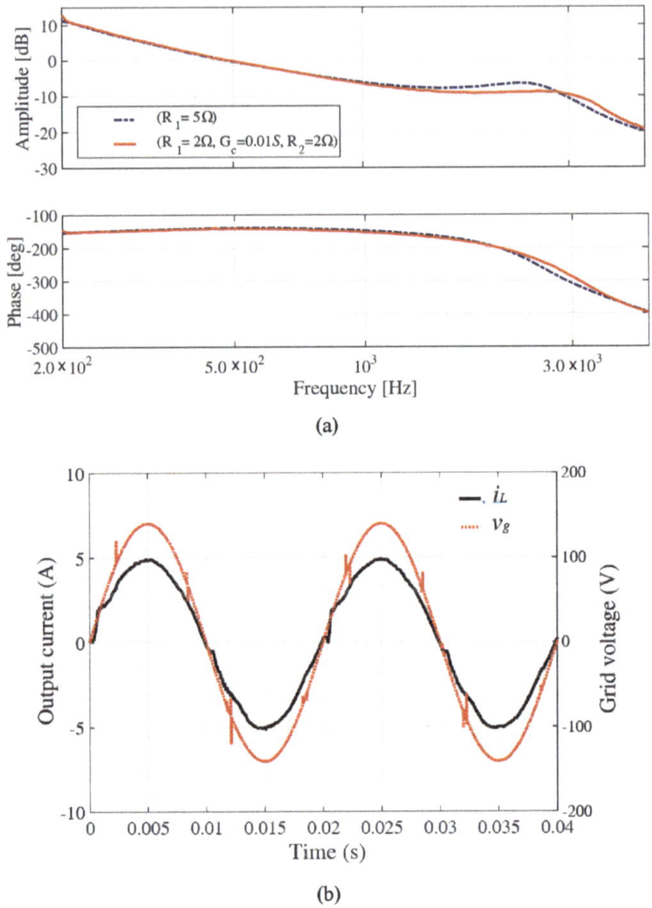

Figure 7.12 (A) Frequency response of the closed-loop system using one and three virtual resistors, (B) the GCC output current and the grid voltage waveforms with thee virtual resistors.

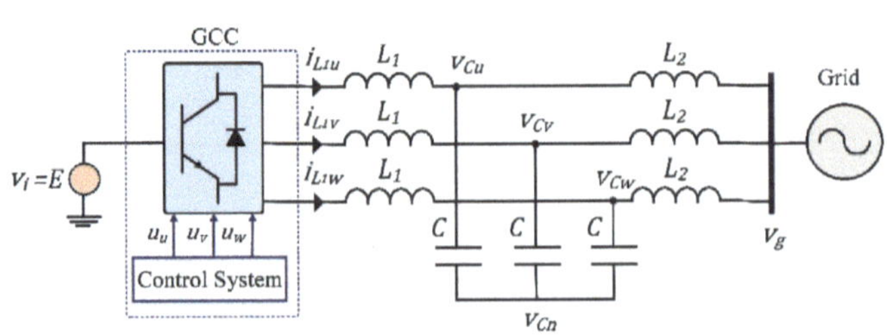

Figure 7.13 A three-phase GCC.

This idea has been already used in Chapter 4 to perform the control system of a three-phase GCC (Fig. 4.12). Here, using the same idea, the overall control system with delay compensation blocks for a three-phase GCC is realized as shown in Fig. 7.14. The upper and lower parts correspond to blocks of phases u and w respectively, and the middle part corresponds to a block of phase v.

According to the given discussion on computing input $\widehat{u}[i,m]$ based on Fig. 7.8, in a three-phase GCC, a phase voltage pulse of the output is produced by the difference between the phase output voltage and the common-mode voltage v_{Cn} (Fig. 7.13), which has a pulse waveform that is classified into six pattern modes depending on the pulse widths of the three phases. An example mode when a case of $u_u[i] > u_w[i] > u_v[i]$ is used, and the waveforms for phase u are shown in Fig. 7.15, where the control system has a time delay of one sample ($T_d = T$).

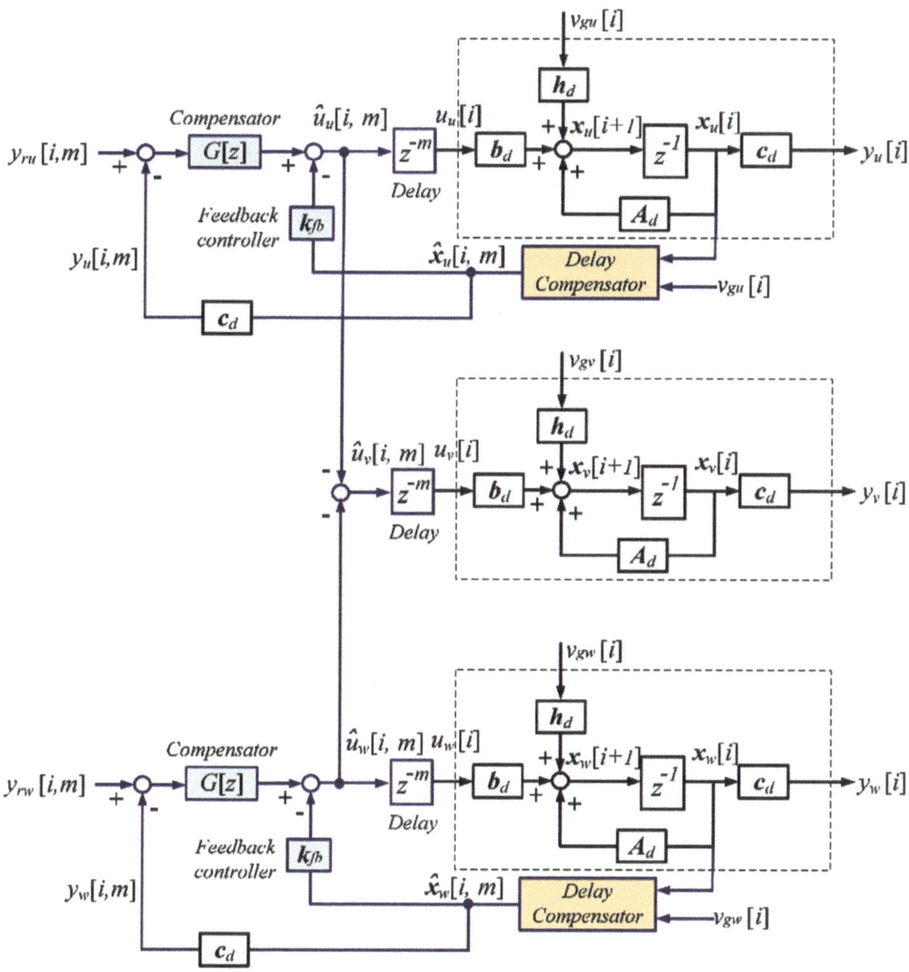

Figure 7.14 Overall control system with delay compensation blocks.

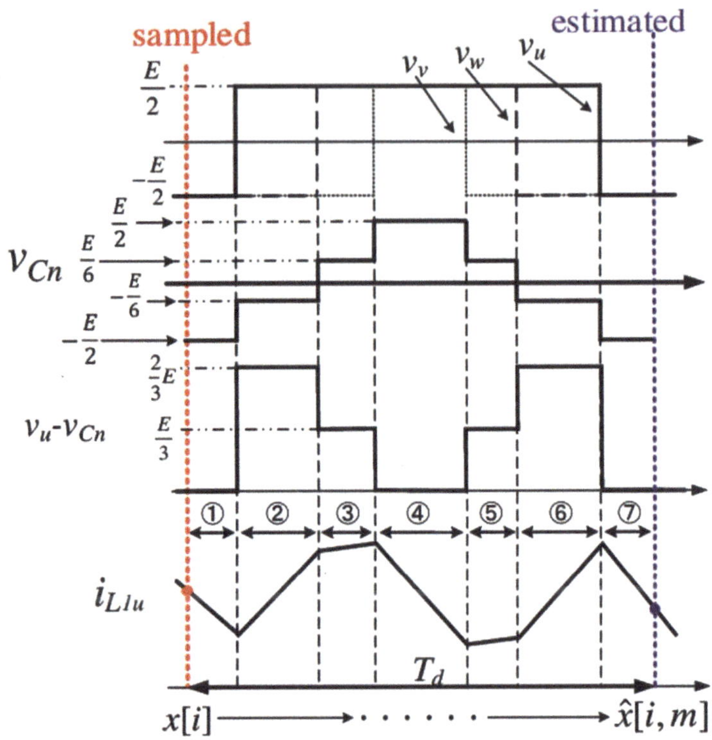

Figure 7.15 Precise prediction of the state-variables, considering waveform ripples.

Using this example, the addressed compensation principle can be simply understood. In Fig. 7.15, from top to bottom, phase u voltage v_u, common-mode voltage v_{Cn}, and inductor current i_{L_1u} are illustrated. The common-mode voltage is a stepwise waveform and it consists of seven time intervals. In this case, the voltage level of $v_u - v_{Cn}$ is switched to 0, $E/3$, or $2E/3$. Considering the following transients during seven time intervals for the delay $T_d = mT$, the state vector $x[i,m]$ of the phase u can be estimated as follows:

$$u_v[i] = -(u_u[i] + u_w[i]) \tag{7.29}$$

$$\widehat{\boldsymbol{x}}_j[i] = \boldsymbol{A}_j\widehat{\boldsymbol{x}}_{j-1}[i] + \boldsymbol{b}_j u_j + \boldsymbol{h}_j v_g[i] \ (j = 1, 2, \cdots, 7) \tag{7.30}$$

where

$$\widehat{\boldsymbol{x}}_0[i] = \boldsymbol{x}[i],\ \widehat{\boldsymbol{x}}[i,m] = \widehat{\boldsymbol{x}}_7[i] \tag{7.31}$$

These coefficient matrices and vectors are described in Eq. (7.32) and time T_n is the time length of the n-th interval.

$$\boldsymbol{A}_j = e^{AT_j},\ \boldsymbol{b}_j = \int_0^{T_j} e^{A\tau} d\tau \boldsymbol{b},\ \boldsymbol{h}_j = \int_0^{T_j} e^{A\tau} d\tau \boldsymbol{h}\quad (\mathrm{j} = 1, 2, \cdots, 7). \tag{7.32}$$

From magnitude relationship among $u_u[i], u_v[i]$, and $u_w[i]$, the output pattern mode is determined. Then seven time lengths are determined from $u_u[i], u_v[i]$, and $u_w[i]$. The symbol u_j stand for a pulse height and $u_j = 0(j = 1, 4, 7)$. The other intervals $u_j(j = 2, 3, 5, 6)$ are symmetrical, and $u_2 = u_6$ and $u_3 = u_5$. Their values are determined according to the six pattern modes from I to VI given in Eq. (7.33). For example, the state variable prediction shown in Fig. 7.15 belongs to mode I.

$$\left.\begin{array}{llll}
\textbf{I}: & u_u > u_w > u_v & u_2 = u_6 = \frac{2}{3} & u_3 = u_5 = \frac{1}{3} \\
\textbf{II}: & u_u > u_v > u_w & u_2 = u_6 = \frac{2}{3} & u_3 = u_5 = \frac{1}{3} \\
\textbf{III}: & u_v > u_u > u_w & u_2 = u_6 = \frac{1}{3} & u_3 = u_5 = -\frac{1}{3} \\
\textbf{IV}: & u_v > u_w > u_u & u_2 = u_6 = -\frac{1}{3} & u_3 = u_5 = -\frac{2}{3} \\
\textbf{V}: & u_w > u_v > u_u & u_2 = u_6 = -\frac{1}{3} & u_3 = u_5 = -\frac{2}{3} \\
\textbf{VI}: & u_w > u_u > u_v & u_2 = u_6 = \frac{1}{3} & u_3 = u_5 = -\frac{1}{3}
\end{array}\right\} \tag{7.33}$$

and

$$u_j = 0\ (j = 1, 4, 7) \tag{7.34}$$

The above explained algorithm can be summarized in the following steps:

Step 1: Sampling the state variables $x[i]$ and $v_g[i]$ at $t = iT$;

Step 2: Recall the previous control inputs $\widehat{u}_j[i-1, m](j = u, v, w)$;

Step 3: Determine the pattern mode among I-IV and u_2, u_6, u_3, u_5 according to Eq. (7.33);

Step 4: Compute $T_j(j = 1, \cdots, 7)$ from $\widehat{u}_j[i-1, m](j = u, v, w)$;

Step 5: Calculate $\widehat{x}[i,m]$ based on the transient equations, (7.30) and (7.31) if $T_j > 0(j = 1,\cdots,7)$;

Step 6: Compute and memorize the next control inputs $\widehat{u}_j[i,m](j = u,w)$ according to the given control principle and compute third phase control input using $\widehat{u}_v[i,m] = -(\widehat{u}_u[i,m] + \widehat{u}_w[i,m])$;

Step 7: Go back to the next sampling step (*Step 1*).

The extended compensation method is applied to estimate the inductor current i_{L1} of the 3-phase GCC for $T_d = T$. The estimated current value $\widehat{i_{L1}}[i,m]$ is assessed to determine if it is close enough to the actual value $i_{L1}[i,m]$. The measured results are shown in Fig. 7.16A where the presampling time is negligible. The oscillatory gray waveform represents the current ripples, and the red curve represents its sampled values, and the orange curve represents the estimated values. The delay is considered as one sampling time and the actual values of $i_{L1}[i,m]$ are almost located in the center of the ripples. A magnified view is shown in Fig. 7.16B. The applied voltage $v_u - v_{Cn}$ is denoted by the purple curve (long—amplitude pulses). The predicted $\widehat{i_{L1}}[i,m]$ from $i_{L1}[i]$ crosses the actual $i_{L1}(t)$ waveform at the predicted time. Similar results are also obtained for phase w. These results validate the effectiveness of the proposed estimation method.

7.3 Passivity-based stabilization

In Section 7.2, to satisfy a desirable set of performance characteristics such as tracking property, noise rejection, and ripple attenuation; a control system containing digital compensator, static state FB controller, and a delay compensator is used. However, as discussed in Chapter 6, to ensure the overall stability of the GCC system in all operating frequency ranges, an additional controller such as a feedforward (FF) may be required.

In this section, a two steps design method is introduced. In the first step, which has been fully discussed and accomplished in the previous section, a state FB control loop is designed to have a desirable sinusoidal tracking characteristic. The addressed control loop with compensators satisfies the performance objectives. Here, as the second step, a conductive characteristic function is added to the system by an FF control loop which makes the overall system passive and ensures system stability.

Let's redraw the updated GCC and its digital control system depicted in Fig. 7.7, by adding an FF controller, as shown in Fig. 7.17A. Adding of the FF controller, $H_{ff}[z[$, to the system, as discussed in Chapter 4 (Section 4.3), increases design freedom regarding the system stabilizing over a wide range of operations. The sample-and-hold (S/H) operator is used to convert the analog to digital signals. The transfer function of the S/H and delay (z^{-m}) blocks in Fig. 7.17A can be described as Eq. (7.35), which T is the sampling time.

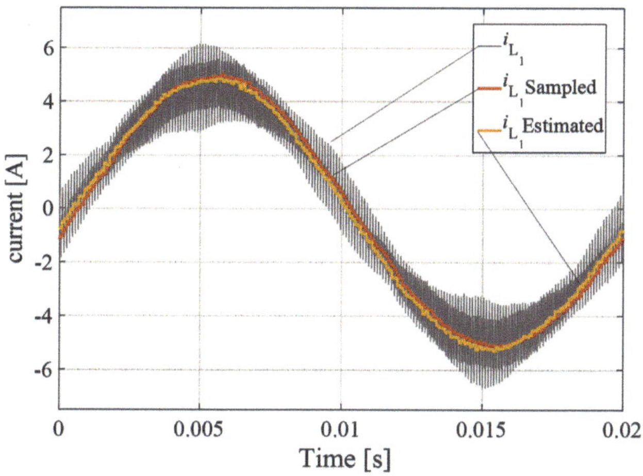

(a)

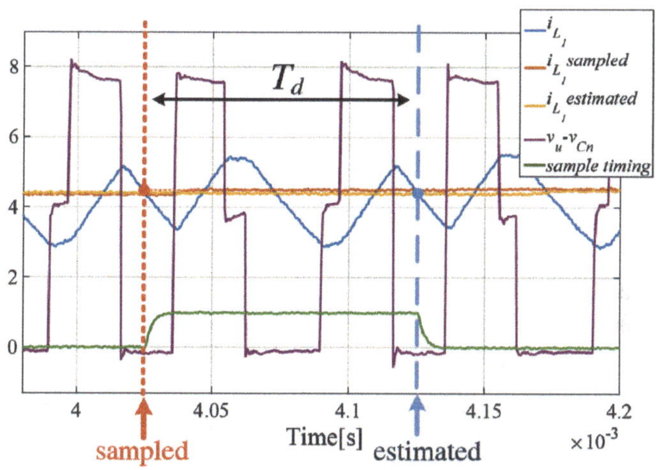

(b)

Figure 7.16 Sampled and estimated waveforms for current i_{L1}; (A) one cycle, and (B) a zoomed view.

$$S_m(s) = \frac{1 - \varepsilon^{-sT}}{sT}; \; z^{-\mathrm{m}} = \varepsilon^{-smT} \tag{7.35}$$

To analyze the dynamic impacts of FB and FF control loops, one may formulate their overall impacts in the GCC output admittance, which shapes the system dynamic characteristics, as shown in Fig. 7.17B. These impacts that already discussed in Chapters 4 and 6 are denoted by Y_{fb} and Y_{ff} which approximately determine the GCC output admittance Y_o.

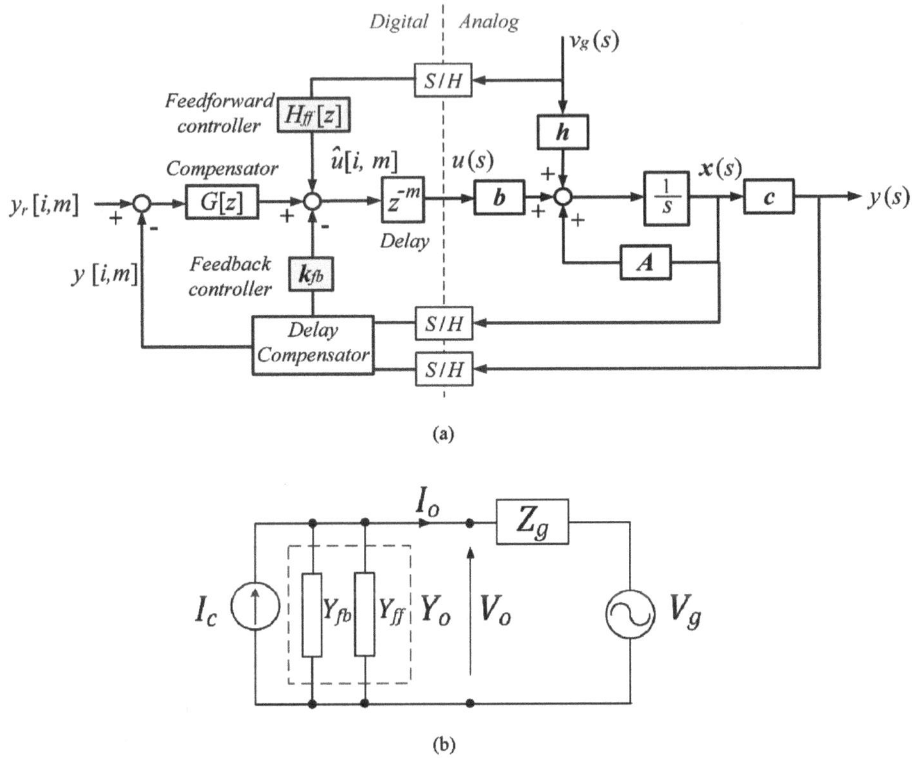

Figure 7.17 (A) The GCC with digital feedforward and feedback control loops; a) block diagram representation, and b) equivalent circuit.

In fact, a conductive characteristic Y_{ff} is added by the FF control loop to ensure stability, because the designed characteristic of the output admittance Y_{fb} may not be passive over the interested frequency band.

7.3.1 Passivity-based stabilizing

As discussed in Chapter 4, the Nyquist criterion in the impedance method essentially checks if the cross point of Nyquist curve of $Z_g Y_o$ at the negative real axis exceeds -1 or not. The line impedance Z_g is passive and its phase stays between $\pm$ 90°. If the phase of the output admittance Y_o of the positive sequence (and the negative sequence) can be kept between ± 90 degree, there is no cross point at the negative real axis and the system is always stable even if the admittance becomes large. It is sufficient when the real part of the admittance of the converter is positive and passive to ensure stability [30], that is:

$$\text{Re}\{Y_o(s)\} > 0. \tag{7.36}$$

The FB controller is designed in the previous section to satisfy the performance objective. However, the system may become unstable due to possible movement of the output admittance phase over ±90 degrees. To fix this probable problem, an FF controller can be designed to satisfy criteria (Eq. 7.36) [29]. Here, to explain the design methodology, the given passivity-based stability analysis formulation given in Chapter 4 is recalled.

According to Fig. 7.17A, the state equations of the closed-loop GCC system, considering both FB and FF controllers and ignoring the dynamic of time delay compensator can be obtained as follows:

$$s\boldsymbol{x}(s) = \boldsymbol{A}\boldsymbol{x}(s) + \boldsymbol{b}u(s) + \boldsymbol{h}_c v_g(s), \tag{7.37}$$

$$u(s) = -\ \boldsymbol{k}_{fb}\, S_m(s)\boldsymbol{x}(s) + G(s)(y_r(s) - \boldsymbol{c}\, S_m(s)\boldsymbol{x}(s)) + H_{ff}(s)S_m(s)v_g(s). \tag{7.38}$$

From Eqs. (7.37 and 7.38), the following equation is achieved.

$$\begin{aligned}\left(s\boldsymbol{I} - \boldsymbol{A} + \boldsymbol{b}\,\boldsymbol{k}_{fb}\, S_m(s) + \boldsymbol{b}G(s)\boldsymbol{c}\, S_m(s)\right)^{-1}\boldsymbol{x}\,(s) &= \boldsymbol{h}v_g(s) + \boldsymbol{b}H_{ff}(s)S_m(s)v_g(s)\\ &+ \boldsymbol{b}G(s)y_r(s)\end{aligned} \tag{7.39}$$

Thus, the GCC output admittance $Y_o(s)$ becomes as follows:

$$\begin{aligned}Y_o(s) = -\frac{I_{L2}}{v_g} &= -\boldsymbol{c}\left(s\boldsymbol{I} - \boldsymbol{A} + \boldsymbol{b}\boldsymbol{k}_{fb}\, S_m(s) + \boldsymbol{b}G(s)\boldsymbol{c}\, S_m(s)\right)^{-1}\boldsymbol{h}\\ &-\boldsymbol{c}\left(s\boldsymbol{I} - \boldsymbol{A} + \boldsymbol{b}\boldsymbol{k}_{fb}\, S_m(s) + \boldsymbol{b}G(s)\boldsymbol{c}\, S_m(s)\right)^{-1}\boldsymbol{b}H_{ff}(s)S_m(s)\end{aligned} \tag{7.40}$$

The output admittance $Y_o(s)$ is divided into an FB component $Y_{fb}(s)$ and an FF component $Y_{ff}(s)$ as follows.

$$\begin{aligned}Y_{fb}(s) &= -\boldsymbol{c}\left(s\boldsymbol{I} - \boldsymbol{A} + \boldsymbol{b}\boldsymbol{k}_{fb}\, S_m(s) + \boldsymbol{b}G(s)\boldsymbol{c}\, S_m(s)\right)^{-1}\boldsymbol{h}\\ Y_{ff}(s) &= -\boldsymbol{c}\left(s\boldsymbol{I} - \boldsymbol{A} + \boldsymbol{b}\boldsymbol{k}_{fb}\, S_m(s) + \boldsymbol{b}G(s)\boldsymbol{c}\, S_m(s)\right)^{-1}\boldsymbol{b}H_{ff}(s)S_m(s) \equiv \overline{Y}_{ff}(s)H_{ff}(s)\end{aligned} \tag{7.41}$$

A conductive characteristic $Y_{ff}(s)$ is added around the nonpassive frequency band by the $H_{ff}(s)$ to make the GCC output admittance characteristic passive. Considering Eq. (7.41), the frequency characteristic of the $Y_{ff}(s)$ near the resonant frequency can be determined such that

$$Y_{ff}(s) = \overline{Y}_{ff}(s)H_{ff}(s) \simeq \widehat{Y}_{ff}(s). \tag{7.42}$$

The transfer function of $Y_{ff}(s)$ or $\widehat{Y}_{ff}(s)$ can be considered as Eq. (7.43), which its conductive characteristic to be R_d at the resonant frequency [31].

$$\widehat{Y}_{ff}(s) \simeq \frac{1}{R_d + sL_d + \dfrac{1}{sC_d}} \tag{7.43}$$

Then, the $H_{ff}[z]$ can be calculated in the following form

$$H_{ff}[z] \simeq \frac{\widehat{Y}_{ff}[z]}{\overline{Y}_{ff}[z]} = \frac{a_0 + a_1 z^{-1} + a_2 z^{-2}}{1 + b_1 z^{-1} + b_2 z^{-2}}. \tag{7.44}$$

The coefficients of $H_{ff}[z]$, that is, a_0, a_1, a_2, b_1, and b_2, must be optimally fitted so that Y_{ff} to be stable and close to $\widehat{Y}_{ff}$. It is not possible to fit Y_{ff} over a wide range of frequency and Y_{ff} must be optimally fitted to $\widehat{Y}_{ff}$ for a limited band over $f_1 \leq f \leq f_2$. An additional filter such as $F[z]$, in (7.45) may be required to reduce the FF side effects outside this fitting band. The coefficients of $F[z]$ (c_0, c_1, and d_1) are determined to satisfy this objective. In this case, the $H_{ff}[z]$ can be determined as (7.46).

$$F[z] = \frac{c_0 + c_1 z^{-1}}{1 + d_1 z^{-1}} \tag{7.45}$$

$$H_{ff}[z] = \frac{a_0 + a_1 z^{-1} + a_2 z^{-2}}{1 + b_1 z^{-1} + b_2 z^{-2}} \cdot \frac{c_0 + c_1 z^{-1}}{1 + d_1 z^{-1}} \tag{7.46}$$

7.3.2 Design examples

7.3.2.1 A single-phase GCC example

As first example, a single-phase GCC (Fig. 7.1) is examined in the presence of a digital control system with sinusoidal and time delay compensators, as well as FB and FF controllers. Circuit parameters are listed in Table 7.1 and the designed FB, FF, and compensator gains are considered as Eq. (7.47) [32].

$$\begin{aligned}
&\boldsymbol{k}_{fb} = [0.0402 \quad -0.0004 \quad 0.0248]; \\
&G[z]:\ k_1 = -0.0185,\ k_2 = 0.0216 \\
&H_{ff}[z]: \\
&a_0 = 0.7075,\ a_1 = -1.2068,\ a_2 = 0.5533, \\
&b_1 = -1.9393,\ b_2 = 0.9407.
\end{aligned} \tag{7.47}$$

Table 7.1 Circuit and control parameters for the single-phase GCC.

Parameters	Value
Input DC voltage	200 V
Output voltage amplitude	141 V
Output frequency	50 Hz
Filter inductance, L_1	0.76 mH
Filter inductance, L_2	0.76 mH
Filter capacitance, C	9.3 μF
Switching and sample frequency	10 kHz
FF damping resistor, R_d	10 Ω
FF equivalent inductor, L_d	1 mH
FF equivalent capacitor, C_d	158 μF

A digital control system with the described control scheme is used and realized on a DSP-based experimental setup and its performances and stability are investigated. Measured frequency response characteristics of the GCC output admittance Y_o with only FB controller are shown in Fig. 7.18A, and as can be seen, there is a cross-frequency around 410 Hz. According to the impedance method-based Nyquist stability criterion, the system is stable for $L_g = 3$ mH because $|Y_o|\langle|Y_g|$, while it is unstable for a small $Y_g = 1/Z_g$, e.g., $L_g = 5$ mH ($|Y_o|\rangle|Y_g|$). This check result for stability is investigated by the actual waveforms of the GCC output currents in Fig. 7.18B. The system is stable and the current has small oscillations of 410 Hz when $L_g = 1$ mH. The oscillations become larger when $L_g = 3$ mH, and it becomes unstable for $L_g = 5$ mH.

In the second scenario, the GCC response is evaluated with both FB and FF controllers. The FF is expected to provide passivity property and thus ensures the system stability. For this purpose, it is necessary to make the phase characteristic of Y_o within the difference of 180 degrees from that of Y_g. According to the proposed method, the location of the resonant frequency of Y_{ff} is selected at 400Hz. R_d is selected as 10 Ω which corresponds to the amplitude of the FF characteristic. The FF admittance Y_{ff} is optimally fitted to $\widehat{Y}_{ff}$ for a limited band over $100Hz \leq f \leq 600\ Hz$. The fitted characteristics are shown in Fig. 7.19. The given $\widehat{Y}_{ff}$ is a smooth curve and Y_{ff} is fitted to $\widehat{Y}_{ff}$ closely over the band which includes the resonant frequency.

Measured characteristics of the output admittance Y_o with the designed FF control (without time delay and considering $L_g = 3$ mH) are shown in Fig. 7.20A. However, due to the control time delay, the measured characteristics have unnecessary phase-shift especially around the filter resonant frequency. Applying the proposed delay compensation method explained in Section 7.2, the modified frequency response is shown in Fig. 7.20B. The phase characteristic varies only within ± 90 degrees, and thus, the system is stable. The GCC output current waveforms for $L_g = 3$ mH and $L_g = 5$ mH are shown in Fig. 7.21.

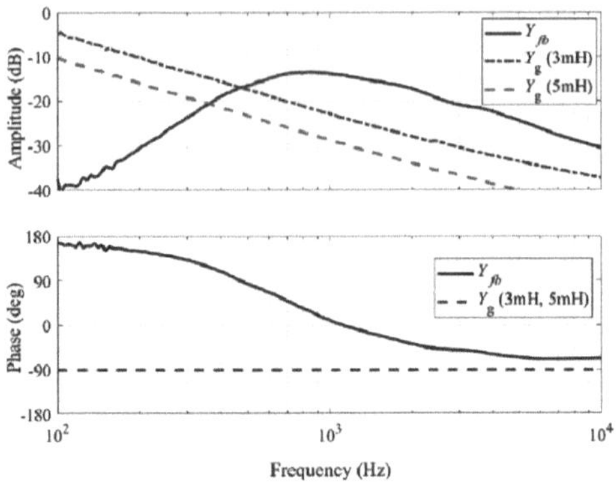

(a)

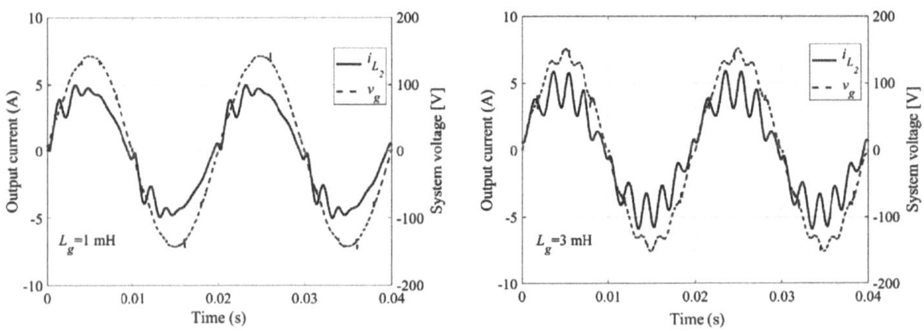

(b)

Figure 7.18 (A) Experimental GCC system response; a) with only feedback controller; b) measured output current and grid voltage for $L_g = 1$ mH and $L_g = 3$ mH.

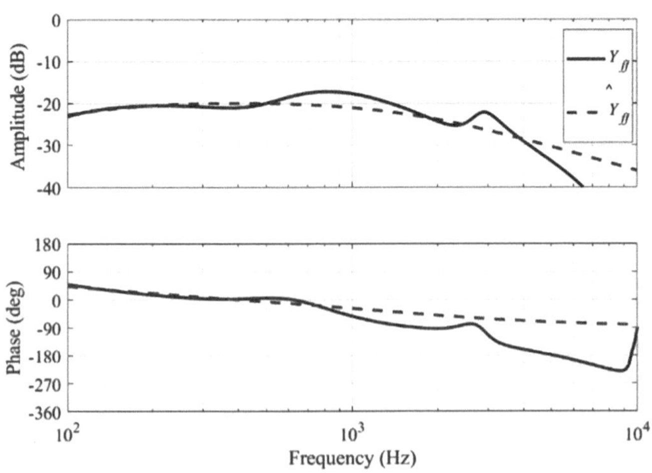

Figure 7.19 The feedforward controller frequency response and the fitted curve.

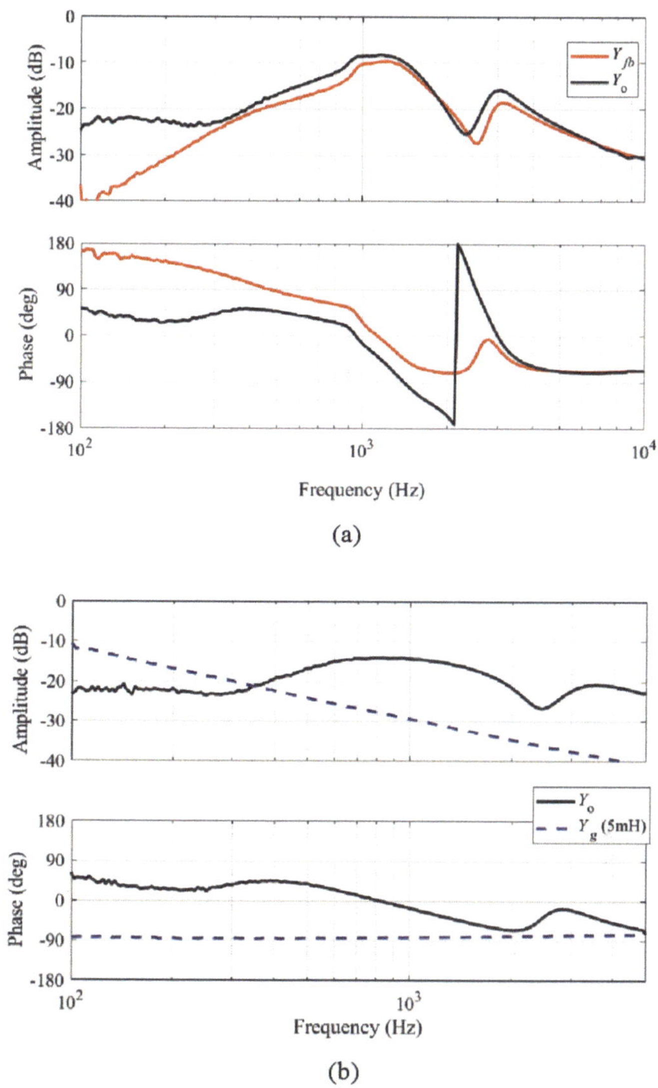

Figure 7.20 System frequency response with both feedback and feedforward controllers; (A) without delay compensation, and (B) with delay compensation.

7.3.2.2 A 3-phase GCC example

As second example, a three-phase GCC (Fig. 7.13) stability and performance are examined in the presence of a digital control system with sinusoidal and time delay compensators, as well as FB and FF controllers [33]. In the first step, only the state FB is applied and then the obtained results in simulation and experimental environments are compared with the results of using both FB and FF controllers. Circuit

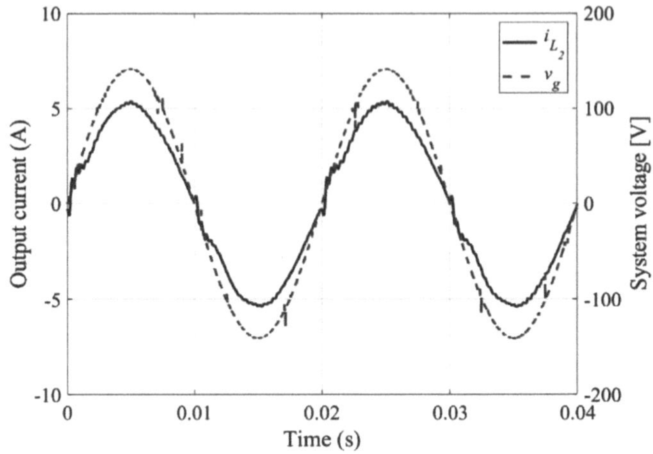

(a)

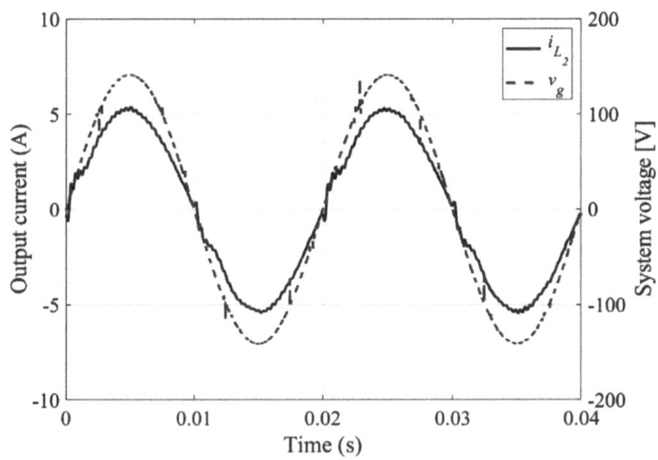

(b)

Figure 7.21 Experimental output current measurements for (A) $L_g = 3$ mH, and (B) $L_g = 5$ mH.

parameters are listed in Table 7.2, and the designed FB and compensator gains are considered as Eq. (7.48).

$$\boldsymbol{k_{fb}} = [0.03941 \quad -0.00004 \quad 0.02624], \; k_1 = -0.02025, \; k_2 = 0.02347. \tag{7.48}$$

- ***With only FB controller***

As first scenario, only the designed state FB controller is used. As discussed in Chapter 4, according to the Nyquist stability criterion used in the impedance-based

Table 7.2 Circuit and control parameters for three-phase inverter.

Parameters	Value
Input DC voltage	400 V
Output voltage amplitude	141 V
Output frequency	50 Hz
Filter inductance L_1	0.76 mH
Filter inductance L_2	0.76 mH
Filter capacitance C	9.3 μ F
Switching and sample frequency	10 kHz
Control time delay T_d	100 μs
FF damping resistor R_d	7.5 Ω
FF equivalent inductor L_d	1 mH
FF equivalent capacitor C_d	158 μF

stability analysis method, the system is stable if the Nyquist graph of $Z_g Y_o$ crosses the right side of $(-1, 0)$ at the cross-frequency of the phase difference of 180 degrees between $Y_g = 1/Z_g$ and Y_o. In the frequency response in simulation environment which is shown in Fig. 7.22A, it can be seen that the cross-frequency exists around 400 Hz. The system is stable at $L_g = 1$ mH because Y_o/Y_g crosses the right side of $(-1, 0)$ at the cross-frequency. However, it is unstable at $L_g = 5$ mH because Y_o/Y_g crosses left side of $(-1, 0)$. The experimental results for the same system in a laboratory environment are shown in Fig. 7.22B. The measured frequency response is close to the obtained one from the simulation.

Moreover, experimental results for the GCC output current waveforms in two different grid impedances are shown in Fig. 7. 23. Fig. 7.23A shows the system is stable for $L_g = 1$ mH. Increasing the line inductance leads the system to an unstable state. The current waveforms for a critical case for $L_g = 2.8$ mH are shown in Fig. 7.23B, where oscillations of 400 Hz which corresponds to the cross-frequency are superposed.

- ***With both FB and FF controllers***

The GCC system designed with the FB control is not passive because it has a negative conductance region around the cross-frequency which limits stable operations. Therefore, the explained FF control design method is to make the system passive and stable. For this purpose, it is necessary to keep the phase variation of Y_o respect to Y_g, within 180 degrees. The design resonant frequency of Y_{ff} is selected at 400 Hz. The coefficients of $H_{ff}[z]$ are optimally selected in a fitting process, such that to be resistive, stable, and close to Y_{ff}. It is difficult to fit Y_{ff} over the wide frequency range, then Y_{ff} is optimally fitted to $\widehat{Y}_{ff}$ for a limited band over 100 Hz $\leq f \leq$ 600 Hz. An additional LPF ($F[z]$) is used to eliminate the FF side effects outside this fitting band. The $H_{ff}[z]$ and $F[z]$ design parameters are given in Ref. [33].

The simulation and experimental results of the synthesized admittance characteristic are shown in Fig. 7.24A and B, respectively.

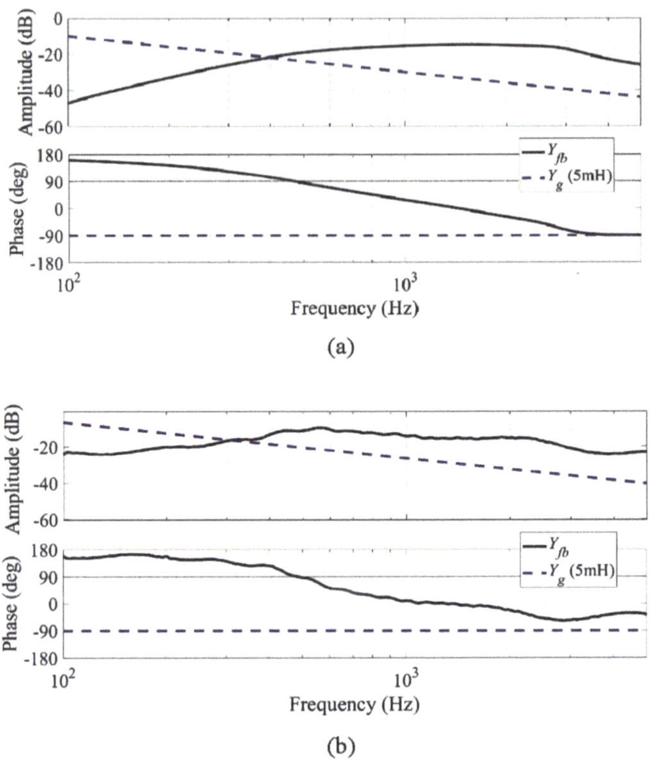

Figure 7.22 System frequency response; (A) simulation, and (B) experimental.

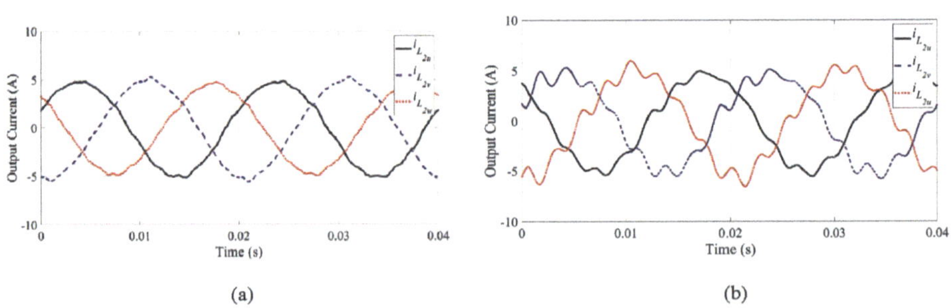

Figure 7.23 Experimental output current measurements for (A) $L_g = 1$ mH, and (B) $L_g = 2.8$ mH.

Fig. 7.24 shows that the phase characteristic varies only within $\pm$ 90 degrees and there is no cross-frequency because the phase difference from any Y_g is within 180 degrees. The GCC by the designed FB and FF controllers is passive and stable for any Y_g. The experimentally measured output current waveforms for $L_g = 5$ mH are also shown in Fig. 7.25.

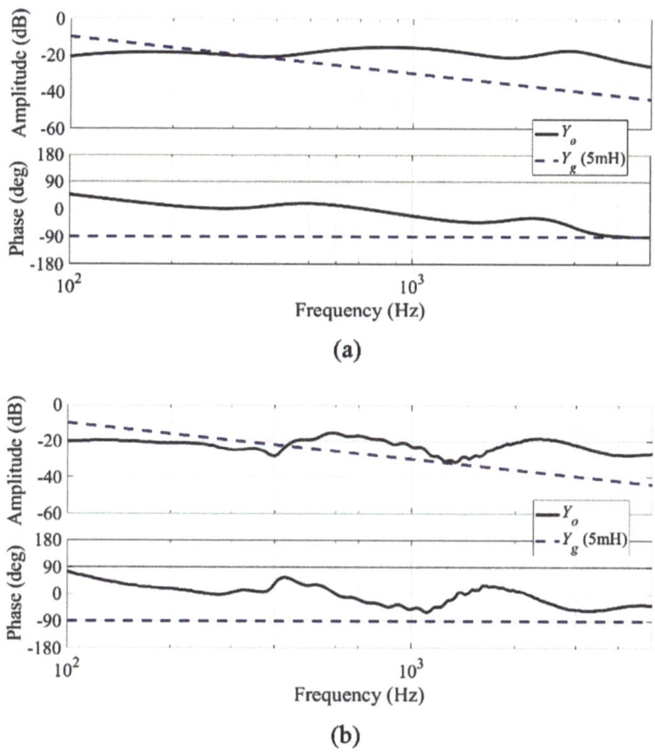

Figure 7.24 System frequency response for both FB and FF controllers; (A) simulation, and (B) experimental.

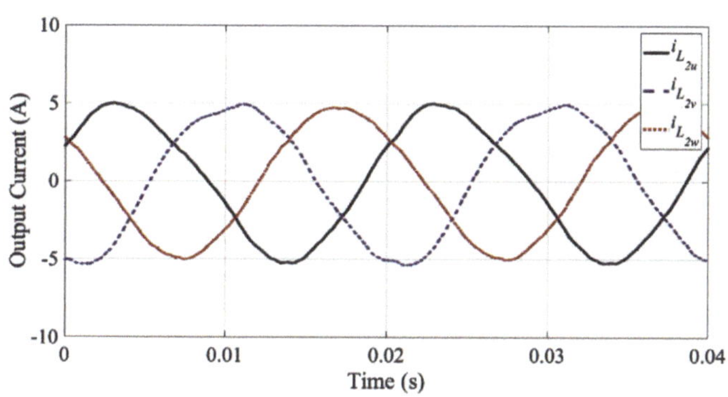

Figure 7.25 Experimental output current measurements for $L_g = 5$ mH.

7.4 Summary

Dynamic stability and performance requirements of the GCC systems contain several GCC needs such as ensuring stability over a wide range of operations, improving damping control and power quality, attenuation of ripples and rejection of noises, providing desirable frequency characteristics and reference tracking, and control time delay compensation.

In this chapter, it is investigated that time delay in control signal communication (due to A/D conversion and computation processes) decreases the GCC performances and frequency characteristics, especially around the resonant frequency. Then, a simple and effective delay compensation method based on a predictive state estimation is addressed. Virtual resistors are used to improve the oscillation damping characteristic. Finally, a combination of FB and FF control schemes is designed to ensure GCC stability and a good tracking characteristic over a wide range of operating frequencies. Particularly, the FF control loop provides an additional admittance term at the output of GCC to satisfy the passivity property.

The explained design methodologies are experimentally applied to a single- and three-phase GCC through a DSP-based digital control system.

References

[1] H. Bevrani, B. Francois, T. Ise, Microgrid Dynamics and Control, Wiley, NJ, USA, July 2017.

[2] M. Lindgren, J. Svensson, Control of a voltage-source converter connected to the grid through an LCL-filter-application to active filtering, IEEE PESC (1998) 229–235.

[3] E. Twining, D.G. Holmes, Grid current regulation of a three-phase voltage source inverter with an LCL input filter, IEEE Trans. Power Electron. 18 (3) (2003) 888–895.

[4] F. Blaabjerg, R. Teodorescu, M. Liserre, A. Timbus, Overview of control and grid synchronization for distributed power generation systems, IEEE Trans. Ind. Electron. 53 (5) (2006) 1398–1409.

[5] M. Malinowski, S. Bernet, A simple voltage sensorless active damping scheme for three-phase PWM converters with an LCL filter, IEEE Trans. Ind. Electron. 55 (4) (2008) 1876–1880.

[6] J. Dannehl, C. Wessels, F.W. Fuchs, Limitations of voltage oriented PI current control of grid-connected PWM rectifiers with LCL filters, IEEE Trans. Ind. Electron. 56 (2) (2009) 380–388.

[7] J. Dannehl, F.W. Fuchs, S. Hansen, P.B. Thogersen, Investigation of active damping approaches for PI-based current control of grid-connected pulse width modulation converters with LCL filters, IEEE Trans. Ind. Appl. 46 (4) (2010) 1509–1517.

[8] M. Huang, X. Wang, P.C. Loh, F. Blaabjerg, W. Wu, Stability analysis and active damping for llcl-filter-based grid-connected inverters, IEEJ J. Ind. Appl. 4 (3) (2015) 187–195.

[9] Y. Tang, P.C. Loh, P. Wang, F.H. Choo, F. Gao, Exploring inherent damping characteristics of LCL-filters for three-phase grid-connected voltage source inverters, IEEE Trans. Power Electron. 27 (3) (2012) 1433–1443.

[10] J. He, Y.W. Li, D. Bosnjak, B. Harris, Investigation and active damping of multiple resonances in a parallel-inverter-based microgrid, IEEE Trans. Power Electron. 28 (1) (2013) 234–246.

[11] M. Semasa, T. Kato, K. Inoue, A simple and effective time delay compensation method for grid-connected inverter with an LCL filter: application to active damping method, in: 2017 IEEE 18th Workshop on Control and Modeling for Power Electronics (COMPEL), 2017, pp. 1–7.

[12] J. Liu, Y. Miura, T. Ise, Fixed-parameter damping methods of virtual synchronous generator control using state feedback, IEEE Access 7 (2019) 99177−99190.
[13] M. Semasa, T. Kato, K. Inoue, Simple and effective time delay compensation method for active damping control of grid-connected inverter with an LCL filter, IEEJ J. Ind. Appl. 7 (6) (2018) 454−461.
[14] T. Kato, K. Inoue, Y. Donomoto, Fast current-tracking control for grid-connected inverter with an LCL filter by sinusoidal compensation, in: Energy Conversion Congress and Exposition (ECCE), 2011, pp. 2543−2548.
[15] D. Pan, X. Ruan, C. Bao, W. Li, X. Wang, Capacitor-current-feedback active damping with reduced computation delay for improving robustness of LCL-type grid-connected inverter, IEEE Trans. Power Electron. 29 (7) (2014) 3414−3427.
[16] C. Zou, B. Liu, S. Duan, R. Li, Influence of delay on system stability and delay optimization of grid-connected inverters with LCL filter, IEEE Trans. Ind. Inf. 10 (3) (2014) 1775−1784.
[17] Y. Lyu, H. Lin, Y. Cui, Stability analysis of digitally controlled LCL type grid-connected inverter considering the delay effect, IET Power Electron. 8 (9) (2015) 1651−1660.
[18] D. Yang, X. Ruan, H. Wu, A real-time computation method with dual sampling modes to improve the current control performance of the LCL-type grid-connected inverter, IEEE Trans. Ind. Electron. 62 (7) (2015) 4563−4572.
[19] J. Wang, J.D. Yan, L. Jiang, J. Zou, Delay-dependent stability of single-loop controlled grid-connected Inverters with LCL filters, IEEE Trans. Power Electron. 31 (1) (2016) 743−757.
[20] H. Deng, R. Oruganti, D. Srinivasan, PWM methods to handle time delay in digital control of a UPS inverter, IEEE Trans. Power Electron. 3 (1) (Mar. 2005) 1−6.
[21] P. Mattavelli, F. Polo, F.D. Lago, S. Saggini, Analysis of control delay reduction for the improvement of UPS voltage-loop bandwidth, IEEE Trans. Ind. Electron. 55 (8) (2008) 2903−2911.
[22] Z. Wan, J. Xiong, J. Lei, C. Chen, Analyze and reduce the impact of sampling delay on LCL converter with capacitor current feedback active damping, in: Proc. IEEE 11th Int. Conf. Power Elect. Drive Syst., 2015, pp. 539−545.
[23] C. Chen, J. Xiong, Z. Wan, J. Lei, K. Zhang, A time delay compensation method based on area equivalence for active damping of an LCL-type converter, IEEE Trans. Power Electron. 32 (1) (2017).
[24] S. Hyunl, S. Hongl, C. Wonl, A Compensation Method to Reduce Sampling Delay of Zero Dead-time PWM Using 3-Level NPC PWM Inverter, ITEC Asia-Pacific, 2016.
[25] M.G.F. Gous, H.J. Beukes, Time delay and dead-time compensation for a current controlled four-leg voltage source inverter utilized as a shunt active filter, in: Conference Record of 39th IAS Annual Meeting, 2004, pp. 115−122.
[26] P. Cortes, J. Rodriguez, C. Silva, A. Flores, Delay compensation in model predictive current control of a three-phase Inverter, IEEE Trans. Ind. Electron. 59 (2) (2012) 1323−1325.
[27] K. Ogata, Designing Linear Control Systems with MATLAB, Prentice Hall, NJ, 1994.
[28] T. Kato, K. Inoue, Y. Akiyama, K. Ohashi, Optimum and adjustable damping control of grid-connected inverter with an LCL filter, in: 2015 9th International Conference on Power Electronics and ECCE Asia (ICPE-ECCE Asia), June 2015, https://doi.org/10.1109/ICPE.2015.7167775.

[29] T. Kato, K. Inoue, M. Semasa, N. Minanimo, Effective time delay compensation for control of grid-connected inverter with an LCL filter, in: 2018 IEEE 19th Workshop on Control and Modeling for Power Electronics (COMPEL), Sept. 2018, https://doi.org/10.1109/COMPEL.2018.8459999.
[30] R. Ortega, A. Loría, P.J. Nicklasson, H. Sira-Ramírez, Passivity-based Control of Euler-Lagrange Systems, Springer Verlag, 2010.
[31] T. Messo, A. Aapro, T. Suntio, T. Roinila, Design of grid-voltage feedforward to increase impedance of grid-connected three-phase inverters with LCL-filter, IPEMC-ECCE Asia (2016) 2675–2682.
[32] T. Kato, K. Inoue, Y. Nakajima, Stabilization of grid-connected inverter system with feed-forward control, in: IEEE Energy Conversion Congress and Exposition (ECCE), 2017, pp. 3375–3382.
[33] T. Kato, K. Inoue, Y. Yamamoto, Optimal digital controller design for passive stabilization of a grid-connected three-phase inverter with LCL-filter, in: IEEE Energy Conversion Congress and Exposition (ECCE), September 2019, https://doi.org/10.1109/ECCE.2019.8912644.

Advanced control synthesis methods for grid connected converters

8

8.1 Optimal control design

The grid connected converter (GCC) in form of grid-forming converter is known as a voltage source converter (VSC). Conventionally, the cascaded proportional-integral (PI) controllers (as shown in Fig. 2.8) are applied to the GCCs as the inner loop controllers for the voltage regulation. The PI controllers in cascaded structures are independently tuned in frequency-domain by setting a faster response time for inner loop and slower one for outer loops [1]. However, it is difficult to tune the controller gains because of the modes coupling and the controller's bandwidth. Although some methods have been developed in the literature to tune control gains [2], the power coupling still needs to be improved.

This section addresses a linear quadratic regulator (LQR) as an alternative controller [3], and the control gains are tuned using the participation factors [4,5]. It is shown that the LQR-based controller has better transient characteristics and decoupling effects between voltage and active power than a conventional cascaded PI controller.

8.1.1 Case study and dynamic modeling

Fig. 8.1 shows a three-phase GCC connected with an AC grid through LCL filters and supplied by an ideal DC source, which has been already depicted in Fig. 3.19. The control system uses the dq-frame characteristics. Assume the state variables are the GCC current i_o through the inductor L_f, the voltage v_c across the filter capacitor C_f, and the current i_g through the transformer inductance L_c. The V_{dc}, v_m, and v_g (voltage at point of common coupling-v_{pcc}) are the ideal DC bus voltage, the GCC output voltage and the grid-voltage which is transformed to the capacitor phase, respectively. The following equations from Kirchhoff's laws can be obtained for the equivalent circuit of the GCC in Fig. 8.1:

$$\begin{cases} \boldsymbol{v}_m = \left(R_f + L_f\dfrac{d}{dt}\right)\boldsymbol{i}_o + \boldsymbol{v}_c \\ C_f\dfrac{d}{dt}\boldsymbol{v}_c = \boldsymbol{i}_o - \boldsymbol{i}_g \\ \boldsymbol{v}_c = \left(R_c + L_c\dfrac{d}{dt}\right)\boldsymbol{i}_g + \boldsymbol{v}_g \end{cases} \tag{8.1}$$

Grid Connected Converters. https://doi.org/10.1016/B978-0-323-99902-1.00002-1

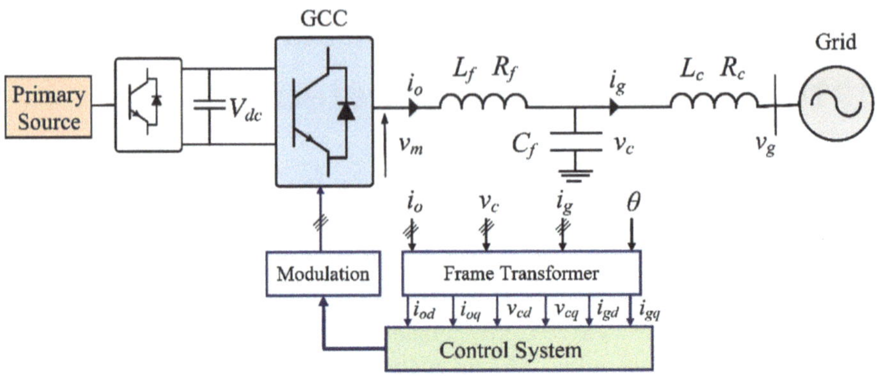

Figure 8.1 A GCC with *dq*-frame based control system.

where

$$\boldsymbol{v}_m^T = [v_{ma} \quad v_{mb} \quad v_{mc}],$$
$$\boldsymbol{i}_o^T = [i_{oa} \quad i_{ob} \quad i_{oc}],$$
$$\boldsymbol{v}_c^T = [v_{ca} \quad v_{cb} \quad v_{cc}],$$

$$\boldsymbol{i}_g^T = [i_{ga} \quad i_{gb} \quad i_{gc}],$$
$$\boldsymbol{v}_g^T = [v_{ga} \quad v_{gb} \quad v_{gc}].$$

To have a uniform representation and in order to absorb large deviations in absolute values into base relationships, the per-unit system can be used [6]. Therefore, by defining a base rated apparent power and impedance as:

$$S_{\text{base}} = \frac{P_{\text{base}}}{\cos\varphi}; \; Z_{\text{base}} = \frac{U_{ac}^2}{S_{\text{base}}}$$

the per-unit parameters of the GCC can be calculated as follows:

$$R_{fpu} = \frac{R_f}{Z_{\text{base}}}; \; \frac{L_{fpu}}{\omega_b} = \frac{L_f}{Z_{\text{base}}}; \; \frac{C_{fpu}}{\omega_b} = C_f Z_{\text{base}} \tag{8.2}$$

The term of ω_b is the base value for the angular frequency in rad/s, corresponding to the rated frequency of the grid. Park transformation in the per-unit system transfers Eqs. (8.1) and (8.2) in *d-q* frame.

$$\begin{cases} \dfrac{L_{fpu}}{\omega_b}\dfrac{d}{dt}\begin{bmatrix} i_{od} \\ i_{oq} \end{bmatrix} = -R_f\begin{bmatrix} i_{od} \\ i_{oq} \end{bmatrix} + \omega L_{fpu}\begin{bmatrix} i_{oq} \\ -i_{od} \end{bmatrix} - \begin{bmatrix} v_{cd} \\ v_{cq} \end{bmatrix} + \begin{bmatrix} v_{md} \\ v_{mq} \end{bmatrix} \\ \dfrac{C_{fpu}}{\omega_b}\dfrac{d}{dt}\begin{bmatrix} v_{cd} \\ v_{cq} \end{bmatrix} = \omega C_{fpu}\begin{bmatrix} v_{cq} \\ -v_{cd} \end{bmatrix} + \begin{bmatrix} i_{od} \\ i_{oq} \end{bmatrix} - \begin{bmatrix} i_{gd} \\ i_{gq} \end{bmatrix} \\ \dfrac{L_{cpu}}{\omega_b}\dfrac{d}{dt}\begin{bmatrix} i_{gd} \\ i_{gq} \end{bmatrix} = \omega\begin{bmatrix} i_{gq} \\ -i_{gd} \end{bmatrix} - R_{cpu}\begin{bmatrix} i_{gd} \\ i_{gq} \end{bmatrix} + \begin{bmatrix} v_{cd} \\ v_{cq} \end{bmatrix} - \begin{bmatrix} v_{gd} \\ v_{gq} \end{bmatrix} \end{cases} \tag{8.3}$$

The state variables are transformed in dq-frame where θ is the instantaneous phase of the capacitor voltage deduced by the droop control.

Thus, the state space model for the three-phase GCC system in dq-frame can be obtained as Eq. (8.4). For simplicity, the subscript of "pu" is removed from the variables.

$$\begin{cases} \dot{x} = \boldsymbol{A}\boldsymbol{x} + \boldsymbol{B}\boldsymbol{u} + \boldsymbol{E}\boldsymbol{w} \\ y = \boldsymbol{C}\boldsymbol{x} \end{cases} \tag{8.4}$$

where,

$$\boldsymbol{x}^T = \begin{bmatrix} i_{od} & i_{oq} & v_{cd} & v_{cq} & i_{gd} & i_{gq} \end{bmatrix}$$

$$\boldsymbol{u}^T = \begin{bmatrix} v_{md} & v_{mq} \end{bmatrix}$$

$$\boldsymbol{w}^T = \begin{bmatrix} v_{gd} & v_{gq} \end{bmatrix}$$

$$\boldsymbol{A} = \begin{bmatrix} -\dfrac{R_f\omega_b}{L_f} & \omega\omega_b & -\dfrac{\omega_b}{L_f} & 0 & 0 & 0 \\ -\omega\omega_b & -\dfrac{R_f\omega_b}{L_f} & 0 & -\dfrac{\omega_b}{L_f} & 0 & 0 \\ \dfrac{\omega_b}{C_f} & 0 & 0 & \omega\omega_b & -\dfrac{\omega_b}{C_f} & 0 \\ 0 & \dfrac{\omega_b}{C_f} & -\omega\omega_b & 0 & 0 & -\dfrac{\omega_b}{C_f} \\ 0 & 0 & \dfrac{\omega_b}{L_c} & 0 & -\dfrac{R_c\omega_b}{L_c} & \omega\omega_b \\ 0 & 0 & 0 & \dfrac{\omega_b}{L_c} & -\omega\omega_b & -\dfrac{R_c\omega_b}{L_c} \end{bmatrix}$$

$$B^T = \begin{bmatrix} \frac{\omega_b}{L_f} & 0 & 0 & 0 & 0 & 0 \\ 0 & \frac{\omega_b}{L_f} & 0 & 0 & 0 & 0 \end{bmatrix}$$

$$C = \begin{bmatrix} 0 & 0 & 1 & 0 & 0 & 0 \\ 0 & 0 & 0 & 1 & 0 & 0 \end{bmatrix}$$

$$E^T = \begin{bmatrix} 0 & 0 & 0 & 0 & -\frac{\omega_b}{L_c} & 0 \\ 0 & 0 & 0 & 0 & 0 & -\frac{\omega_b}{L_c} \end{bmatrix}$$

The vectors $\boldsymbol{x}$ and $\boldsymbol{u}$ are the state variables and the system control inputs, respectively, and $\boldsymbol{w}$ is the grid-voltage considered as the disturbance input. Here, the output $\boldsymbol{y}$ is the voltage across the capacitor (v_c).

For contribution to the frequency regulation of the grid, the GCCs need to operate at a power lower than the maximum power point tracking (MPPT) power. The remaining power is considered as a regulation power reserve for frequency control support. The supported power can be shared according to the corresponding droop control gains as shown in Fig. 8.2. The droop control system contains the $P-\omega$ and $Q-E_g$ droops since the nature of the grid is inductive enough ($X_g \gg R_g$). The droop control system determines the amplitude, phase, and frequency of the capacitor voltage reference, thus grid-forming GCC can be synchronized with the grid and can share power without any detection of the grid-phase.

The injected instantaneous three-phase active and reactive power components by the GCC, p_{mes} and q_{mes}, are given by the following equations

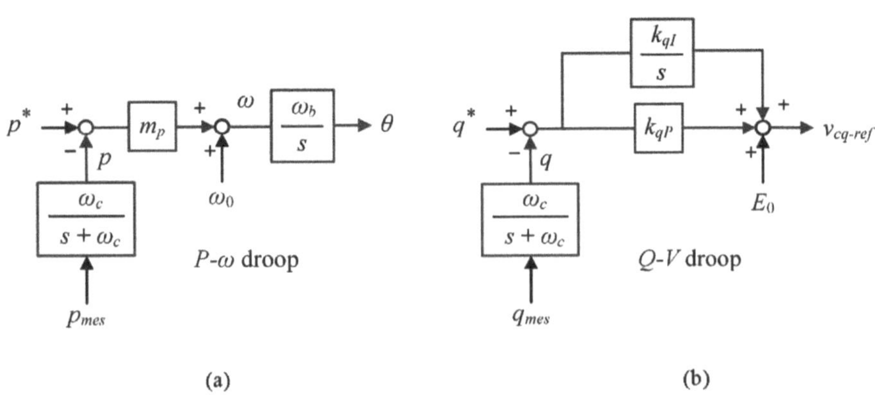

Figure 8.2 Block diagram of droop controls; (A) $P-\omega$ droop, and (B) $Q-V$ droop.

$$p_{mes} = v_{cd}i_{gd} + v_{cq}i_{gq}\ ;\ q_{mes} = v_{cd}i_{gd} - v_{cq}i_{gq} \tag{8.5}$$

To avoid power noises, a low-pass filter (LPF), as given in Eq. (8.6), is added on the way of active and reactive power measurements.

$$p = \frac{\omega_c}{s + \omega_c} p_{mes};\ q = \frac{\omega_c}{s + \omega_c} q_{mes} \tag{8.6}$$

where ω_c is the cut-off frequency.

8.1.2 Optimal voltage controller design

Fig. 8.3 shows a GCC with a control system containing the droop and the optimal state feedback controller. The amplitude and frequency of the output voltage (v_{abc}) are determined by a nominal amplitude E_0 and nominal frequency ω_0. The GCC generate a three-phase balanced sinusoidal AC voltage using the cascaded PI controllers in the *dq* reference frame. The phase angle of the AC voltage is formed by the droop control system. Therefore, the GCC with its primary source can operate as an ideal voltage source, and hence, the power sharing among different grid-forming GCCs is a function of their output and tie-line impedances [1].

This section presents an LQR controller composed of a state-feedback and an integral compensator instead of the conventional cascaded voltage and current controllers. An integral compensator is sufficient to control the voltage across the capacitor without any steady-state error in the synchronous rotating frame. The block diagram of the LQR system for the voltage regulation is shown in Fig. 8.4. $\boldsymbol{K}_{fb}$ and $\boldsymbol{K}_i$ are the state-feedback gain and the integral compensator matrices, respectively. The vector $\boldsymbol{r}$ is defined as the voltage reference across the capacitor from the Q-V droop operation. The vector $\dot{\boldsymbol{e}}$ is the derivative of the error between the reference and the output voltage; these are expressed in Eq. (8.7).

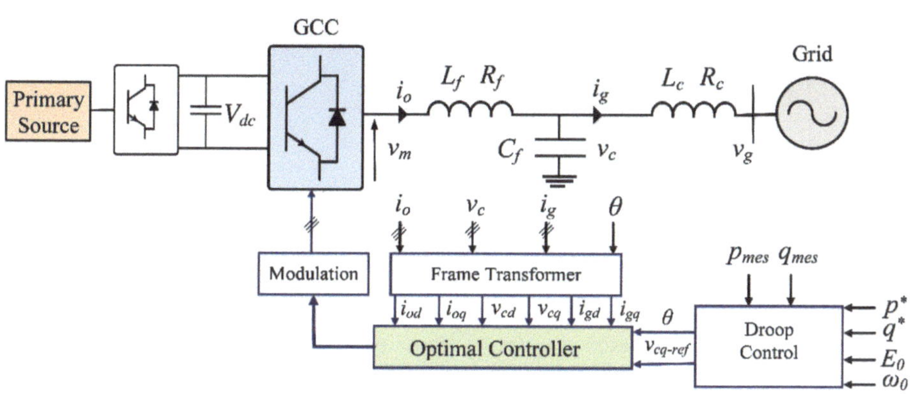

Figure 8.3 Grid-forming GCC example with droop and optimal controls.

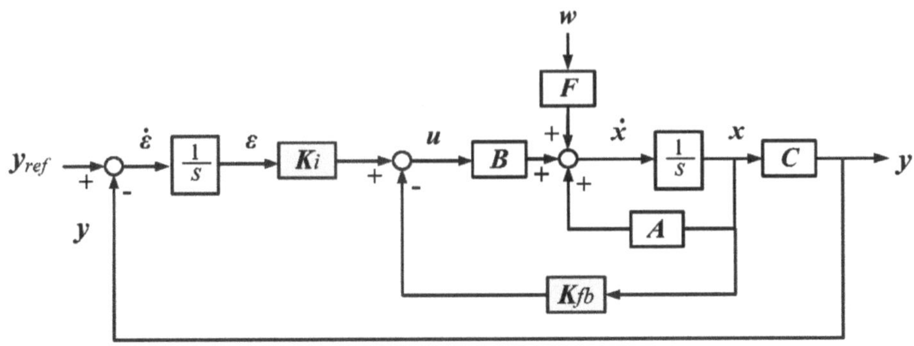

Figure 8.4 Block diagram of LQR for voltage regulation.

$$\boldsymbol{r} = \begin{bmatrix} v_{cd\_ref} & v_{cq\_ref} \end{bmatrix}^T; \ \dot{\boldsymbol{\varepsilon}} = \boldsymbol{r} - \boldsymbol{y} = \boldsymbol{r} - \boldsymbol{Cx} \tag{8.7}$$

The augmented state-space model of the closed-loop system can be represented as

$$\begin{cases} \begin{bmatrix} \dot{\boldsymbol{x}} \\ \dot{\boldsymbol{\varepsilon}} \end{bmatrix} = \begin{bmatrix} \boldsymbol{A} & \boldsymbol{0} \\ -\boldsymbol{C} & \boldsymbol{0} \end{bmatrix} \begin{bmatrix} \boldsymbol{x} \\ \boldsymbol{\varepsilon} \end{bmatrix} + \begin{bmatrix} \boldsymbol{B} \\ \boldsymbol{0} \end{bmatrix} u + \begin{bmatrix} \boldsymbol{0} \\ \boldsymbol{I} \end{bmatrix} r + \begin{bmatrix} \boldsymbol{F} \\ \boldsymbol{0} \end{bmatrix} w \\ y = \begin{bmatrix} \boldsymbol{C} & \boldsymbol{0} \end{bmatrix} \begin{bmatrix} \boldsymbol{x} \\ \boldsymbol{\varepsilon} \end{bmatrix} \\ u = -\boldsymbol{K}_{fb} x + \boldsymbol{K}_i \varepsilon = -\begin{bmatrix} \boldsymbol{K}_{fb} & -\boldsymbol{K}_i \end{bmatrix} \begin{bmatrix} \boldsymbol{x} \\ \boldsymbol{\varepsilon} \end{bmatrix} \end{cases} \tag{8.8}$$

In addition, in order to utilize the LQR, the augmented matrices are defined as

$$\widetilde{\boldsymbol{x}} = \begin{bmatrix} \boldsymbol{x} \\ \boldsymbol{\varepsilon} \end{bmatrix} \in R^{8\times 1}, \ \widetilde{\boldsymbol{K}} = \begin{bmatrix} \boldsymbol{K}_{fb} & -\boldsymbol{K}_i \end{bmatrix} \in R^{2\times 8} \tag{8.9}$$

$$\widetilde{\boldsymbol{A}} = \begin{bmatrix} \boldsymbol{A} & \boldsymbol{0} \\ -\boldsymbol{C} & \boldsymbol{0} \end{bmatrix} \in R^{8\times 8}, \widetilde{\boldsymbol{B}} = \begin{bmatrix} \boldsymbol{B} \\ \boldsymbol{0} \end{bmatrix}, \widetilde{\boldsymbol{F}} = \begin{bmatrix} \boldsymbol{F} \\ \boldsymbol{0} \end{bmatrix} \in R^{8\times 2}$$

This controller has 16 gains in order to regulate the voltage, meaning that the feedback controller has a greater degree of freedom. Moreover, the coupling effect between the *d*- and *q*-axes can be taken into account when the gain matrices are determined.

The quadratic optimal regulator problem is used to determine the control gain matrix $\widetilde{\boldsymbol{K}}$ such that minimizes the performance index of

$$J = \int_0^{\infty} \left(\widetilde{\boldsymbol{x}}^T \boldsymbol{Q} \widetilde{\boldsymbol{x}} + \boldsymbol{u}^T \boldsymbol{R} \boldsymbol{u} \right) dt \tag{8.10}$$

where $\boldsymbol{Q}$ and $\boldsymbol{R}$ are positive-definite (or positive-semidefinite) Hermitian or real symmetric matrices. Note that the first term on the right side of the equation is related to the convergence speed of each of the state variables, and the second term is related to the expenditure of the energy of the control signals. The matrices $\boldsymbol{Q}$ and $\boldsymbol{R}$ determine the relative importance of the convergence speed and the expenditure of this energy, respectively. An advantage of the quadratic optimal control method over the pole-placement method is that it provides a systematic way of computing the state-feedback control gain matrix [7].

As mentioned, controller $\widetilde{\boldsymbol{K}}$ must be determined such that the performance index J (Eq. 8.10) to be minimized. In this case, the control law $\boldsymbol{u} = -\widetilde{\boldsymbol{K}}\widetilde{\boldsymbol{x}}$ is optimal for any initial state $\widetilde{\boldsymbol{x}}(0)$, and considering Eq. (8.11), the eigenvalues of $\widetilde{\boldsymbol{A}} - \widetilde{\boldsymbol{B}}\widetilde{\boldsymbol{K}}$ have negative real parts, and Eq. (8.10) can be rewritten as Eq. (8.12).

$$\dot{\widetilde{\boldsymbol{x}}} = \widetilde{\boldsymbol{A}}\widetilde{\boldsymbol{x}} - \widetilde{\boldsymbol{B}}\widetilde{\boldsymbol{K}}\widetilde{\boldsymbol{x}} = \left(\widetilde{\boldsymbol{A}} - \widetilde{\boldsymbol{B}}\widetilde{\boldsymbol{K}} \right) \widetilde{\boldsymbol{x}} \tag{8.11}$$

$$\boldsymbol{J} = \int_0^{\infty} \left(\widetilde{\boldsymbol{x}}^T \boldsymbol{Q} \widetilde{\boldsymbol{x}} + \widetilde{\boldsymbol{x}}^T \widetilde{\boldsymbol{K}}^T \boldsymbol{R} \widetilde{\boldsymbol{K}} \widetilde{\boldsymbol{x}} \right) dt = \int_0^{\infty} \widetilde{\boldsymbol{x}}^T \left(\boldsymbol{Q} + \widetilde{\boldsymbol{K}}^T \boldsymbol{R} \widetilde{\boldsymbol{K}} \right) \widetilde{\boldsymbol{x}} dt \tag{8.12}$$

Let us set

$$\widetilde{\boldsymbol{x}}^T \left(\boldsymbol{Q} + \widetilde{\boldsymbol{K}}^T \boldsymbol{R} \widetilde{\boldsymbol{K}} \right) \widetilde{\boldsymbol{x}} = -\frac{d}{dt} \left(\widetilde{\boldsymbol{x}}^T \boldsymbol{P} \widetilde{\boldsymbol{x}} \right) \tag{8.13}$$

where $\boldsymbol{P}$ is a positive-definite Hermitian (real symmetric) matrix. Then, one can obtain

$$\widetilde{\boldsymbol{x}}^T \left(\boldsymbol{Q} + \widetilde{\boldsymbol{K}}^T \boldsymbol{R} \widetilde{\boldsymbol{K}} \right) \widetilde{\boldsymbol{x}} = -\dot{\widetilde{\boldsymbol{x}}}^T \boldsymbol{P} \widetilde{\boldsymbol{x}} - \widetilde{\boldsymbol{x}}^T \boldsymbol{P} \dot{\widetilde{\boldsymbol{x}}} = -\widetilde{\boldsymbol{x}}^T \left[\left(\widetilde{\boldsymbol{A}} - \widetilde{\boldsymbol{B}}\widetilde{\boldsymbol{K}} \right)^T \boldsymbol{P} + \boldsymbol{P} \left(\widetilde{\boldsymbol{A}} - \widetilde{\boldsymbol{B}}\widetilde{\boldsymbol{K}} \right) \right] \widetilde{\boldsymbol{x}} \tag{8.14}$$

Comparing both sides of Eq. (8.14), it is required that

$$\left(\widetilde{A}-\widetilde{B}\widetilde{K}\right)^{T}P+P\left(\widetilde{A}-\widetilde{B}\widetilde{K}\right)=-\left(Q+\widetilde{K}^{T}R\widetilde{K}\right) \tag{8.15}$$

It can be proven that if $\widetilde{A}-\widetilde{B}\widetilde{K}$ is stable, there exists a positive-definite matrix P that satisfies Eq. (8.15) [7,8]. This means if one solves the above equation and finds a positive-definite matrix P, then the system is stable. Since all eigenvalues of $\widetilde{A}-\widetilde{B}\widetilde{K}$ are assumed to have negative real parts, then $\widetilde{x}(t)$ must be bounded, and the performance index J based on Eq. (8.13) can be evaluated as

$$J=\widetilde{x}^{T}(0)P\widetilde{x}(0)-\widetilde{x}^{T}(\infty)P\widetilde{x}(\infty) \tag{8.16}$$

Note that this system has references such that $\widetilde{x}(\infty)$ is not equal to 0. Thus, the performance index J can be obtained in terms of the initial condition $\widetilde{x}(0)$, P, and the final condition $\widetilde{x}(\infty)$, which is deduced from reference r. To obtain the solution to the quadratic optimal control problem, first assume a nonsingular matrix T, such that $R=T^{T}T$, where R is a positive-definite Hermitian or real symmetric matrix. Then Eq. (8.15) can be written as

$$\left(\widetilde{A}-\widetilde{B}\widetilde{K}\right)^{T}P+P\left(\widetilde{A}-\widetilde{B}\widetilde{K}\right)+Q+\widetilde{K}^{T}T^{T}T\widetilde{K}=0 \tag{8.17}$$

which can be rewritten as

$$t\widetilde{A}^{T}P+P\widetilde{A}+\left[T\widetilde{K}-\left(T^{T}\right)^{-1}\widetilde{B}^{T}P\right]^{T}\left[T\widetilde{K}-\left(T^{T}\right)^{-1}\widetilde{B}^{T}P\right]+Q-P\widetilde{B}R^{-1}\widetilde{B}^{T}P=0 \tag{8.18}$$

The minimization of the performance index J with respect to $\widetilde{K}$ requires the minimization of

$$\widetilde{x}^{T}\left[T\widetilde{K}-\left(T^{T}\right)^{-1}\widetilde{B}^{T}P\right]^{T}\left[T\widetilde{K}-\left(T^{T}\right)^{-1}\widetilde{B}^{T}P\right]\widetilde{x}^{T}.$$

Since this expression is nonnegative, the minimum occurs when it is zero, or

$$T\widetilde{K}=\left(T^{T}\right)^{-1}\widetilde{B}^{T}P. \tag{8.19}$$

Hence

$$\widetilde{K}=T^{-1}\left(T^{T}\right)^{-1}\widetilde{B}^{T}P=R^{-1}\widetilde{B}^{T}P \tag{8.20}$$

Eq. (8.20) gives the optimal matrix $\widetilde{\boldsymbol{K}}$. Thus, the optimal control law for the quadratic optimal control problem when the performance index J is given by Eq. (8.20) is given by

$$\boldsymbol{u}(t) = -\widetilde{\boldsymbol{K}}\widetilde{\boldsymbol{x}}(t) = -\boldsymbol{R}^{-1}\widetilde{\boldsymbol{B}}^{T}\boldsymbol{P}\widetilde{\boldsymbol{x}}(t) \tag{8.21}$$

The matrix $\boldsymbol{P}$ in Eq. (8.21) must satisfy the following reduced equation, which is called the *reduced-matrix Riccati equation*.

$$\widetilde{\boldsymbol{A}}^{T}\boldsymbol{P} + \boldsymbol{P}\widetilde{\boldsymbol{A}} + \boldsymbol{Q} - \boldsymbol{P}\widetilde{\boldsymbol{B}}\boldsymbol{R}^{-1}\widetilde{\boldsymbol{B}}^{T}\boldsymbol{P} = 0 \tag{8.22}$$

Therefore, assuming that the system is controllable, the design steps of the optimal controller can be summarized as follows: (1) solve Eq. (8.22), the reduced-matrix Riccati equation, for the matrix $\boldsymbol{P}$. If a positive-definite matrix $\boldsymbol{P}$ exists, the system is stable and (2) substitute matrix $\boldsymbol{P}$ into Eq. (8.20). The resulting matrix $\widetilde{\boldsymbol{K}}$ is the optimal matrix.

In MATLAB, the command *lqr* solves the continuous-time, quadratic regulator problem, and the associated Riccati equation which is mentioned above [8].

8.1.3 Application example

Consider the GCC system shown in Fig. 8.1, which its parameters at the nominal operating point are given in Table 8.1. Firstly, the controllability of the system considering $\boldsymbol{A}$ and $\boldsymbol{B}$ matrices in Eq. (8.4) must be examined. The controllability matrix can be defined as follows

Table 8.1 System parameters and operating point.

Parameter	Value	Parameter	Value
P	0.5 [pu]	ω_c	31.4 [rad/s]
f	50 [Hz]	R_f	0.005 [pu]
ω_{ref}	1 [pu]	L_f	0.15 [pu]
e_{gdref}	1 [pu]	R_c	0.005 [pu]
e_{gqref}	0 [pu]	L_c	0.15 [pu]
ω_b	$2\pi f$ [rad/s]	C_f	0.066 [pu]
m_p	0.02 [pu]	n_{qi}	−10 [pu]
		n_{qp}	0.1 [pu]
Operating point			
θ_s	0.316 [rad]	I_{gd}	0.5 [pu]
$\cos\theta_s$	0.95	I_{gq}	0 [pu]
$\sin\theta_s$	0.3123	V_{gd}	1 [pu]
E_{gd}	1 [pu]	V_{gq}	0 [pu]
E_{gq}	0 [pu]		

$$\phi_C = \left[\boldsymbol{B}\ \boldsymbol{AB}\ \boldsymbol{A}^2\boldsymbol{B}\ \boldsymbol{A}^3\boldsymbol{B}\ \boldsymbol{A}^4\boldsymbol{B}\ \boldsymbol{A}^5\boldsymbol{B} \right] \tag{8.23}$$

The rank of the controllability matrix ϕ_C is 6, which is the same dimension of matrix $\boldsymbol{A}$, so this system has controllable, and the optimal state feedback controller can be designed [7].

The servo-system for voltage regulation is determined for given matrices $\boldsymbol{Q}$ and $\boldsymbol{R}$ values in the above equations. However, the phase of the grid-voltage and the droop control that is nonlinear should be taken into account for the parameter determination. Moreover, the feedback gains should be tuned carefully to decouple the active power control from the reactive power control, which is related to voltage regulation. A tuning method using the participation factors for the same case study is given in Ref. [3]. The participation factor shows a sensitivity of an eigenvalue to a state variable.

A comparison between the cascaded PI control [9] and the addressed control methodology in this section is investigated in MATLAB/Simulink environment. The voltage step response is applied to the system with $E_0 = 1.1$ pu at $t = 0.1$ s. This case is chosen in order to check the coupling between the active power and controlled voltage. The results are shown in Fig. 8.5. The proposed method generates a faster response without overshoot than the cascaded PI controller does. The proposed method presents a small transient coupling effect between the active and reactive powers in contrast to the cascaded PI controller.

8.2 Digital optimal control design

This section addresses an optimal control methodology to design a digital sinusoidal compensator and state feedback gains to provide a desirable tracking and active damping effects for a GCC with an LCL filter. The introduced design methodology is validated through simulation and experiment by a digital signal processor-based control system. The compensator and feedback gains of the GCC are easily determined by an optimal control technique.

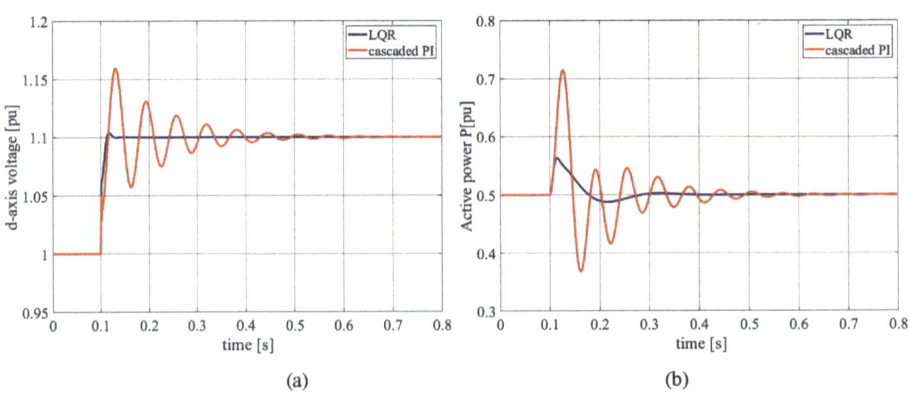

Figure 8.5 Proposed control performance compared to conventional one; (A) d-axis voltage and (B) active power.

8.2.1 Case study and discrete dynamic model

Reconsider a single-phase full-bridge GCC configuration shown in Fig. 7.1, which is redrawn as Fig. 8.6. The converter is connected to a grid with voltage v_g. The GCC works as a VSC that uses a digital control scheme to provide the pulse-width modulation (PWM) switching signals. An LCL filter of L_1, C, L_2 is employed at the converter output. The state vector $\boldsymbol{x}(t)$ contains i_{L_1}, v_C, i_{L_2} and its sampled form with sampling time of T is shown as $\boldsymbol{x}[i]$. It is assumed that the input voltage $v_i[i]$ is provided by an ideal DC voltage. As discussed in Section 7.1, the discrete state-space model of the GCC can be obtained as Eq. (8.24).

$$\begin{aligned} \boldsymbol{x}[i+1] &= \boldsymbol{A}_d x[i] + \boldsymbol{b}_d u[i] + \boldsymbol{h}_d v_g[i] \\ y[i] &= \boldsymbol{c}_d x[i] \end{aligned} \tag{8.24}$$

where

$$\boldsymbol{A}_d = e^{AT_S};\ \boldsymbol{b}_d = \int_0^{T_s} e^{A\ \tau} d\tau \boldsymbol{b};\ \boldsymbol{h}_d = \int_0^{T_s} e^{A\ \tau} d\tau \boldsymbol{h};\ \boldsymbol{c}_d = \boldsymbol{c}. \tag{8.25}$$

and the $\boldsymbol{A}$, $\boldsymbol{b}$, $\boldsymbol{c}$, and $\boldsymbol{h}$ are the matrix coefficients of the continuous form of state-space model and are

$$t\boldsymbol{A} = \begin{bmatrix} 0 & -1/L_1 & 0 \\ 1/C & 0 & -1/C \\ 0 & 1/L_2 & 0 \end{bmatrix};\ \boldsymbol{b} = \begin{bmatrix} E/L_1 \\ 0 \\ 0 \end{bmatrix};\ \boldsymbol{c} = [0\ \ 0\ \ 1];\ \boldsymbol{h} = \begin{bmatrix} 0 \\ 0 \\ -1/L_2 \end{bmatrix}.$$

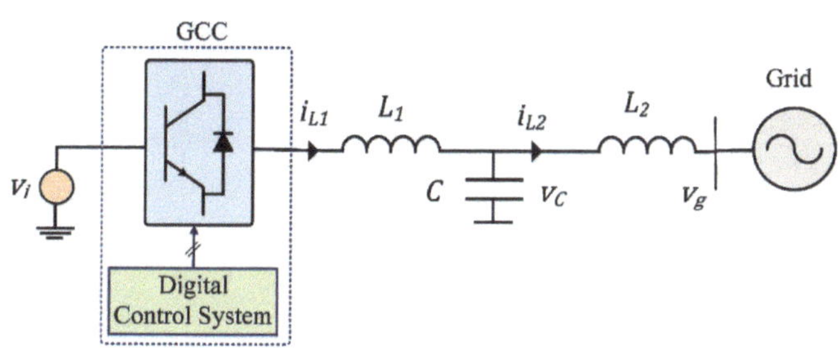

Figure 8.6 A single-phase full-bridge GCC with a digital control system.

8.2.2 Design of sinusoidal compensator and state-feedback gains

As shown in Fig. 8.7A, conventionally in a GCC control system, a state feedback control with an integral compensator $G[z]$ $(= G_I[z])$ is often used to reduce steady-state error between output y and its reference y_{ref}. Then, using an augmented variable $w[i]$, the control problem is reduced to find an integral gain k_1, and a feedback gain vector $\boldsymbol{k}_{fb}$, which are described in the following equation:

$$G[z] = G_I[z] = \frac{k_1}{z-1} \tag{8.26}$$

$$\begin{aligned} u[i] &= -\boldsymbol{k}_{fb}\,\boldsymbol{x}[i] + k_1 w[i] \\ w[i+1] &= w[i] + \left(y_{ref}[i] - y[i]\right) \end{aligned} \tag{8.27}$$

The above equation is a general form and some elements of the feedback gains $\boldsymbol{k}_{fb}$ can be zero. A different type of compensator, which is called proportional-resonance (PR) compensator is given in Eq. (8.28).

$$G[z] = C_{PR}[z] = k_1\left(1 + \frac{\sin\omega T}{\omega_0 T_r}\frac{z-1}{z^2 - 2z\cos\omega_0 T + 1}\right) \tag{8.28}$$

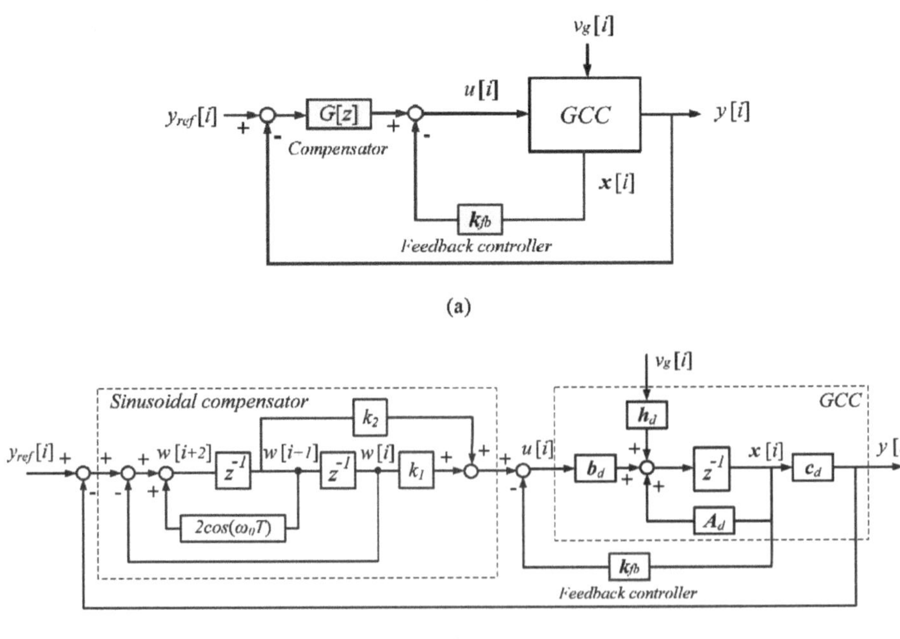

Figure 8.7 Closed-loop system with compensator and state feedback controller; (A) overall framework and (B) detailed block diagram.

However, the integral compensator may add an additional time delay. Moreover, tuning only one parameter (k_1) provides a minimum control freedom. Therefore, the sinusoidal compensator with the following transfer function in the z-domain, which were used in Chapters 4 and 7, addresses a better solution.

$$G[z] = C_S[z] = \frac{k_2 z + k_1}{z^2 - 2z\cos\omega T + 1} \tag{8.29}$$

Using Eq. (8.29), the control principle can be written as follows

$$\begin{aligned} u[i] &= -\boldsymbol{k}_{fb}\,\boldsymbol{x}[i] + k_1 w[i] + k_2 w[i+1] \\ w[i+2] &= -w[i] + 2\cos\omega T w[i+1] + \left(y_{ref}[i] - y[i]\right) \end{aligned} \tag{8.30}$$

Its block diagram is shown in Fig. 8.7B, where the system order is five because both the compensator and the converter are second and third orders. Using the auxiliary variables $w[i], w[i+1]$ for the compensator, the final form of the closed-loop control system is:

$$\left.\begin{aligned} &\begin{bmatrix} \boldsymbol{x}[i+1] \\ w[i+1] \\ w[i+2] \end{bmatrix} = \begin{bmatrix} \boldsymbol{A}_d & \boldsymbol{0} & \boldsymbol{0} \\ 0 & 0 & 1 \\ -\boldsymbol{c}_d & -1 & 2\cos\omega T \end{bmatrix} \begin{bmatrix} \boldsymbol{x}[i] \\ w[i] \\ w[i+1] \end{bmatrix} \\ &+ \begin{bmatrix} \boldsymbol{b}_d \\ 0 \\ 0 \end{bmatrix} \begin{bmatrix} -\boldsymbol{k}_{fb} & k_1 & k_2 \end{bmatrix} \begin{bmatrix} \boldsymbol{x}[i] \\ w[i] \\ w[i+1] \end{bmatrix} \\ &+ \begin{bmatrix} \boldsymbol{0} \\ 0 \\ 1 \end{bmatrix} y_{ref}[i] + \begin{bmatrix} \boldsymbol{h}_d \\ 0 \\ 0 \end{bmatrix} v_g[i] = \begin{bmatrix} \boldsymbol{A}_d - \boldsymbol{b}_d\boldsymbol{k}_{fb} & \boldsymbol{b}_d k_1 & \boldsymbol{b}_d k_2 \\ 0 & 0 & 1 \\ -\boldsymbol{c}_d & -1 & 2\cos\omega T \end{bmatrix} \begin{bmatrix} \boldsymbol{x}[i] \\ w[i] \\ w[i+1] \end{bmatrix} \\ &+ \begin{bmatrix} \boldsymbol{0} \\ 0 \\ 1 \end{bmatrix} y_{ref}[i] + \begin{bmatrix} \boldsymbol{h}_d \\ 0 \\ 0 \end{bmatrix} v_g[i] \end{aligned}\right\} \tag{8.31}$$

or

$$\widehat{\boldsymbol{x}}[i+1]=\widehat{\boldsymbol{A}}\widehat{\boldsymbol{x}}\,[i]+\widehat{\boldsymbol{g}}\,y_r[i]+\widehat{\boldsymbol{h}}\,v_g[i]. \tag{8.32}$$

Based on the above augmented system matrix $\widehat{\boldsymbol{A}}$ and the input vector $\widehat{\boldsymbol{h}}$, three feedback gains $\boldsymbol{k}_{fb}$ and two compensator gains k_1, k_2 must be determined by the designer. The last term $v_g[i]$ works as a disturbance whose behavior is sinusoidal, and it is expected to be attenuated by the introduced sinusoidal compensator.

8.2.3 Optimal tuning of control system parameters

The control gains of $\boldsymbol{K} = \left[\boldsymbol{k}_{fb}\ -k_1\ -k_2\right]$ can be easily computed using an optimal control method, such as the one introduced in the previous section. The $\boldsymbol{K}$ is determined to minimize the following second-order index function with weighting constants $\boldsymbol{Q}$ and r.

$$J=\sum_{i=0}^{\infty}\left\{\widehat{\boldsymbol{x}}[i]^2\boldsymbol{Q}\widehat{\boldsymbol{x}}[i]+ru[i]^2\right\} \tag{8.33}$$

Then, the optimum gains vector $\boldsymbol{K}$ can be found as follows:

$$\boldsymbol{K}=\left(\boldsymbol{k}_{fb}+\widetilde{\boldsymbol{b}}^{\,T}\boldsymbol{P}\widetilde{\boldsymbol{b}}\right)^{-1}\widetilde{\boldsymbol{b}}^{\,T}\boldsymbol{P}\widetilde{\boldsymbol{A}} \tag{8.34}$$

where

$$\widetilde{\boldsymbol{A}}=\begin{bmatrix}\boldsymbol{A}_d & \boldsymbol{0} & \boldsymbol{0}\\ 0 & 0 & 1\\ -\boldsymbol{c}_d & -1 & 2\cos\omega T\end{bmatrix};\ \widetilde{\boldsymbol{b}}=\left[\boldsymbol{b}^{\,T}\ \ 0\ \ 0\right]^T$$

and $\boldsymbol{P}$ is the solution of the following Ricatti equation.

$$\boldsymbol{P}=\boldsymbol{Q}+\widetilde{\boldsymbol{A}}^{\,T}\boldsymbol{P}\widetilde{\boldsymbol{A}}-\widetilde{\boldsymbol{A}}^{\,T}\boldsymbol{P}\widetilde{\boldsymbol{b}}\left(r+\widetilde{\boldsymbol{b}}^{\,T}\boldsymbol{P}\widetilde{\boldsymbol{b}}\right)^{-1}\widetilde{\boldsymbol{b}}^{\,T}\boldsymbol{P}\widetilde{\boldsymbol{A}} \tag{8.35}$$

8.2.4 Estimation of state variables by the minimum-order observer

The state feedback control basically needs as many sensors as number of state variables. In this case, three measured values for i_{L_1}, v_C, i_{L_2} are necessary. The number can be reduced to one for i_{L2} which is necessary to track the target and the others are estimated by the minimum-order observer [10,11].

Dividing the system equations into the measured variable $y[i] = i_{L_2}[i]$ and estimated variables $\boldsymbol{\eta}[i] = [i_{L_1}[i]\ v_c[i]]^T$, they can be rewritten as follows.

$$\begin{bmatrix} y[i+1] \\ \boldsymbol{\eta}[i+1] \end{bmatrix} = \begin{bmatrix} a_{11} & \boldsymbol{a}_{12} \\ \boldsymbol{a}_{21} & \boldsymbol{A}_{22} \end{bmatrix} \begin{bmatrix} y[i] \\ \boldsymbol{\eta}[i] \end{bmatrix} + \begin{bmatrix} b_1 \\ \boldsymbol{b}_2 \end{bmatrix} u[i] + \begin{bmatrix} h_1 \\ \boldsymbol{h}_2 \end{bmatrix} v_g[i]$$
$$y[i] = [1 \quad \boldsymbol{0}] \begin{bmatrix} y[i] \\ \boldsymbol{\eta}[i] \end{bmatrix} \tag{8.36}$$

The grid voltage v_g and the output current state variable i_{L_2} can be measured (accessible), and $v[i] = [i_{L_1}, v_C]^T$ must be estimated by the minimum-order observer as follows where $\boldsymbol{\eta}[i]$ and the observer gains as $\boldsymbol{l}$.

$$\boldsymbol{v}[i] = \boldsymbol{\eta}[i] - \boldsymbol{l}y[i] \tag{8.37}$$

The observer state equation is as follows.

$$\boldsymbol{v}[i+1] = \widehat{\boldsymbol{A}}\boldsymbol{v}[i] + \widehat{\boldsymbol{B}}\,y[i] + \widehat{\boldsymbol{J}}u[i] + \widehat{\boldsymbol{h}}v_g[i] \tag{8.38}$$

$$\widehat{\boldsymbol{x}}[i] = \widehat{\boldsymbol{C}}\widehat{\boldsymbol{v}}[i] + \widehat{\boldsymbol{D}}y[i]$$

where

$$\widehat{\boldsymbol{A}} = \boldsymbol{A}_{22} - \boldsymbol{la}_{12},\ \widehat{\boldsymbol{B}} = \boldsymbol{a}_{21} - \boldsymbol{l}a_{11} + \widehat{\boldsymbol{A}}\boldsymbol{l},\ \widehat{\boldsymbol{C}} = \begin{bmatrix} 0 \\ \boldsymbol{I} \end{bmatrix},\ \widehat{\boldsymbol{D}} = \begin{bmatrix} 1 \\ \boldsymbol{l} \end{bmatrix}$$

$$\widehat{\boldsymbol{J}} = \boldsymbol{b}_2 - \boldsymbol{l}b_1,\ \widehat{\boldsymbol{h}} = \boldsymbol{h}_2 - \boldsymbol{l}h_1$$

Only from measured i_{L_2}, it is possible to estimate the other two state variables (i_{L_1}, v_C) with the minimum-order observer. The closed-loop system frequency responses with the minimum-order observer from both simulation and experiment are shown in Fig. 8.8A. It is shown that while the observer reduces the number of sensors, the responses almost coincide well; the experimental result is almost same as the simulation one. The experimental output current signal in the presence of the addressed optimal digital controller is also shown in Fig. 8. 8B.

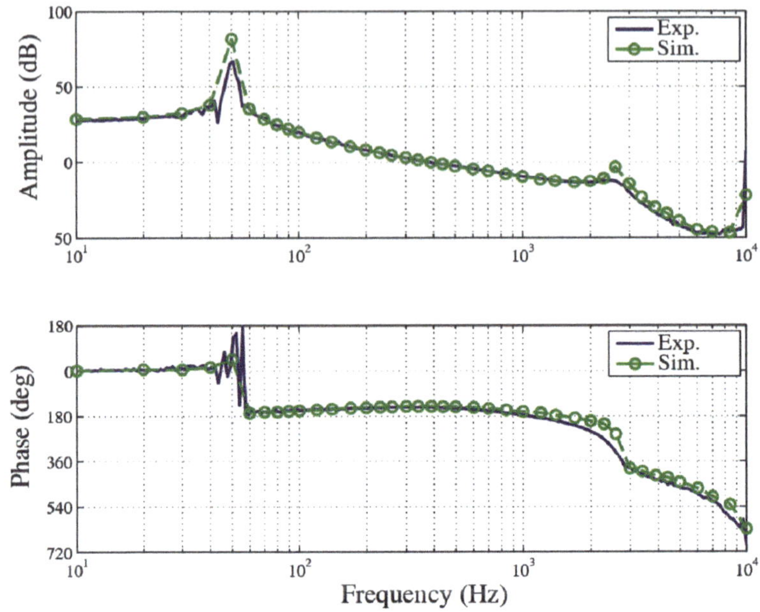

(a)

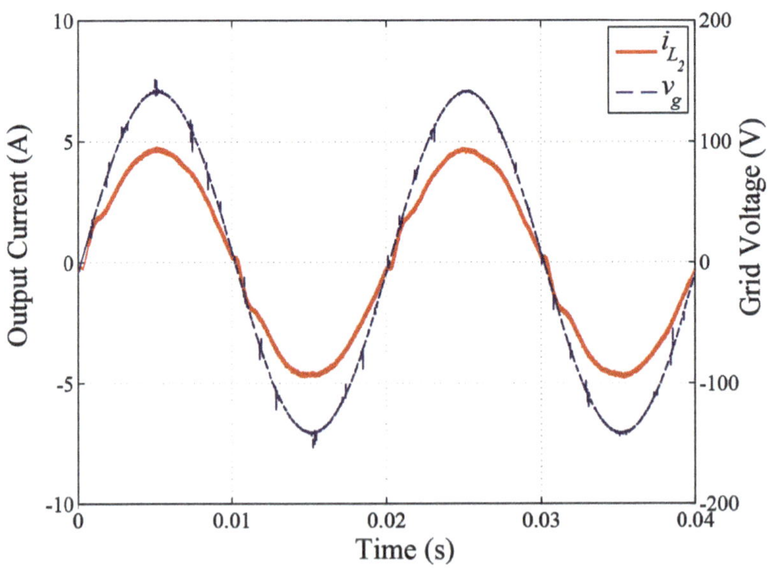

(b)

Figure 8.8 Closed-loop system response; (A) frequency response with and without the minimum-order observer, and (B) output current and grid voltage signals.

8.3 Lyapunov-based digital control design

The GCC are usually controlled with compensators and state feedback gains after linearization at designed operating points. The gains are determined using a linear control method such as optimal control technique which is discussed in previous sections. Energy function-based methods provide other effective solutions for the GCC control system design, specifically in the case of more complex structure and interconnected dynamics such as mutual interactions among multiple parallel converters. Lyapunov function is one of these energy-based functions, and this section addresses a Lyapunov-based digital control system synthesis with several functionalities for a GCC system [12,13].

8.3.1 Lyapunov-based control synthesis methodology

Reconsider a single-phase full-bridge GCC configuration shown in Fig. 8.6 with the dynamic model given in Eq. (8.1). The state equations of the desired operation are described by reference variables of $\boldsymbol{x}_r[i]$ and $u_r[i]$, with the subscript r, as follows:

$$\boldsymbol{x}_r[i+1] = \boldsymbol{A}_d\boldsymbol{x}_r[i] + \boldsymbol{b}_d u_r[i] + \boldsymbol{h}_d v_g[i] \tag{8.39}$$

Deviation between Eqs. (8.24) and (8.39) is defined as follows.

$$\widetilde{\boldsymbol{x}}[i+1] = \boldsymbol{A}_d\widetilde{\boldsymbol{x}}[i] + \boldsymbol{b}_d\widetilde{u}[i] \tag{8.40}$$

where

$$\widetilde{\boldsymbol{x}}[i] = \boldsymbol{x}[i] - \boldsymbol{x}_r[i];\ \widetilde{u}[i] = u[i] - u_r[i]$$

Its Lyapunov function $\boldsymbol{V}\left(\widetilde{\boldsymbol{x}}[i]\right)$ becomes as follows.

$$\boldsymbol{V}\left(\widetilde{\boldsymbol{x}}[i]\right) = \frac{1}{2}\widetilde{\boldsymbol{x}}^T[i]\boldsymbol{Q}\widetilde{\boldsymbol{x}}[i] \tag{8.41}$$

where $\boldsymbol{Q}$ is a positive definite matrix. The convergence condition for above function can be derived as follows (for simplicity in presentation, the subscript "d" is removed from $\boldsymbol{A}_d$ and $\boldsymbol{b}_d$.

$$\begin{aligned}
&t\boldsymbol{V}\left(\widetilde{\boldsymbol{x}}[i+1]\right) - \boldsymbol{V}\left(\widetilde{\boldsymbol{x}}[i]\right)\\
&t = \frac{1}{2}\widetilde{\boldsymbol{x}}^T[i+1]\boldsymbol{Q}\widetilde{\boldsymbol{x}}[i+1] - \frac{1}{2}\widetilde{\boldsymbol{x}}^T[i]\boldsymbol{Q}\widetilde{\boldsymbol{x}}[i]\\
&t = \frac{1}{2}\left(\boldsymbol{A}\widetilde{\boldsymbol{x}}[i] + \boldsymbol{b}\widetilde{u}[i]\right)^T\boldsymbol{Q}\left(\boldsymbol{A}\widetilde{\boldsymbol{x}}[i] + \boldsymbol{b}\widetilde{u}[i]\right) - \frac{1}{2}\widetilde{\boldsymbol{x}}^T[i]\boldsymbol{Q}\widetilde{\boldsymbol{x}}[i]\\
&= \frac{1}{2}\widetilde{\boldsymbol{x}}^T[i](\boldsymbol{A}^T\boldsymbol{Q}\boldsymbol{A} - \boldsymbol{Q})\widetilde{\boldsymbol{x}}[i] + \frac{1}{2}\left(\widetilde{u}[i]\boldsymbol{b}^T\boldsymbol{Q}\boldsymbol{A}\widetilde{\boldsymbol{x}}[i] + \widetilde{\boldsymbol{x}}^T[i]\boldsymbol{A}^T\boldsymbol{Q}\boldsymbol{b}\widetilde{u}[i] + \widetilde{u}[i]\boldsymbol{b}^T\boldsymbol{Q}\boldsymbol{b}\widetilde{u}[i]\right) < 0
\end{aligned} \tag{8.42}$$

The convergence condition is that the above two terms Eq. (8.42) of the discrete Lyapunov function become negative. Due to the passivity of the converter, the first term is negative. Therefore, it requires the second term to be negative. This condition is satisfied with the input $\widetilde{u}[i]$ in the form of Eq. (8.43), where α is a positive design parameter which determines attenuation rate.

$$\widetilde{u}[i] = -\alpha \boldsymbol{b}^T \boldsymbol{Q} \boldsymbol{A} \widetilde{\boldsymbol{x}}[i] = -\alpha \widetilde{\boldsymbol{x}}^T[i] \boldsymbol{A}^T \boldsymbol{Q} \boldsymbol{b} = -\alpha \rho \tag{8.43}$$

In order to determine the parameter α, one can use Eq. (8.43) in the second term of Eq. (8.42) to find when it to be negative.

$$\begin{aligned} &t\frac{1}{2}\left(\widetilde{u}[i]\boldsymbol{b}^T\boldsymbol{Q}\boldsymbol{A}\widetilde{\boldsymbol{x}}[i] + \widetilde{\boldsymbol{x}}^T[i]\boldsymbol{A}^T\boldsymbol{Q}\boldsymbol{b}\widetilde{u}[i] + \widetilde{u}[i]\boldsymbol{b}^T\boldsymbol{Q}\boldsymbol{b}\widetilde{u}[i]\right) \\ &= \frac{1}{2}\left(-\alpha\rho^2 - \alpha\rho^2 + \alpha^2\rho^2\boldsymbol{b}^T\boldsymbol{Q}\boldsymbol{b}\right) = -\alpha\rho^2\left(1 - \frac{\alpha}{2}\boldsymbol{b}^T\boldsymbol{Q}\boldsymbol{b}\right) \end{aligned} \tag{8.44}$$

The positive parameter α must be selected to satisfy the negative condition of Eq. (8.20). The condition is as follows.

$$0 < \alpha < \frac{2}{\boldsymbol{b}^T \boldsymbol{Q} \boldsymbol{b}} \tag{8.45}$$

It is noteworthy that possible α is bounded and cannot to be too large. For example, it can be set to $\alpha = 1/\left(\boldsymbol{b}^T\boldsymbol{Q}\boldsymbol{b}\right)$ under its allowable range ($u_{\min} \le u \le u_{\max}$) to minimize the difference and maximize the convergence rate.

8.3.2 Reference generator

The addressed Lyapunov-based control needs to have reference waveforms of $\boldsymbol{x}_r[i]$ and $u_r[i]$ as given in Eq. (8.39). Conventionally reference variables can be obtained by analytical calculations. However, they have to be calculated implicitly at the same time from a given output reference y_r. This makes the analytically calculation very complex and sometimes almost impossible. To avoid this challenge, here a numerical generation method for the desirable/reference variables is introduced [12]. The method is based on the proposed digital control scheme (with state feedback and compensator) discussed in Chapter 7, and its overall scheme is shown in Fig. 8.9A.

When the target output waveform is a sinusoidal waveform, which is common in practice, the sinusoidal type $G_s[z]$ could be useful as the compensator in the reference generator as shown in Fig. 8.9B. It has a denominator of $D_G = z^2 - 2z\cos\omega T + 1$ which is identical to the z transform of the sinusoidal function $Y_r[z]$ in Eq. (8.46). It has a numerator of a simple first order with gains k_1, k_2 in Eq. (8.47).

$$Y_r[z] = \frac{N_R[z]}{D_R[z]} = \frac{z\sin\omega T}{z^2 - 2z\cos\omega T + 1} \tag{8.46}$$

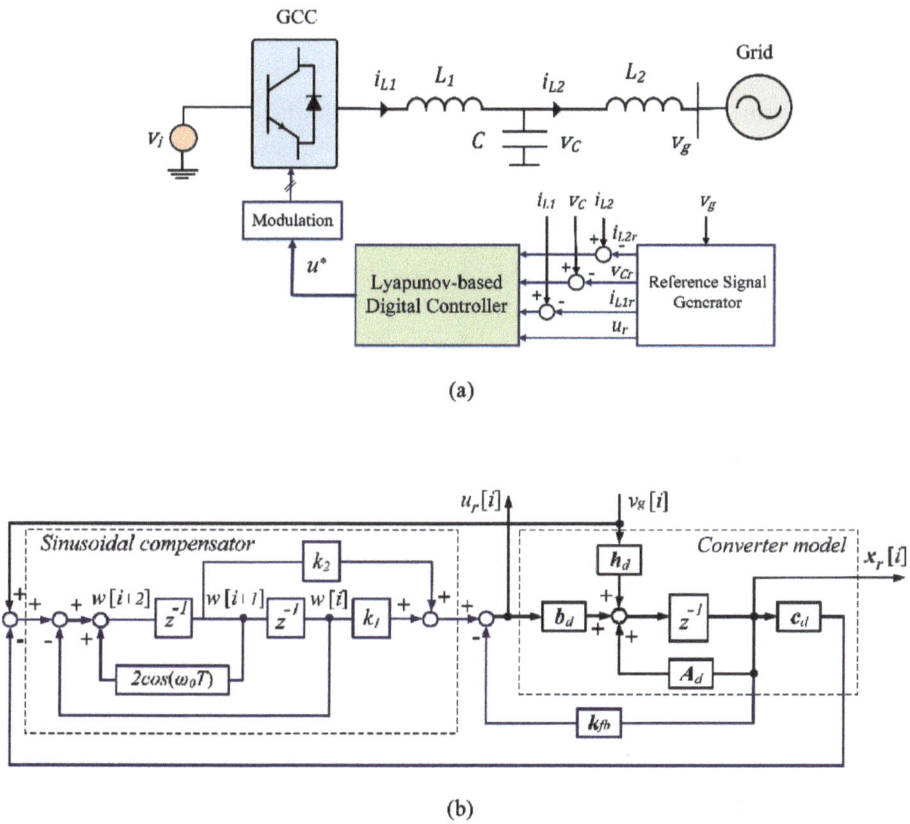

Figure 8.9 The GCC closed-loop control system; (A) overall digital control scheme, and (B) reference signal generator.

$$G_s[z] = \frac{N_G[z]}{D_G[z]} = \frac{k_2 z + k_1}{z^2 - 2z\cos\omega T + 1} \tag{8.47}$$

The structure of digital control system with the sinusoidal compensator is similar to the structure of reference generator which is shown in Fig. 8.9B. Its control law becomes as Eq. (8.30) with auxiliary variables $w[i], w[i+1]$ and feedback gains $\boldsymbol{k}_{fb}$. The total system, as given by Eq. (8.31) becomes fifth order because the converter system is third order and the compensator is second order. Similar to previous section, the feedback and compensator gains, $\boldsymbol{k}_{fb}$ and k_1, k_2 can be determined by the optimal control.

8.3.3 Simulation and experimental results

The proposed Lyapunov-based digital control is applied to the single-phase GCC shown in Fig. 8.1. Circuit parameters are the input DC voltage $E = 200$ V, the system voltage amplitude $V_g = 141$ V, and filter parameters $L_1 = L_2 = 1$ mH, $C = 10\mu$ F. Gain parameters are $f = [0.0191, 0.0008, 0.0231]$, $k_1 = -0.0217$, $k_2 = 0.0249$, and $\alpha = 0.4$.

The experimental result is shown in Fig. 8.10A, where the output current is controlled sinusoidally according to the grid voltage phase and has a few high-frequency oscillations whose frequency corresponds to the filter resonance characteristic. As can be seen, the output current is well controlled with the exception of some minor deformation due to small oscillation and time delay/deadtime effects. The oscillations can be effectively damped using the virtual impedance/resistor method, and the time delay can be compensated using time delay compensator that are both discussed in Chapter 7. The application results for two parallel GCCs using the addressed Lyapunov-based digital control methodology are given in Ref. [13].

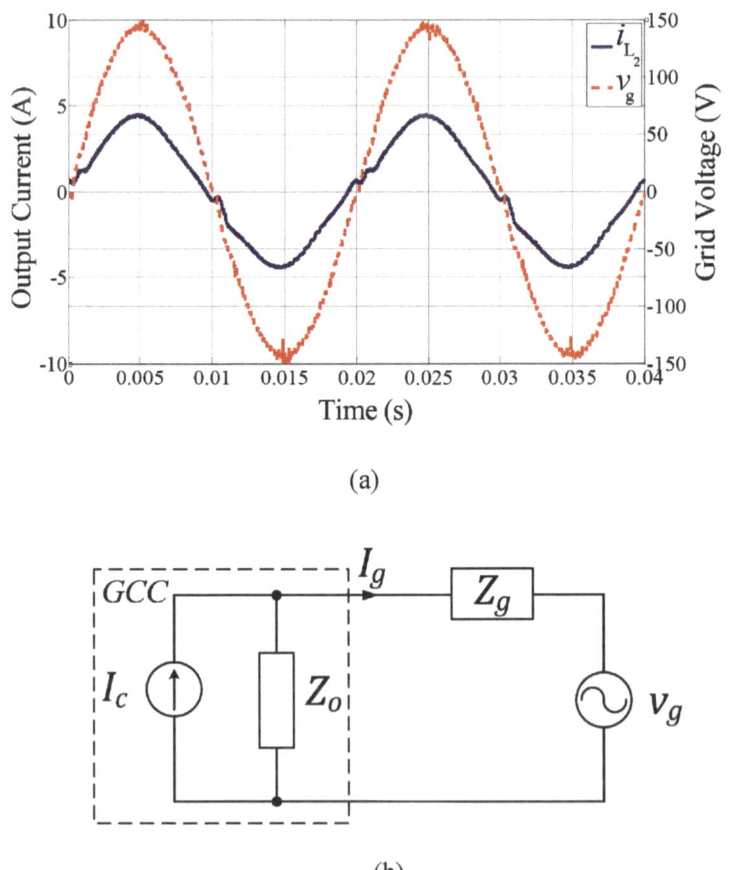

Figure 8.10 (A) The GCC experimental response for the addressed digital control system, and (B) equivalent circuit of a GCC with an LCL filter.

8.3.4 Stability analysis

The GCC system stability is analyzed according to a simple equivalent model which is shown in Fig. 8.10B. The inverter is expressed with the Norton's equivalent circuit of Z_o, I_c and the grid system is expressed with the Thevenin's equivalent circuit of v_g, Z_g in the frequency domain. The current to the grid system I_g can be obtained as follows:

$$I_c = I_g + \frac{Z_g I_g + v_g}{Z_o}$$
$$\frac{Z_o + Z_g}{Z_o} I_g = I_c - \frac{v_g}{Z_o} \tag{8.48}$$
$$I_g = \left(I_c - \frac{v_g}{Z_o}\right) \frac{1}{1 + Z_g/Z_o}$$

According to this criterion, the system becomes unstable when the phase difference is 180 degree and $1 + Z_g/Z_o \leq 0$, or $|Z_o| \langle |Z_g|$. It is possible to investigate the system stabilities from frequency characteristics of two impedances Z_o and Z_g, which are measured from response currents for a perturbation voltage signal Δv_g whose amplitude is about 5% of the system voltage and the frequency is swept between target ranges.

First, the conventional linear state-feedback control for a pure inductive line case is analyzed and investigated for its frequency characteristics to compare them with those of the proposed Lyapunov control. The system is stable because the amplitude of the line impedance is smaller than that of the GCC output impedance at the frequency, where phase difference between Z_o and Z_g is 180 degrees (simulation result shown in Fig. 8.11A). When the line inductor is larger than 3 mH, the current becomes diverged and the system becomes unstable. This is because the amplitude of the line impedance is larger than that of the converter output impedance at the frequency where phase difference is 180 degrees.

Then, the proposed Lyapunov control for a pure inductive line case is also analyzed and investigated. The Z_o and Z_g simulation and experimental frequency responses are shown in Fig. 8.11B, where the line impedance is assumed to be an inductor of $L_g =$ 10 mH. The inductor value is larger than that of the conventional feedback case. However, the controlled system is stable. Because the phase characteristic of the converter output impedance is between 0 and 90 degrees and never the phase difference between Z_o and Z_g becomes 180 degrees.

8.4 Model predictive control-based controller design

8.4.1 An overview

The MPC addresses the idea of precalculation of a system behavior via mathematical model and using these values to determine optimum actuating variables. Therefore, the

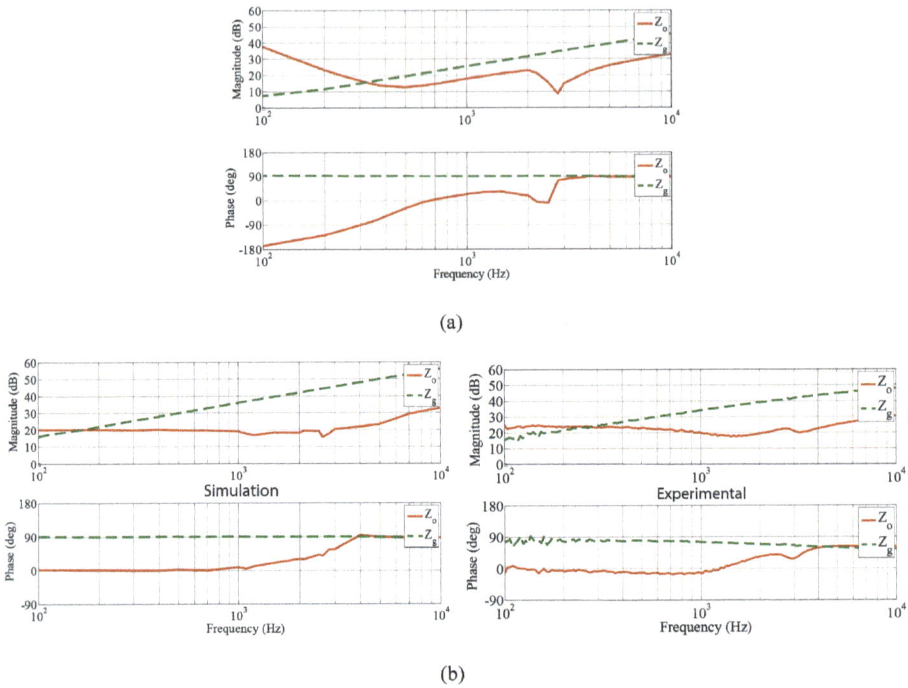

Figure 8.11 The frequency responses of Z_o and Z_g; (A) simulation results for application of conventional state feedback control in the presence of $L_g = 3$ mH, and (B) simulation and experimental results for the proposed digital control system, where $L_g = 10$ mH.

common characteristic of all these controllers is the principle of determining an optimum value for the actuating variable by using an explicit model of the system. The basic structure of the MPC can be summarized as follows [14]:

(i) Predict the future behavior of the process state/output with a system model.
(ii) Compute the future control input signals on line at each step by minimizing a cost function under inequality constraints on the manipulated (control) and/or controlled variables.
(iii) Apply vector control variable on the controlled plant and repeat the previous step with new measured input/state/output variables.

In the area of GCC, due to its high sampling frequency, an intuitive approach to reducing the computational load for solving optimal control problems is required, for instance, by using a fixed number of iterations in each sampling step. This is possible due to the discrete nature of power converters, which means the number of actuations of the system is limited to the number of possible switching states of the converter. Alternatively, the computational load can be reduced by using the previous solution for the next step of an optimization algorithm in order to successively reduce the suboptimality of the predicted trajectories.

Regarding application in GCCs, three important classes of MPC are *finite control set model predictive control (FCS-MPC)*, *two-configuration predictive control*

(*2PC*), and *PWM Model Predictive Control* (*PPC*) [15]. The FCS-MPC takes advantage of the inherent features of the GCC, which can only produce eight different switching states by turning on and off the gates for each leg of the converter. This control approach uses the GCC switching states as control variables directly [16]. Thus, it is also known as *direct model predictive control.*

The 2PC consists of applying two configurations of the power converter during a computation step; one of these configurations leads to null voltages. First, the future values of load currents are calculated, assuming that null voltages are applied during the whole next sampling period (the free response). Then, an active state of the converter (and the response based on this state) is selected, for instance, the values that the controlled values would reach if this converter configuration were applied for the whole sampling period will be computed (the force response). Finally, from these calculated values, one can acquire the time in which the null voltage and active configuration should be applied in one period so that the actual actuating values are as close as possible to the desired values [16].

The PPC calculates the voltages required to reach the desired point after one period and remains there. Then, a PWM is used to translate these desired voltages into switching orders. This method is also noted as "dead-beat control" by some works. The PPC operates with six configurations per sampling period. Hence, the required actuating values could be precisely reached [17].

Furthermore, a number of novel MPC strategies are proposed in the literature, which combines the concepts of above three classes of MPCs, such as "continuous control set MPC" or "modulated MPC" [18,19]. Although these methods offer clear advantages over FCS-MPC, for instance, the reduction in ripple and a fixed switching frequency, similar to 2PC and PPC MPC, the tuning of cost function's weighting parameters of such methods are more complicated than FCS-MPC when multiple variables are considered in the controllers.

The FCS-MPC provides a flexible control scheme that allows easy inclusion of multiple control variables, system constraints, and nonlinearities. A number of control solutions based on FCS-MPC for GCC application appeared in the literature [17,20,21]. Compared with the classic linear PI and PWM controller, the MPC offers many advantages, including minimizing steady-state errors over a wide target frequency range, minimal harmonic distortion, fast dynamic response, and maximum dc-link utilization.

The FCS-MPC allows several control objectives to be fulfilled simultaneously by utilizing the system model and the real-time values of the system to determine the optimized converter's physical switching states directly. This grants the predictive controller significant advantages over the existing PWM-based methods, as it is capable of regulating the amplitudes of output values without employing zero current vectors.

A combination of an FCS-MPC and a virtual synchronous generator (VSG) schemes are used to address an effective GCC control system [22], in this section. Considering the operation of VSG control-based GCC, several controls and regulation issues could be satisfied. Some of these are essential to ensure proper operation of GCCs and to maintain power grid stability, like the voltage and frequency control, while others are imposed for security reasons, like current limitations and voltage restrictions. This characteristic converges in a natural way to the utilization of FCS-MPC, as it is multiinput multioutput (MIMO)-based control that has the ability to control several control variables and constraints at the same time via the inclusion of those terms in a single cost function. In addition, since the design of any control system must take into account the model of the system for adjusting the controller parameters, the employment of model predicting control will only simplify the control system design. This is beneficial for complex systems such as GCCs. Furthermore, with practically all current control strategies being implemented in discrete time, the rapid advancement of the control platform, and the discrete nature of GCC that require a discrete signal to command the turn-on and turn-off of their switching devices, the compatibility of FCS-MPC for GCCs are clearly highlighted.

8.4.2 Finite control set model predictive control

The basic ideas of FCS-MPC are based on the use of a system model to predict the future behavior of the system variable. The system model is a discrete-time model which can be expressed in the state-space model form with the sates of $\boldsymbol{x}(k)$ and control inputs of $\boldsymbol{u}(k)$. The optimal actuation will be selected among these predictions based on a predefined cost function that represents the desired behavior of the system. The cost function is a function of the references, future states, and the future actuation, which can be defined as follows:

$$J(k) = f(\boldsymbol{x}(k+1),\ \boldsymbol{u}(k)) \tag{8.49}$$

In fact, the MPC is an optimization problem that consists of minimizing the cost function J. For each sampling instant, the controller will employ the optimal actuator in the GCC. Then at the next sampling instant, the new measured data will be utilized to obtain a new optimal actuation. This is called a receding horizon strategy. Giving the fact that the number of actuators (switching states that a converter can switch into) is finite, the number of forecasts that need to be made is limited to this number, and thus, the algorithm is called finite control set MPC. The operation principle of FCS-MPC is summarized in Fig. 8.12 [23].

In order to FCS-MPC design, the discrete-time model of the GCC system, and the modeling of the power converter with consideration of its possible switching states are required. The fundamental elements of a GCC are the semiconductor switches, which can be simplified as ideal switches with only on and off states. Hence, the total number of switching states of a GCC is equal to the number of possible combination of states of each switch, which can be described with $N = x^y$, when x and y are the number of possible states of each leg and the number of legs of the GCC (converter), respectively.

Nevertheless, the states that resulted in undesired system conditions such as short-circuits must be removed from the set of possible switching stats. Utilizing the corresponding voltage or current, which is produced by each possible switching state and the system model, a prediction for the future system behavior can be conducted.

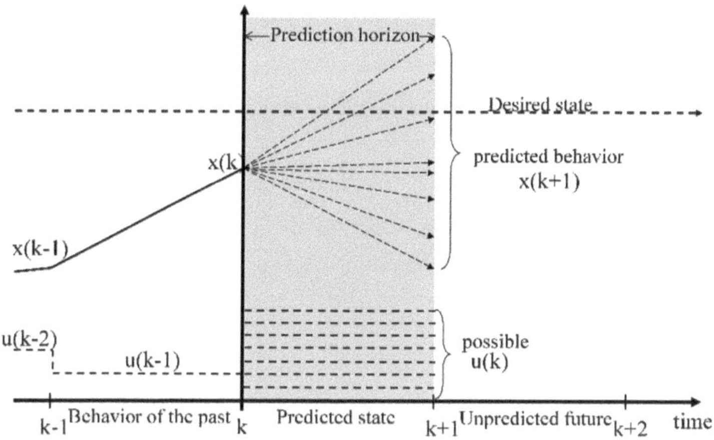

Figure 8.12 Functional principle of the FCS-MPC.

The circuit configuration of a three-phase VSC is shown in Fig. 8.13A. As seen, either one of the power switches of each leg (for instance either $S1$ or $S4$) is allowed to turn on at any time to prevent the short circuit of the input DC-link. Thus, taking into account that the number of possible states of each leg is 2 and the number of legs is 3, the converter can generate 2^3 equal to eight possible different switching states as listed in Table 8.2, where the switching signal Sa, Sb, and Sc can be defined as 1 if an upper switch on and a lower switch off, and as 0 if an upper switch off and a lower switch on. The converter voltage corresponding to these switching signals is determined by Clarke's transformation.

$$v_{\alpha\beta} = \frac{2}{3}\begin{bmatrix} 1 & -\frac{1}{2} & -\frac{1}{2} \\ 0 & \frac{\sqrt{3}}{2} & -\frac{\sqrt{3}}{2} \end{bmatrix}\begin{bmatrix} v_{an} \\ v_{bn} \\ v_{cn} \end{bmatrix} = \frac{2}{3}\begin{bmatrix} v_{an} - \frac{v_{bn}}{2} - \frac{v_{cn}}{2} \\ \frac{\sqrt{3}}{2}v_{bn} - \frac{\sqrt{3}}{2}v_{cn} \end{bmatrix} \tag{8.50}$$

Considering Fig. 8.13B, it can be concluded that if we converse the values of v_{aN}, v_{bN}, and v_{cN}; and v_{an}, v_{bn}, and v_{cn} with Clarke's transformation, their values in $\alpha\beta$-frame are actually equivalent. Hence Eq. (8.50) can be written as follows.

$$v_{\alpha\beta} = \frac{2}{3}\begin{bmatrix} S_a V_{dc} - \frac{S_b V_{dc}}{2} - \frac{S_c V_{dc}}{2} \\ \frac{\sqrt{3}}{2}S_b V_{dc} - \frac{\sqrt{3}}{2}S_c V_{dc} \end{bmatrix} \tag{8.51}$$

when V_{dc} is the dc-link voltage and v_{aN}, v_{bN}, and v_{cN} are the phase-to-neutral voltage and can be described as a function of S_a, S_b and S_c (Eq. 8.52).

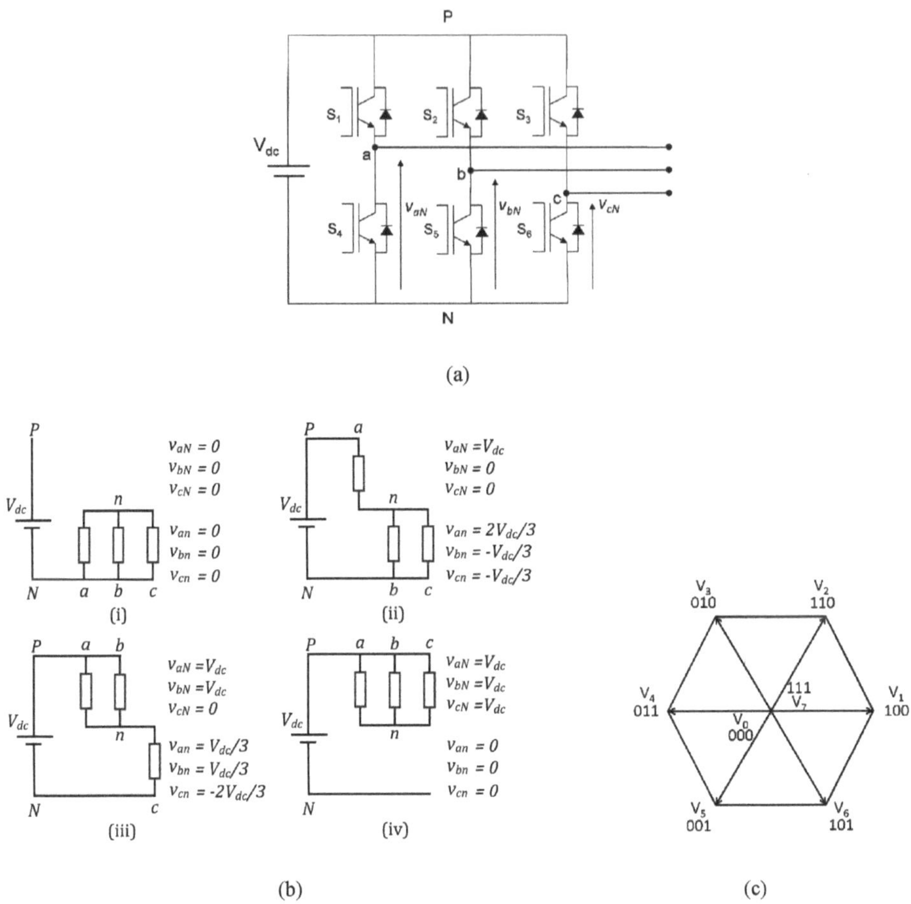

Figure 8.13 A 3-phase voltage source converter; (A) topology, (B) equivalent load configuration, and (C) space voltage vectors representation.

Table 8.2 Switching states and voltage vectors.

ptS_a	ptS_b	ptS_c	**Voltage vector**
0	0	0	V_0
1	0	0	V_1
1	1	0	V_2
0	1	0	V_3
0	1	1	V_4
0	0	1	V_5
1	0	1	V_6
1	1	1	V_7

$$v_{aN} = S_a V_{dc}\ ;\ v_{bN} = S_b V_{dc}\ ;\ v_{cN} = S_c V_{dc} \tag{8.52}$$

Considering the configuration in Fig. 8.12, Eq. (8.50), and the possible switching states in Table 8.2, using Clarke's transformation, the converter voltage ($\boldsymbol{v}_{vs}$) in αβ-coordination for each switching state can be determined with Eq. (8.53).

$$\boldsymbol{v}_{vs,\alpha\beta} = \frac{2}{3}\left(v_{aN} + a v_{bN} + a^2 v_{cN}\right) \tag{8.53}$$

Where a is the unitary vector, which represents the 120 degree phase displacement between the phase. Therefore, the converter voltage produced by the eight possible switching states of the converter can be represented in αβ-coordination with space voltage vector $V_0 - V_7$, as shown in Fig. 8.13C. This representation is known as space vector representation, where $(V_1,\ V_2,\ V_3,\ V_4,\ V_5,\ V_6)$ represent six active switching states of the rectifier, while V_0 and V_7 indicate the cases when all three upper power transistor switches are all turned on, or all turned off, which this causes a short circuit in three legs of the converter. Hence, these two voltage vectors are called zero voltage vectors. Each of these voltage vectors will be used as input actuators for prediction of the MPC.

8.4.2.1 Types of cost function

A cost function is a tool to analyze performance or to define the operating range of the controller. It allows multiple control objectives to be considered at the same time, which is a unique advantage of MPC. This cost function may consist of several types of control terms. The most common terms that can be included in a cost function are introduced as follows:

(i) *Reference following*: The most common expressions in a cost function are the ones that represent references following. These terms can be expressed generally as the error between the predicted value and its reference, as can be seen in Eq. (8.54).

$$g_1 = (\boldsymbol{x}^* - \boldsymbol{x}(k+1))^2 \tag{8.54}$$

where g_1 is reference following cost function, $\boldsymbol{x}^*$ is the reference value and $\boldsymbol{x}(k+1)$ is the predicted value. An absolute error can be used instead of a squared error. However, when two or more different terms are included in the cost function, a squared error represents a better reference following. Alternatively, an integral value of error for one sampling period is also a promising choice, as cost function considers the trajectory of the variable between the instant k and the instant $k+1$, which leads to more accurate reference tracking. A reference following cost function can be used alone or together with other types of control terms.

(ii) *Actuation constraints*: The actuation constraint is a term that represents the control effort of a GCC, which is related to the voltage or current variations, the switching frequency, or the switching losses. For example, if one wants to minimize the magnitude of the difference between the previously applied voltage vector $v(k+1)$ and the voltage vector to be applied $v(k)$, the actuation constraints are often used together with the reference following cost function as shown in Eq. (8.55).

$$g_2 = g_1 + \gamma_2 \|v(k+1) - v(k)\| \tag{8.55}$$

where g_2 is the actuation constraints and γ_2 is a weighting factor that allows the importance of control effort to be adjusted.

(iii) *Hard constraints*: Although model predictive control can achieve direct control of output variables without the need for inner control loops, the uncontrolled internal variables can reach values that are outside their allowed range. In order to prevent this problem, some limitations can be included as an additional term in the cost function. For example, a hard constraint to ensure that DC-voltage is always positive can be defined as Eq. (8.56), where g_3 is the actuation constraints.

$$g_3 = \begin{cases} 10^6, & u_{DC}(k+1) < 0 \\ 0, & u_{DC}(k+1) \geq 0 \end{cases} \tag{8.56}$$

8.4.2.2 Weighting factor design

Although no weighting factors are necessary for the cost function when only one type of variable is controlled, to fully utilize the advantage of MPC, multiple objectives control is preferable. This can be done simply by introducing the additional control terms in the cost function. However, as the number of terms in the cost function increases, the adjustment of the weight parameters is not a straightforward task, especially when they are of a different nature. Each additional term in the cost function has a specific weighting factor, which is used to tune the importance of that term related in relation to the other control targets. These parameters have to be appropriately designed in order to achieve the desired performance [20].

In a *cost function with a second (less important) term*, which is the easiest case for weighting factor adjustment, one can simply neglect the secondary terms by fixing the weighting factor at zero, $\gamma_2 = 0$. In general, in order to establish the tradeoff between the first term and the second term, it is necessary to establish measurements or figures of merit that will be used to evaluate the quality achieved by adjusting the weighting factors. Once the measures are defined, evaluation of the system behavior with simulations starts with $\gamma_2 = 0$ and increases the value gradually.

In a *cost functions with equally important terms*, the first step is to normalize the cost function. Once normalized, all the terms will be equally important, and now, $\gamma_2 = 1$ can be considered as a starting point since the second term is as important as the first term in this control system, thus $\gamma_2 = 0$ is not allowed. More explanation regarding this tuning method can be found in Ref. [20].

8.4.3 The proposed FCS-MPC-based VSG control

In recent years, the FCS-MPC has been presented as an attractive alternative for the control of GCCs because it has a flexible control scheme that allows easy inclusion of multiple control variables, system constraints, and nonlinearities. Some control solutions based on FCS-MPC for DGs are given in the literature; for instance, FCS-MPC strategies for a grid connected operation are proposed in Refs. [22–26]. Conclusively, several benefits of the MPC scheme can be observed, including being simple to apply in multivariable systems, the possible inclusion of constraints, straightforward control law, and fast dynamic response.

Among these, the effective MIMO control of FCS-MPC and the fast dynamic response is highly attractive for achieving overcurrent prevention ability. The multi-variable system of the controller allows the inclusion of both voltage and current control into a single control loop. This grants the controller with the ability to limit overcurrent without a transition between normal and overcurrent mode. The fast dynamic response of FCS-MPC can also be considered as the main advantage over other PWM-based methods. Additionally, the existence of voltage outer loop control in $\alpha\beta$-axis can assist in dealing with nonlinearity and distortion in current and voltage waveform under the effect of transient disturbance [27].

8.4.3.1 Overall framework

The overall control diagram of the proposed control is shown in Fig. 8.14A. It is composed of the FCS-MPC, the VSG control, and the current command parts. The FCS-MPC uses a mathematic model of the system to predict the system variables and to evaluate the optimal switching states for GCC. Before describing the FCS-MPC part, let discuss on the VSG control part first. In this part, the voltage command variables of the VSG are determined. In order to accomplish the FCS-MPC scheme with both voltage and current control, a current command need to be provided. This command is generated from the current command part [28].

8.4.3.2 VSG control

The objective of VSG is to track the dispatch power command while providing virtual inertia to slow down any deviation of the grid frequency. This can be realized with the structure of a VSG control scheme as shown in Fig. 8.14B. It consists of "Swing equation," "Governor model," and "Q droop" parts. P_o, P_{out} , P_{gov}, Q_o, Q_{out} V_{out}, ω_m, $\widehat{\omega}_g$, E^* and θ^* represent the output active power command, the output active power, the shaft power from governor, the output reactive power command, the output reactive power, the output voltage amplitude, the mechanical frequency of rotors, the output frequency measured by phase-locked loop (PLL), the voltage reference magnitude, and the power angle reference, respectively. Virtual inertia is emulated in the "Swing equation" block with the help of the well-known swing Eq. (8.57).

$$P_{gov} - P_{\text{out}} + D(\omega_m - \widehat{\omega}_g) = J\omega_m \frac{\mathrm{d}\omega_m}{\mathrm{d}t} \tag{8.57}$$

where J is the moment of inertia, D is the damping factor produced by the damper windings, P_{out} is the output active power, ω_m is the virtual rotor angular frequency, and $\widehat{\omega}_g$ is the measured grid frequency. The function of the shaft power P_{gov} regulated by a governor is shown in Eq. (8.58).

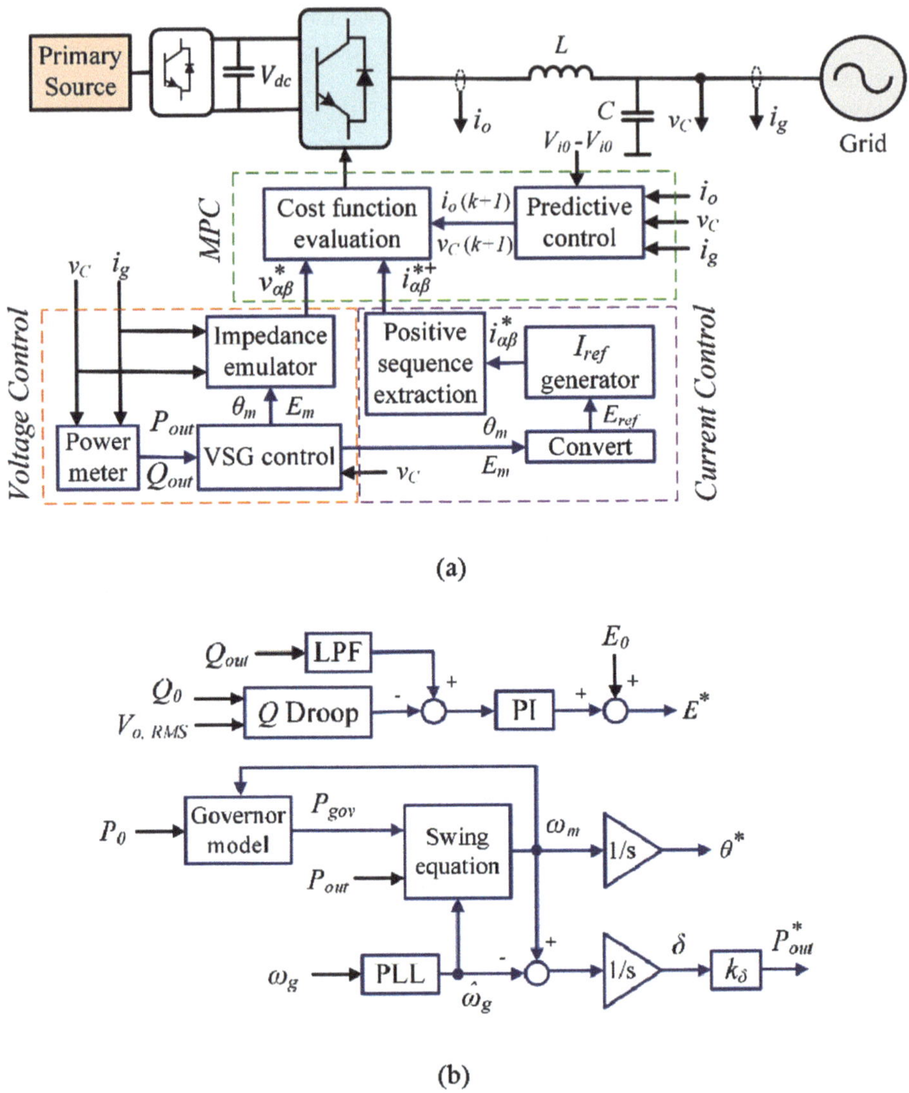

Figure 8.14 Control Diagram of the purposed FCS-MPC-Based VSG Control; (A) overall structure and (B) VSG control scheme.

$$P_{gov} = P_0 - k_p(\omega_m - \omega_0) \tag{8.58}$$

where k_p is the droop coefficient and ω_0 is the nominal grid frequency. The block "Q droop," as well as the design details of other blocks of VSG control system, and the tuning of VSG parameters, are all explained in Ref. [29].

Here, the power command of VSG control is determined with the help of the equation of the active power flow in a two-bus bar system [4] as follows,

$$P_{\text{out}} = \frac{V_o V_c}{X} \sin\delta \tag{8.59}$$

where V_o is the output voltage amplitude, V_c is the output filter's capacitor voltage amplitude, X is the output filter reactance, and δ is the phase angle difference between the GCC and grid. Assuming V_o, V_c and X to be constant and $\sin\delta \approx \delta$ for a small phase angle, P_{out} is equal to $K\delta$ when $K = V_c V_o / X$. However, since the value K of the proposed control is not limited by the physical values of the system like real SGs and can be arbitrarily selected to optimize the system performance, P_{out} of the proposed VSG control can be rewritten as:

$$P_{\text{out}} = k_\delta \delta \tag{8.60}$$

where k_δ is the optimized synchronizing constant, and δ can be determined with Eq. (8.61).

$$\delta = \int (\omega_m - \widehat{\omega}_g) dt \tag{8.61}$$

8.4.3.3 Current command generation for the FCS-MPC

As mentioned, the voltage command for the FCS-MPC is obtained with the VSG control scheme. However, to achieve resonance damping, both voltage and current control must be embedded into a single control loop, as shown in Fig. 8.14A. Therefore, the current command must be provided for the FCS-MPC. The current command can be determined by considering the relationship between the GCC output voltage, the output current, and the output capacitor phase voltage. It can be described in a stationary ($\alpha\beta$) frame as

$$\boldsymbol{v}_{c,\alpha\beta} = \boldsymbol{v}_{o,\alpha\beta} + \boldsymbol{i}_{o,\alpha\beta}(R + j(X)). \tag{8.62}$$

where $\boldsymbol{v}_{c,\alpha\beta}$ is the capacitor voltages, $\boldsymbol{i}_{o,\alpha\beta}$ the output current and $\boldsymbol{v}_{o,\alpha\beta}$ is the output voltage of the GCC. X denotes the filter reactance and R denotes filter resistance. If we consider that the $\boldsymbol{v}_{c,\alpha\beta}$ in Eq. (8.62) equals to the voltage command described by E^* and θ^* from the VSG control, the currents references i_α^* and i_β^* can be expressed as illustrated in Eq. (8.63), and they will be used as the current command of the FCS-MPC.

$$\begin{bmatrix} i_\alpha^* \\ i_\beta^* \end{bmatrix} = \boldsymbol{Y} \left\{ \begin{bmatrix} E^* \cos\theta^* \\ E^* \sin\theta^* \end{bmatrix} - \begin{bmatrix} v_{o,\alpha} \\ v_{o,\beta} \end{bmatrix} \right\} \tag{8.63}$$

where $\boldsymbol{Y} = \frac{1}{R^2 + X^2} \begin{bmatrix} R & X \\ -X & R \end{bmatrix}$.

8.4.3.4 FCS-MPC synthesis

In this work, the FCS-MPC is chosen for the implementation of VSG control instead of PWM-methods. The FCS-MPC utilizes the system model and the real-time values of the system to determine the optimized GCC's switching states directly. Thus it enables proper tracking of voltage and current commands even without the use of zero current vectors of the converter.

The FCS-MPC is also applied in the (primary energy) source-side converter because of the faster dynamic response of the FCS-MPC compared to the PWM-based methods. This is crucial for the proposed VSG control, as the inertial power is provided from the power control of the source side converter [28].

- GCC model for Implementation of FCS-MPC

To realize a GCC that can operate in both grid connected and islanded modes, the converter output voltage and current controls should be the objectives of MPC. In order to use the MPC to forecast the future values of system variables, first, a simple mathematical model of the system is required. An equivalent circuit of the GCC with inductor and filter capacitor of an *LC*-filter is depicted in Fig. 8.15. This circuit provides a model for forecasting the voltage and current output of the system, which can be described in $\alpha\beta$-frame with capacitor and inductance dynamics Eq. (8.64).

$$\begin{aligned} C\frac{d\boldsymbol{v}_c}{dt} &= \boldsymbol{i}_o - \boldsymbol{i}_g \\ L\frac{d\boldsymbol{i}_g}{dt} &= \boldsymbol{v}_c - \boldsymbol{v}_o \end{aligned} \tag{8.64}$$

where $\boldsymbol{v_c} = \left[v_{c,\alpha}\ v_{c,\beta}\right]^T$, $\boldsymbol{v}_o = \left[v_{o,\alpha}\ v_{o,\beta}\right]^T$, $\boldsymbol{i}_o = \left[i_{o,\alpha}\ i_{o,\beta}\right]^T$ and $\boldsymbol{i}_g = \left[i_{g,\alpha}\ i_{g,\beta}\right]^T$ are capacitor voltage, output voltage, converter current, and the grid current, respectively. These equations can be rewritten as follows:

$$\dot{\boldsymbol{x}} = \boldsymbol{A}\boldsymbol{x} + \boldsymbol{B}\boldsymbol{v}_o + \boldsymbol{B}_w\boldsymbol{i}_g \tag{8.65}$$

where

Figure 8.15 Control diagram of the purposed FCS-MPC-based GCC control.

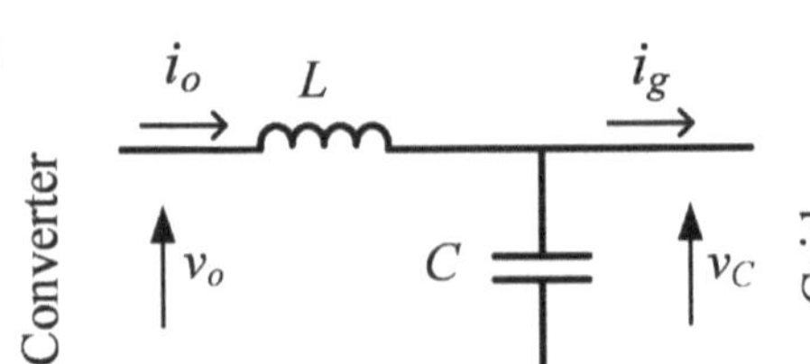

$$x = \begin{bmatrix} i_{g,\alpha} \\ i_{g,\beta} \\ v_{c,\alpha} \\ v_{c,\beta} \end{bmatrix}, A = \begin{bmatrix} 0 & 0 & \frac{1}{L} & 0 \\ 0 & 0 & 0 & \frac{1}{L} \\ \frac{1}{C} & 0 & 0 & 0 \\ 0 & \frac{1}{C} & 0 & 0 \end{bmatrix}, B = \begin{bmatrix} -\frac{1}{L} & 0 \\ 0 & -\frac{1}{L} \\ 0 & 0 \\ 0 & 0 \end{bmatrix}, B_w = \begin{bmatrix} 0 & 0 \\ 0 & 0 \\ -\frac{1}{C} & 0 \\ 0 & -\frac{1}{C} \end{bmatrix}.$$

A discrete-time model of Eq. (8.65), using the general forward-difference Euler formula, for a sampling time T_s can be expressed as

$$\boldsymbol{x}(k+1) = \boldsymbol{A}_d\boldsymbol{x}(k) + \boldsymbol{B}_d\boldsymbol{v}_o(k) + \boldsymbol{B}_{wd}\boldsymbol{i}_g(k) \tag{8.66}$$

where

$$\boldsymbol{A}_d = e^{AT_s},\ \boldsymbol{B}_d = \int_0^{T_s} e^{A\tau}\boldsymbol{B}d\tau \text{ and } \boldsymbol{B}_{wd} = \int_0^{T_s} e^{A\tau}\boldsymbol{B}_d d\tau.$$

From Eqs. (8.65 and 8.66), it can be seen that the future values $\boldsymbol{v}_c(k+1)$ and $\boldsymbol{i}_g(k+1)$ can be determined according to converter stage current $\boldsymbol{i}_o(k)$. For each active current vectors of GCC, the converter stage current $\boldsymbol{i}_o(k)$ is defined as follows [22]:

$$\boldsymbol{i}_o(k) = \begin{bmatrix} S_{ap} - S_{an} & S_{bp} - S_{bn} & S_{cp} - S_{cn} \end{bmatrix}^T i_{DC} \tag{8.67}$$

where S_{ap}, S_{bp}, S_{cp} S_{an}, S_{bn}, S_{cn} represent the switching states of the six switches in the converter stage. The value of the switching state consists of 1 and 0, for which it represents the closed and open states, respectively. As shown in Eq. (8.67), although only active current vectors can be used in the GCC, by predicting the converter output variables corresponding to all possible current vectors, the switching state that produces the desired amplitude and phase of the converter output variables can be found. In contrast to the modulation method of the PWM-based control, which only follows the phase angle reference and only utilizes two nearest space vectors to the reference phasor to generate output current, the FCS-MPC certainly has more freedom to select an optimal switching state to fulfill the GCC control goal.

- *Cost Function Definition*

Contrarily to classical control schemes, the presence of the cost function allows the FCS-MPC to take into account a number of control goals, and to control different state variables simultaneously. Therefore, it is highly relevant to define the cost function properly. For the GCC control, important variable quantities must be considered to make the whole system operate correctly. The GCC control system is set to track the voltage and current references, simultaneously. This can be achieved by defining a cost function as follows.

$$g = k_v\left(\boldsymbol{v}_{c,\alpha\beta,pu}(\mathrm{k}+1) - \boldsymbol{v}^*_{\alpha\beta,\,pu}\right)^2 + k_i\left(\boldsymbol{i}_{o,\alpha\beta,\,pu}(\mathrm{k}+1) - \boldsymbol{i}^*_{\alpha\beta,\,pu}\right)^2 + g_m \tag{8.68}$$

where, $\boldsymbol{v}^*_{\alpha\beta} = \left[v^*_\alpha \; v^*_\beta\right]^T$ is the reference vector of the capacitor voltage, $\boldsymbol{i}^*_{\alpha\beta} = \left[i^*_\alpha \; i^*_\beta\right]^T$ is the reference vector of the inductor current, $\boldsymbol{v}_{c,\alpha\beta}\,(k+1) = \left[v_{c,\alpha} \; v_{c,\beta}\right]^T$ is the predicted capacitor voltage and $\boldsymbol{i}_{o,\alpha\beta}\,(k+1) = \left[i_{f,\alpha} \; i_{f,\beta}\right]^T$ is the predicted inductance current, and the subscript *pu* indicates the per-unit value. g_m is a constraint, which is included in the cost function to ensure that dc-link voltage never becomes negative. It can be defined as g_3 in Eq. (8.56).

Although the FCS-MPC is capable of controlling different variables at the same time through the modification of the cost function, the tuning of the weighting parameters is a complex task when a number of terms exist in the cost function. Here, the ratio between the weighting parameters $\boldsymbol{k}_v$ and $\boldsymbol{k}_i$ has been arbitrarily set equal to 1:1.

- *Filter Resonance Damping*

To ensure proper operations, the GCC requires an inductive output filter. However, in order to improve the quality of the waveforms, *LC*-filter is usually preferred. Nevertheless, the presence of resonance frequency has to be taken into account since the FCS-MPC possesses a variable switching frequency. It might generate harmonics components at the resonant frequency of *LC*-filter at a certain operating point. However, this will result in resonance, if only the output inverter current $\boldsymbol{i}_g(s)$ is directly controlled, whereas the capacitor voltage $\boldsymbol{v}_c(s)$ is indirectly controlled by the current and thus highly dependent on filter impedances.

The proposed FCS-MPC utilizes a multivariable system to control both $\boldsymbol{v}_c(s)$ and $\boldsymbol{i}_g(s)$ simultaneously. This is different from conventional cascade control of voltage and current, where settling time of the inner loop must be significantly faster than the settling time of the outer loop. Therefore, at the resonant frequency where $\boldsymbol{v}_c(s)$ and $\boldsymbol{i}_g(s)$ are oscillating with the same dynamics; the control of the outer loop is not fast enough to provide a damping effect. Contrarily, the multivariable control of the FCS-MPC is conducted at the same bandwidth. This ensures that the regulation of $\boldsymbol{v}_c(s)$ can damp the distortion caused by the harmonics of $\boldsymbol{i}_g(s)$ and the control of $\boldsymbol{i}_g(s)$ is able to reduce the disturbance caused by the harmonics of $\boldsymbol{v}_c(s)$. Hence, by controlling both *LC*-filter voltage and current at the same instances, the frequency response of the filter is no longer influenced by its impedance alone. Thus the uncontrolled resonant energy oscillation between the inductances and the capacitance is effectively damped.

8.4.4 Analysis of PLL impact on the system stability and performance

The applied VSG control scheme suggests the use of $\widehat{\omega}_g$ to determine the output power command. However, the dynamic response of the PLL, which is given by Eq. (8.69), is lower than the real ω_g and this might affect the performance of the modified VSG control. In order to study the influence of the PLL toward the stability and transient performance of the VSG control, a state-space model for the proposed VSG control (with PLL) is obtained as given in Eq. (8.70). The model and the transient response of the PLL is well discussed in Ref. [30]. The transient response can be described with the following expression.

$$G_{pll}(s) = \frac{\widehat{\omega}_g(s)}{\omega_g(s)} = \frac{k_{p,pll}s + k_{i,pll}}{s^2 + k_{p,pll}s + k_{i,pll}} \tag{8.69}$$

Since the dynamic response of the FCS-MPC is fast, the output power is assumed to be equal to the power command. Then the state-space model can be expressed as follows.

$$\begin{cases} \dot{x} = Ax + Bu + Ew \\ y = Cx \end{cases} \tag{8.70}$$

where x, u, w, and y are states, input, disturbance, and outputs vectors, respectively, and

$$y = [\Delta\omega_m \quad \Delta P_{\text{out}} \quad \Delta\widehat{\omega}_g]^T; \; x = \left[\Delta\omega_m \quad \Delta P_{\text{out}} \quad \int \Delta\delta_{err,pll} \quad \Delta\,\delta_{err,\,pll}\right]^T$$

$$u = \Delta P_0\,; \; w = \Delta\omega_g$$

$$A = \begin{bmatrix} \frac{-(k_p + D)}{J\omega_o} & \frac{-1}{J\omega_o} & \frac{Dk_{i,pll}}{J\omega_o} & \frac{Dk_{p,pll}}{J\omega_o} \\ k_\delta & 0 & -k_\delta k_{i,pll} & -k_\delta k_{p,pll} \\ 0 & 0 & 0 & 1 \\ 0 & 0 & -k_{i,pll} & -k_{p,pll} \end{bmatrix}; \; B = \begin{bmatrix} \frac{1}{J\omega_o} & 0 & 0 & 0 \end{bmatrix}^T$$

$$\boldsymbol{E} = [0 \quad 0 \quad 0 \quad 1\,]^T;\ \boldsymbol{C} = \begin{bmatrix} 1 & 0 & 0 & 0 \\ 0 & 1 & 0 & 0 \\ 0 & 0 & k_{i,pll} & k_{p,pll} \end{bmatrix}$$

where

$$\delta_{err,pll} = \int (\omega_g - \widehat{\omega}_g)dt.$$

It is clear that the poles of transfer function $Y(s)/W(s)$ are available in the eigenvalues of $\boldsymbol{A}$. The loci of eigenvalues of $\boldsymbol{A}$ with a variation of $k_{i,pll}$ is shown in Fig. 8.16A when $k_{p,pll}$ is selected as given in Ref. [30]. First, it can be seen in Fig. 8.16A that all the poles of the system located on the left half of the s-plane, which implies that the system is stable. It is also shown in Fig. 8.16A that if the PLL has a rather quick dynamic response (if $k_{i,pll}$ is large enough), the effect of the third pole that locates near the dominant poles will be lessened as it is located farther to the left of the two dominant poles. In this case, the response of the system will be mostly defined by the two dominant poles, as can be described with the following expressions [31]:

$$\omega_n = \sqrt{\frac{k_\delta}{J\omega_0}};\ \zeta = \frac{k_p + D}{2\sqrt{k_\delta J\omega_0}} \tag{8.71}$$

where ω_n is the undamped natural frequency, and ζ is the damping ratio. However, as shown in Fig. 8.16, even with a properly designed PLL with a fast dynamic response, the locations of the two additional poles are not far enough to the left of dominant poles, their effect cannot be totally neglected. Hence, the response of P_{out} will be slower than the previous VSG controls, thus finally results in a less inertial

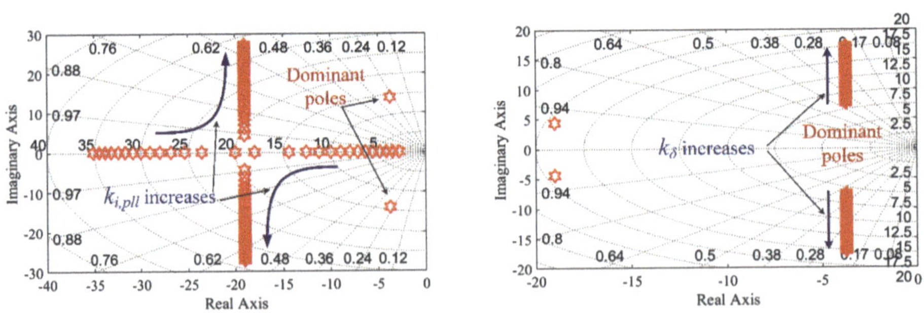

(a) (b)

Figure 8.16 System root locus for changing of (A) $k_{i,pll}$, and (B) k_δ.

response. This can be interpreted as a delay effect of the PLL, because unlike the VSG system with an energy storage element, P_{out} of the proposed control is not dispatched directly from the energy storage element. It is instead regulated by the power command, which depends on the $\widehat{\omega}_g$. This is crucial for VSG control, as P_{out} needs to react quickly to the power imbalance in order to create a proper inertial response.

Since k_δ of the proposed VSG control is not limited by the physical values of the system and can be arbitrarily selected, we can find an optimal value of k_δ by studying the effect of adjusting k_δ with the loci of eigenvalues of $\boldsymbol{A}$, as shown in Fig. 8.16B. It indicates that increasing k_δ leads to an increase in ω_n. This can compensate the delay effect of the PLL. Hence, $k_\delta = 2.5\ V_c V_o / X$ is chosen for the addressed control system.

8.4.5 Simulation and experimental results

In order to evaluate the performance of the proposed control scheme, simulation-based comparative studies between the proposed scheme and the investigated existing method are performed in simulation environment. The simulation parameters are given in Table 8.3.

The performance of the proposed VSG control for the GCC is verified with simulations of the test circuit displayed in Fig. 8.14A. Initially, the reference/command active power for the GCC (P_0) equals 4 kW and reference reactive power (Q_0) equals 1 kVar. The power commands are changed to 5 kW and 0 kVar at $T = 3$ s. The DG system is then set to operate in islanded mode at $T = 5$ s. The load initially connected to the system is 3 kW + 1 kVar, and it is increased to 5.5 kW + 0.05 kVar at $T = 7\ s$.

The output active and reactive powers, the output frequency, and the output voltage amplitude for the whole simulation are illustrated in Fig. 8.17. It can be seen that the system quickly and seamlessly commutates from the grid connected mode into the islanded mode, supplying the required 3 kW + 1 kVar of power to the local load. The system is also able to react to change in load during islanded operation, as evidenced in Fig. 8.17. The output frequency and the output active power of the control system in Fig. 8.17 are illustrating how the system is settling to operate at a different frequency according to the droop characteristic, whereas the output voltage amplitude of the system is maintained around the nominal value (V_{base}) throughout the

Table 8.3 Simulation parameters.

Parameter	Value	Parameter	Value
S_{base}	5 kVA	J	0.28 $\text{kg}\cdot\text{m}^2$
k_p	40 pu	D	200 pu
V_{base}	200 V	k_δ	70,000
$T_{i,pll}$	0.5 s	$k_{p,pll}$	0.1
$\omega_0 = \omega_{\text{grid}}$	376.99 rad/s		

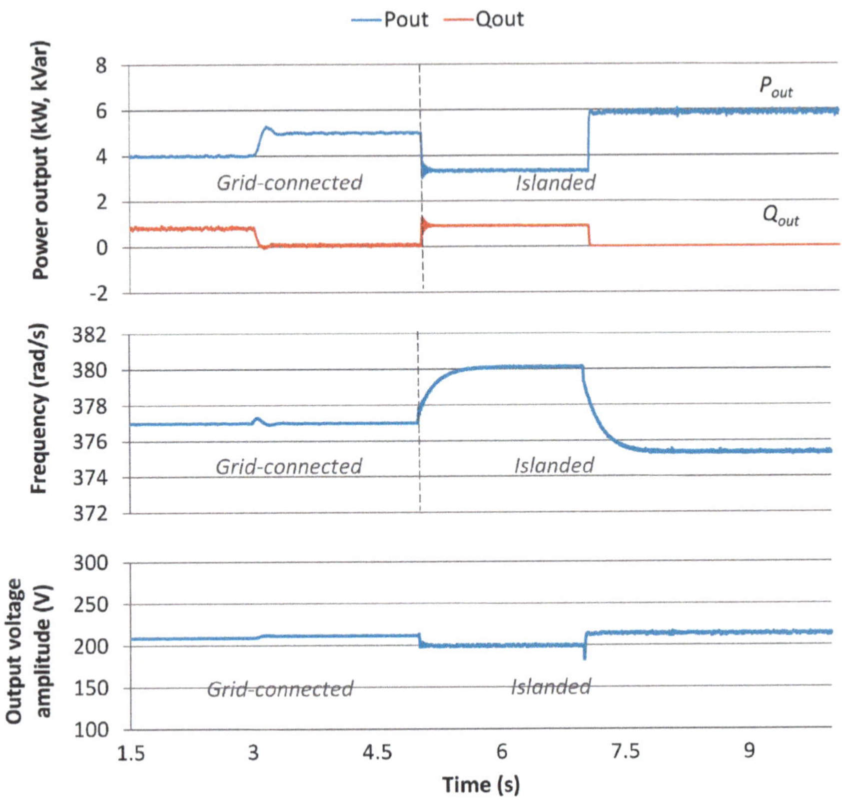

Figure 8.17 Simulation results of the output active and reactive powers, the output frequency, and the output voltage amplitude of the proposed control scheme, from top to bottom, respectively.

operation. Furthermore, the change in output frequency during the load transition in islanded mode demonstrates the inertial response of the controller, as the output frequency slowly decreases from around 378.5 rad/s to around 376 rad/s. The reduction in frequency occurs for around 1 s until the new frequency set point is reached. This is a typical VSGs performance, as analyzed in Refs. [1,29].

To verify the simulation results, a scaled-down version of the simulation circuit depicted in Fig. 8.14A is conducted with the testbed illustrated in Fig. 8.18A. Three-phase 30 Hz/40 V (line-to-line RMS) input ac-source and 60 Hz/60 V (line-to-line RMS) stepped-down grid voltage are considered for the experiment. The system parameters were, respectively, realized as follows: $L_i = 7$ mH, $R_i = 0.1\ \Omega$, $L = 4$ mH, $R = 0.1\ \Omega$, $C = 30$ and 2 μF for fictitious dc-link capacitor. These values are chosen from those available in the laboratory. S_{base} and V_{base} are changed to 150 W and 60 V, respectively. k_δ is set to 2.5 ${V_{\text{base}}}^2/X$, whereas other control parameters are adopted from the simulation study, as concluded in Table 8.3.

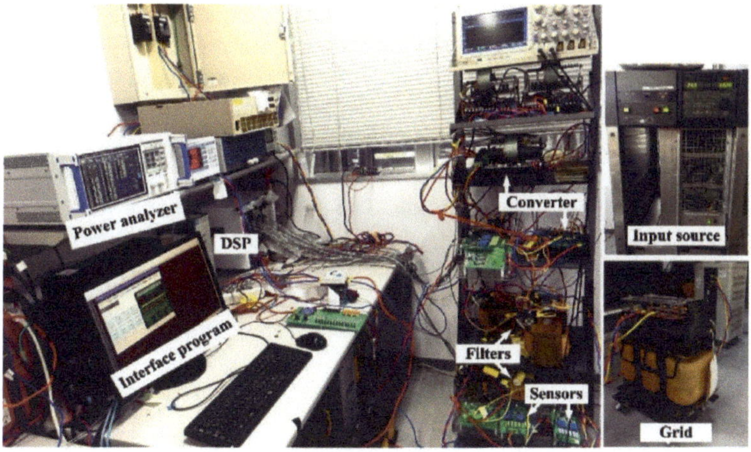

(a)

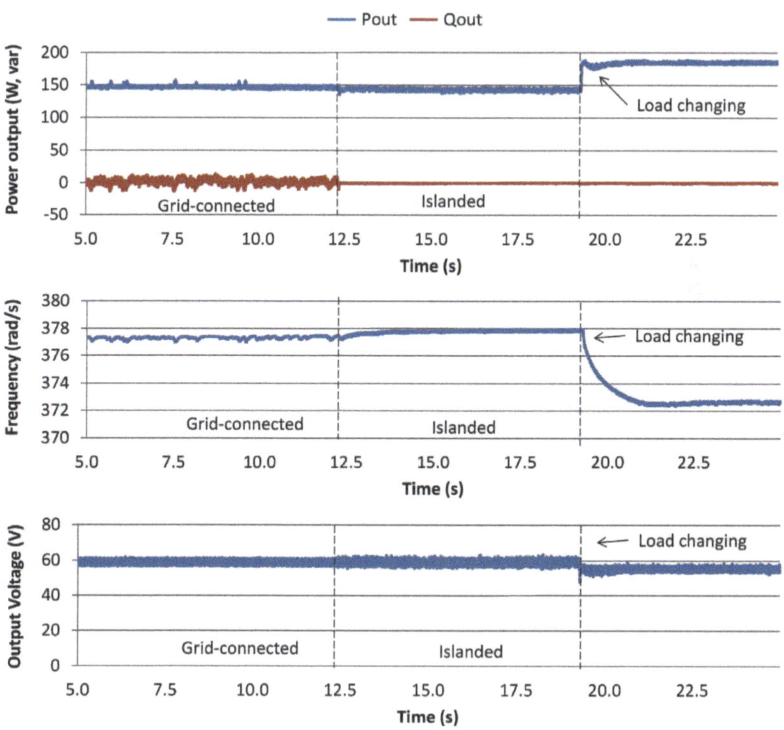

(b)

Figure 8.18 (A) Testbed for experimental verification and (B) Experimental results of the output active and reactive powers, the output frequency, and the output voltage amplitude of the proposed method during the grid connected mode, islanded mode, and load changing transition in islanded mode from top to bottom, respectively.

The output active and reactive powers, the output frequency, and the output voltage amplitude throughout the experiment scenarios are illustrated in Fig. 8.18B. The input active and reactive powers are set to follow the reference values of $P_{\mathrm{in}}^{*} = 150$ W and $Q_{\mathrm{in}}^{*} = 0$ Var. It can be seen that the system quickly and seamlessly transfers from the grid connected mode into the islanded mode, supplying the power to a 20 Ω local load, whereas the output voltage amplitude of the system is maintained around the nominal value (60 V) throughout the operation. The system is also able to react to the load changing during islanded operation when another star-connected 50 Ω local load is connected, as shown in Fig. 8.18B at $T \approx 19$ s.

8.5 Robust damping control

This section addresses a robust damping control for the GCCs. The proposed control strategy calculates the optimal damping and inertia parameters in form of constant gains via an H_∞-based static output feedback (H_∞-SOF) control synthesis [32]. The capability and effectiveness of the addressed control methodology are proved by theoretical analysis, simulation, and experimental results on a typical GCC. It is shown that the designed controller can guarantee the robust performance and improves system stability.

Here, to suppress the undesirable active power fluctuation at GCC output, first the damping control design is transferred to a static output feedback (SOF) control design problem. Then, to obtain the damping control parameters, the problem is solved based on the H_∞-SOF control theory. The simplicity of control structure, and providing desirable inertia and damping characteristics in a software program can be considered as the main advantages of the proposed methodology.

8.5.1 *Case study and dynamic modeling*

A typical grid-forming GCC with its detailed control system is shown in Fig. 8.19. The control system is discussed in Chapter 3. For performing the dynamic characteristic emulation system, several blocks are considered. The "P-droop, inertia and damping emulator" block is the core of GCC system control. In the block "Output impedance emulator," a virtual impedance control loop is applied. The "Q-droop emulator" block, the droop relation between voltage and reactive power is realized [29].

To emulate the steady-state operation of an SG, the swing equation of conventional SGs can be emulated in the "P-droop, inertia and damping emulator" block. The swing equation, presenting the inertia, droop, and damping characteristic is as follows [33].

$$P_{\mathrm{out}} - P_0 = J\omega_0 \frac{\mathrm{d}\omega_m}{\mathrm{d}t} + k_p(\omega_m - \omega_0) + D(\omega_m - \omega_g) \tag{8.72}$$

where P_0 is the reference power determined by the governor, P_{out} is the measured output power, J is the moment of inertia, D is the damping factor, ω_m is the rotor

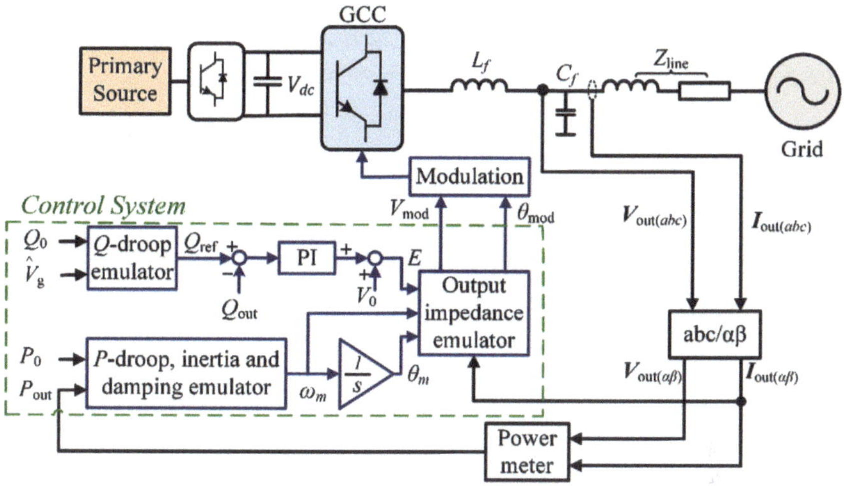

Figure 8.19 Typical control scheme of a grid-forming GCC.

angular frequency, ω_g is the angular frequency of the grid where the voltage sensor is installed, k_p is droop coefficient, and ω_0 is the nominal angular frequency.

In this section, a simple dynamic model is presented for the active power-frequency response of a GCC which could be useful for damping control design. In the literature, with the assumption of the grid is stiff enough, usually a GCC with the grid is considered as a single DG connected to infinite bus system as shown in Fig. 8.20. In the model of this system, ω_g could be considered as a disturbance from the grid side.

It is well-known that the synchronizing power coefficient K_P of the GCC can be expressed as follows [33]:

$$K_P = \frac{\partial P_{out}}{\partial \delta} \approx \frac{E V_g \cos\delta_0}{X_T} \approx \frac{V_{base}^2 \sqrt{1 - X_T^{*2}}}{X_T} \tag{8.73}$$

where the per unit value of output reactance X_T can be calculated as $X_T^* = X_T S_{base}/V_{base}^2$. E is the electromotive force, V_g is the voltage of infinite bus, V_{base} is the rated voltage, and S_{base} is the rated power. As it can be seen in Fig. 8.20, the output reactance

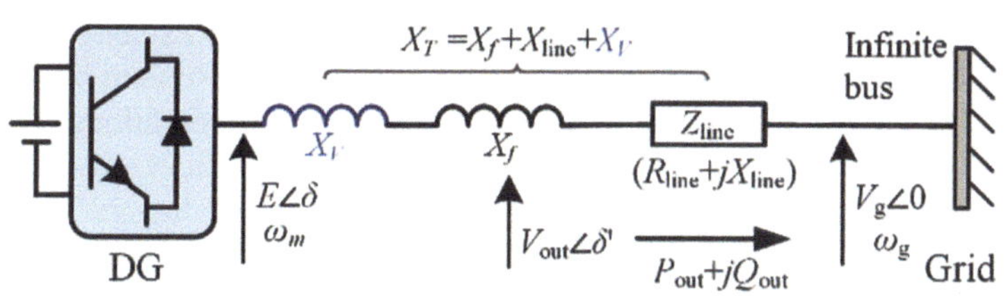

Figure 8.20 The model of GCC connected to an infinite bus.

(X_T) covers three reactance components provided by the line, the output LCL filter, and the virtual reactance created by "Output impedance emulator" block (in Fig. 8.19).

The derivative of power angle δ can be represented as

$$\frac{\mathrm{d}\delta}{\mathrm{d}t} = \omega_m - \omega_g \tag{8.74}$$

Thus, for the GCC, a small-signal state-space model can be derived from Eqs. (8.72–8.74) for the "P-droop, inertia, and damping emulator" block, as given in Eq. (8.75).

$$\begin{aligned} \dot{\boldsymbol{x}} &= \boldsymbol{A}\boldsymbol{x} + \boldsymbol{B}u + \boldsymbol{E}w \\ \boldsymbol{y} &= \boldsymbol{C}\boldsymbol{x} \end{aligned} \tag{8.75}$$

where

$$u = \Delta P_0;\ w = \Delta\omega_g;\ \boldsymbol{y} = \Delta P_{\text{out}};\ \boldsymbol{x} = \begin{bmatrix} \Delta\omega_m & \Delta P_{\text{out}} \end{bmatrix}^{\mathrm{T}}$$

The $\boldsymbol{A}$, $\boldsymbol{B}$, $\boldsymbol{E}$, and $\boldsymbol{C}$ depend on the structure and its dynamics in the "P-droop, inertia, and damping emulator" block. For instance, ignoring the PLL dynamics, the model can be derived as [31].

$$\boldsymbol{A} = \begin{bmatrix} -\dfrac{k_p + D}{J\omega_0} & -\dfrac{1}{J\omega_0} \\ S_P & 0 \end{bmatrix};\ \boldsymbol{B} = \begin{bmatrix} \dfrac{1}{J\omega_0} & 0 \end{bmatrix}^{\mathrm{T}} \boldsymbol{E} = \begin{bmatrix} \dfrac{D}{J\omega_0} & -S_P \end{bmatrix}^{\mathrm{T}};\boldsymbol{C} = \begin{bmatrix} 0 & 1 \end{bmatrix}.$$

It is noteworthy that the presented model does not include the detailed and fast dynamics and only useful for analyzing relatively slow dynamics which are of interest in this paper. Including detailed dynamics will only complicate the model and introduce some nondominant poles, which may marginally influence the GCC dominant poles.

8.5.2 Design methodology

The SOF control problem has received so much attention due to simplicity of structure and applicability for the real-world systems. Furthermore, many existing dynamic control synthesis problems can be transferred to an SOF control problem by well-known system augmentation techniques. Necessary and sufficient conditions for the SOF design can be obtained in terms of two linear matrix inequalities (LMIs) through a bilinear matrix equation [34,35]. Particularly, the problem of finding an SOF controller can be restated as a linear algebra problem, which involves two LMIs.

8.5.2.1 Theoretical background

Consider a linear time invariant (LTI) system with the following state-space realization:

$$\begin{aligned} \dot{x} &= Ax + B_1 w + B_2 u \\ z &= C_1 x + D_{11} w + D_{12} u \\ y &= C_2 x + D_{21} w \end{aligned} \tag{8.76}$$

where x is the state variable vector, w is the disturbance and other external input vector, z is the controlled output vector and y is the measured output vector.

The H_∞-SOF control problem for the LTI system with the state-space realization of (8.76) is to find a given matrix K (SOF law), as shown in Fig. 8.21, such that the resulted closed-loop system is internally stable, and the H_∞ norm for the existing transfer function from w to z ($T_{zw}(s)$) is smaller than γ, a specified positive number, that is,

$$\|T_{zw}(s)_\infty\| < \gamma \tag{8.77}$$

Under certain assumptions on the system matrices, the following theorem can be extendable to the H_∞−SOF control problem.

Theorem 1: it is assumed that is stabilizable and detectable. The matrix K is a dynamic H_∞ controller, if and only if there exists a symmetric matrix such that

$$\begin{bmatrix} A_{cl}^T X + X A_{cl} & X B_{cl} & C_{cl}^T \\ B_{cl}^T X & -\gamma I & D_{cl}^T \\ C_{cl} & D_{cl} & -\gamma I \end{bmatrix} < \gamma \tag{8.78}$$

where

$$A_{cl} = A + B_2 K C_2 \; ; \; B_{cl} = B_1$$

$$C_{cl} = C_1 + D_{12} K C_2 \; ; \; D_{cl} = 0$$

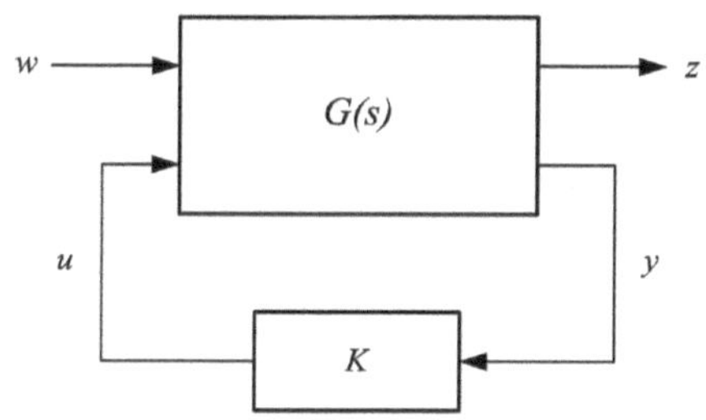

Figure 8.21 Closed-loop system via H_∞ control.

The proof is given in Ref. [34].

It is shown that the H_∞-SOF control problem is reduced to find $\boldsymbol{X} > 0$ and $\boldsymbol{K}$ for a new formulation of above system [36]. Then, a generalized SOF stabilization problem of the system can be solved via Theorem 2.

Theorem 2: the system $(\boldsymbol{A}, \boldsymbol{B}, \boldsymbol{C})$ that may also be identified by the following representation:

$$\begin{aligned} \dot{\boldsymbol{x}} &= \boldsymbol{A}\boldsymbol{x} + \boldsymbol{B}\boldsymbol{u} \\ \boldsymbol{y} &= \boldsymbol{C}\boldsymbol{x} \end{aligned} \tag{8.79}$$

is stabilizable via SOF if and only if there exist $\boldsymbol{P} > 0,\ \boldsymbol{C} > 0$ and $\boldsymbol{K}$ satisfying the following quadratic matrix inequality

$$\begin{bmatrix} \boldsymbol{A}^T\boldsymbol{X} + \boldsymbol{X}\boldsymbol{A} - \boldsymbol{P}\boldsymbol{B}\boldsymbol{B}^T\boldsymbol{X} - \boldsymbol{X}\boldsymbol{B}\boldsymbol{B}^T\boldsymbol{P} + \boldsymbol{P}\boldsymbol{B}\boldsymbol{B}^T\boldsymbol{P} & (\boldsymbol{B}^T\boldsymbol{X} + \boldsymbol{K}\boldsymbol{C})^T \\ \boldsymbol{B}^T\boldsymbol{X} + \boldsymbol{K}\boldsymbol{C} & -\boldsymbol{I} \end{bmatrix} < 0 \tag{8.80}$$

The $\boldsymbol{X}$ and $\boldsymbol{P}$ are symmetric and positive-definite matrices. For this new inequality notation (8.58), the sufficiency and necessity of theorem are already proven [37].

8.5.2.2 Synthesis framework

The overview of the synthesis framework is shown in Fig. 8.22. The robust damping control problem can be obtained and formulated in a dynamic output feedback form as a function of D and J.

$$K_d(s) = f(D,\ J) \tag{8.81}$$

This dynamic control design problem is first transferred to an SOF control problem. The main advantage of this transformation is to use the well-known SOF control techniques to calculate the fixed gains, and once the SOF gain vector is obtained, no additional computation is required. Therefore, by appropriate augmenting the system description the robust optimal damping control problem becomes one of finding an SOF that satisfies the prescribed performance requirements.

$$u = \boldsymbol{K}_s y;\ \boldsymbol{K}_s = [D \quad J] \tag{8.82}$$

The SOF control synthesis is solved through the H_∞ control technique, and then the optimal gains (D^*, J^*) can be carried out using an appropriate solution algorithm. But, it is notable that the $H_\infty-$SOF reformulation generally leads to bilinear matrix inequalities which are nonconvex and its solution cannot be directly achieved by using general LMI technique. Here, in order to solve the mentioned $H_\infty-$SOF, an iterative LMI algorithm is used, which is well explained in Refs. [38,39].

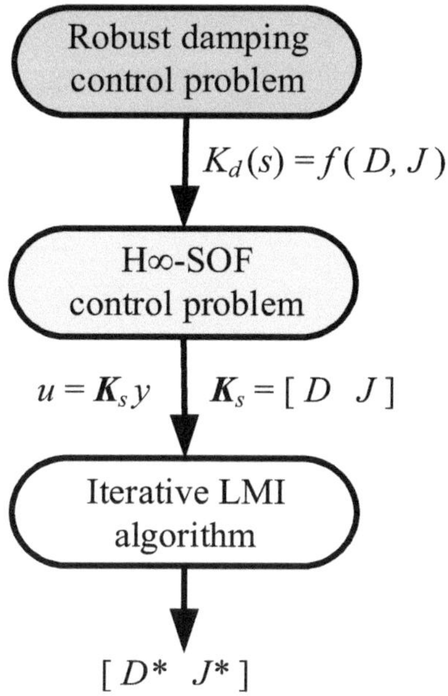

Figure 8.22 Synthesis framework.

The key point is to formulate the H_∞ problem via a generalized stabilizing SOF, such that all eigenvalues of $(\boldsymbol{A} - \boldsymbol{B}\boldsymbol{K}_s\boldsymbol{C})$ shift toward the left half plane through the reduction of a, a real number, to close to feasibility of Eq. (8.80). Theorem 2 gives a family of internally stabilizing SOF gains which can be defined as $\boldsymbol{K}_{SOF}$. A desirable solution can be obtained following tracking an optimal performance index γ that easily can be achieved from the application of H_∞ dynamic output feedback control using *hinflmi* function in MATLAB LMI Control Toolbox [38]. Then the used iterative LMI algorithm determines an admissible SOF law

$$u = \boldsymbol{K}_s y\ ;\ \boldsymbol{K}_s \in \boldsymbol{K}_{SOF}$$

such that

$$\|T_{zw}(s)_\infty\| < \gamma^*\ ;\ |\gamma - \gamma^*| < \varepsilon$$

where ε is a small positive number. The performance index γ^* indicates a lower bound such that the closed-loop system is H_∞ stabilizable.

8.5.3 Simulation and experimental results

To show the effectiveness of the proposed damping control design method, simulation and experimental studies are performed in MATLAB environment and laboratory

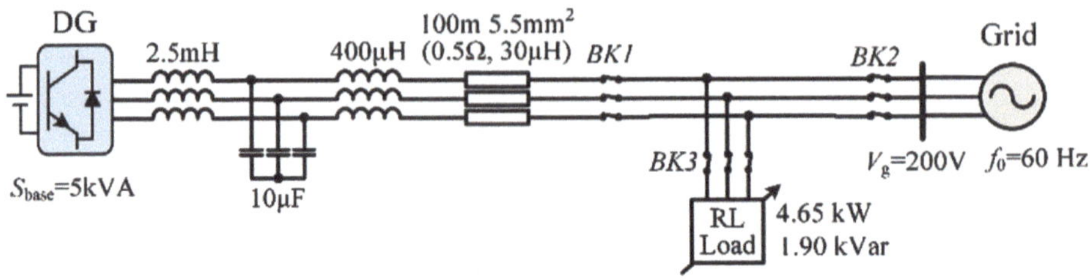

Figure 8.23 The GCC case study.

respectively. An overview of the case study which is used for both simulation, and the experimental system is shown in Fig. 8.23. The circuit parameters are shown in Fig. 8.23. Other system parameters are given in Table 8.4. It is noteworthy that parameters are given in per unit value, and M^* is the inertial constant.

8.5.3.1 Simulation result

To show the capability of the proposed controller several simulation scenarios are performed. Evaluation of the GCC output power P_{out} fluctuation following a step change in the active power reference P_0 is known as the most serious test scenario. In case of having a desirable damping controller, it is expected that P_{out} smoothly follows the P_0. The zero-state response for two cases of with and without damping control is shown in Fig. 8.24. It can be seen that the designed robust optimal damping loop shows a good damping effect compared to the case of no damping characteristic.

8.5.3.2 Experimental verification

To verify the simulation result, a laboratory experiment is conducted for a GCC with the parameters given in Table 8.4. An overview of laboratory setup is shown in Fig. 8.25A. In the performed experiment, the GCC-based DG with a three-phase RL load is connected to the grid. Active power reference P_0 is changed from 0.5 pu to 1 pu. The dynamic response of P_{out} for the cases of with and without damping loop is shown in Fig. 8.25B, which are almost same as the simulation results.

A small overshoot can be seen in the response of system with damping loop, because the PLL is also influenced by V_{out} fluctuation caused by loading transition [31]. This fact is not considered in simulation process (Fig. 8.24). The simulation

Table 8.4 Case study parameters.

Parameter	Value	Parameter	Value
V_{base}	200 V	P_0^*	1 pu
S_{base}	5 kVA	X_T^*	0.3 pu
ω_0	377 rad/s	M^*	8.5 s

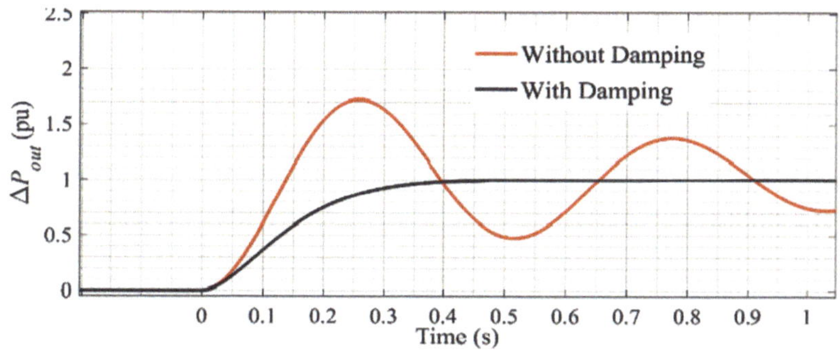

Figure 8.24 Simulation results.

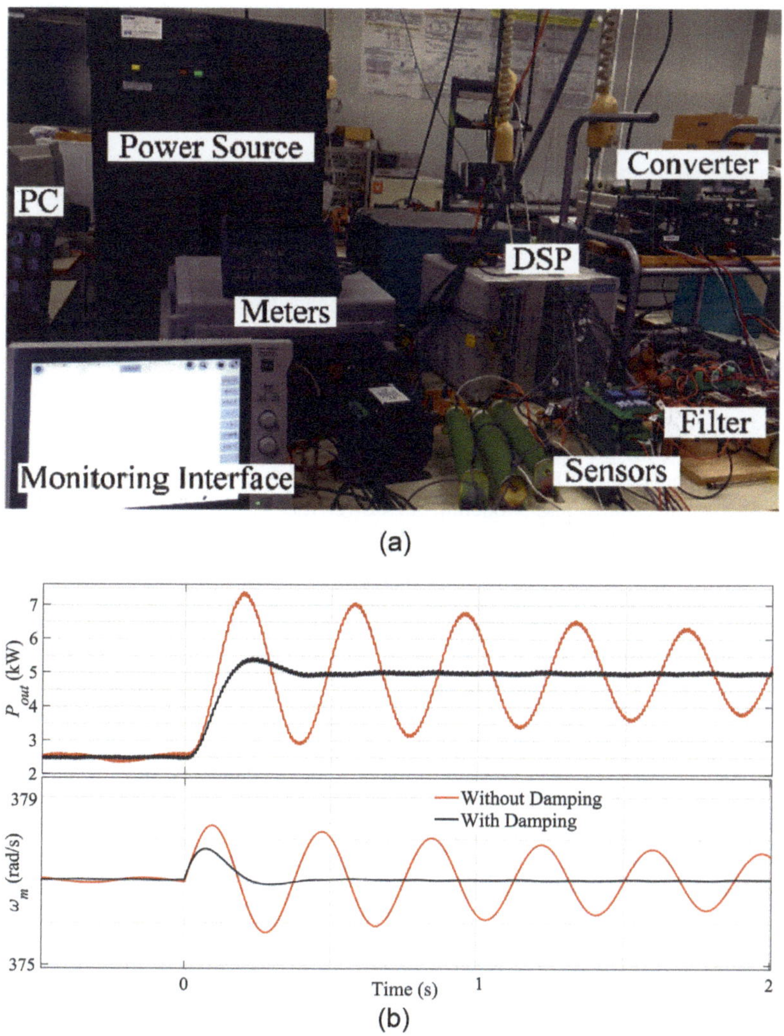

Figure 8.25 Experimental study; (A) laboratory setup and (B) system responses.

and experimental results show that the proposed damping control approach is working as well as the dynamic state feedback with LPF design method discussed in Refs. [31,33].

8.6 Summary

The present chapter addresses five control synthesis methodologies for the GCCs based on continuous and discrete optimal control, Lyapunov energy function, MPC, and robust H_∞ control. The addressed optimal control methods are focused on the output voltage regulation, while Lyapunov-based digital control emphasizes the overall system stability. The H_∞-based robust control provided a damping control by emulating the virtual inertia and damping characteristic, while the MPC-based control methodology addressed both GCC output voltage and current controls using the VSG concept. The effectiveness of the introduced control synthesis methods is supported by several simulations and laboratory experiments.

References

[1] H. Bevrani, B. Francois, T. Ise, Microgrid Dynamics and Control, Wiley, NJ, USA, July 2017.
[2] S. D'Acro, J.-A. Suul, O.-B. Fosso, Automatic tuning of cascaded controllers for power converters using eigenvalue parametric sensitivities, IEEE Trans. Ind. Appl. 51 (2) (2015) 1743−1753.
[3] K. Oue, S. Sano, T. Kato, K. Inoue, Stability analysis of grid-forming inverter in dq frequency domain, in: IEEE Power and Energy Society General Meeting, IEEE 20th Workshop on Control and Modeling for Power Electronics (COMPEL), Paper No. 38, 2019.
[4] H. Bevrani, M. Watanabe, Y. Mitani, Power System Monitoring and Control, Wiley, USA, 2014.
[5] T. Qoria, C. Li, K. Oue, F. Gruson, F. Colas, X. Guillaud, G. Denis, T. Prevost, Tuning of AC voltage-controlled VSC based linear quadratic regulation, IRSEC 18 (2018) 333.
[6] I. Muller, Y.B. Blauth, C.E. Pereira, G. Gabiatti, G. Bonan, Simplifying the design of parallel-connected UPS inverters by means of the per unit system, in: IEEE Brazilian Power Electronics Conference, 2009, pp. 554−550.
[7] K. Ogata, Designing Linear Control Systems with MATLAB, Prentice Hall, NJ, 1994.
[8] K. Ogata, MATLAB for Control Engineers. Upper Saddle River, NJ, 2008.
[9] T. Qoria, F. Gruson, F. Colas, X. Guillaud, M.-S. Debry, T. Prevost, Tuning of cascaded controllers for robust grid-forming voltage source converter, in: IEEE Power Systems Computation Conference (PSCC), 2018.
[10] T. Kato, K. Inoue, Y. Akiyama, K. Ohashi, Optimum and adjustable damping control of grid-connected inverter with an LCL filter, in: 2015 9th International Conference on Power Electronics and ECCE Asia (ICPE-ECCE Asia), June 2015, https://doi.org/10.1109/ICPE.2015.7167775.

[11] T. Kato, K. Inoue, Y. Yamamoto, Optimal digital controller design for passive stabilization of a grid-connected three-phase inverter with LCL-filter, in: IEEE Energy Conversion Congress and Exposition (ECCE), September 2019, https://doi.org/10.1109/ECCE.2019.8912644.
[12] T. Kato, K. Inoue, M. Ishida, Investigation of stabilities of Lyapunov-based digital control for grid-connected inverter, in: IEEE Energy Conversion Congress and Exposition, 2015, pp. 2394–2399.
[13] T. Kato, K. Inoue, M. Ueda, Lyapunov-based digital control of grid-connected inverter with an LCL filter, IEEE J. Emerg. Sel. Topics Power Electron. 2 (4) (2014) 942–948.
[14] T. Zheng, "Model Predictive Control", Sciyo, Chapters published August 18, 2010 under CC BY-NC-SA 3.0 license. www.intechopen.com/books/model-predictive-control.
[15] F. Morel, X. Lin-Shi, J. Retif, B. Allard, C. Buttay, A comparative study of predictive current control schemes for a permanent-magnet synchronous machine drive, IEEE Trans. Ind. Electron. 56 (7) (July 2009) 2715–2728.
[16] J. Rodriguez, J. Pontt, C. Silva, M. Salgado, S. Rees, U. Ammann, P. Lezana, R. Huerta, P. Cortes, Predictive control of three-phase inverter, Electron. Lett. 40 (9) (April 29, 2004) 561–563.
[17] P. Correa, M. Pacas, J. Rodriguez, Predictive torque control for inverter-fed induction machines, IEEE Trans. Ind. Electron. 54 (2) (April 2007) 1073–1079.
[18] A.A. Ahmed, B.K. Koh, Y.I. Lee, A comparison of finite control set and continuous control set model predictive control schemes for speed control of induction motors, IEEE Trans. Ind. Inform. 14 (4) (April 2018) 1334–1346.
[19] F. Donoso, A. Mora, R. Cárdenas, A. Angulo, D. Sáez, M. Rivera, Finite-set model-predictive control strategies for a 3L-NPC inverter operating with fixed switching frequency, IEEE Trans. Ind. Electron. 65 (5) (May 2018) 3954–3965.
[20] P. Cortes, et al., Guidelines for weighting factors design in model predictive control of power converters and drives, in: Proc. IEEE ICIT, 2009, pp. 1–7.
[21] F. Villarroel, et al., Multiobjective switching state selector for finite states model predictive control based on fuzzy decision making in a matrix converter, IEEE Trans. Ind. Electron. 60 (2) (February 2013) 589–599.
[22] J. Jongudomkarn, J. Liu, Y. Yanagisawa, H. Bevrani, T. Ise, Model predictive control for indirect boost matrix converter based on virtual synchronous generator, IEEE Access 8 (2020) 60364–60381.
[23] J. Jongudomkarn, Enhanced Virtual Synchronous Generator Strategy by Model Predictive Multivariable Control: Studies on Voltage Source Inverter and Matrix Converter, PhD Dissertation, Osaka University, 2020.
[24] J. Rodriguez, et al., State of the art of finite control set model predictive control in power electronics, IEEE Trans. Ind. Inform. 9 (2) (May 2013) 1003–1016.
[25] J. Hu, J. Zhu, D.G. Dorrell, Model predictive control of inverters for both islanded and grid-connected operations in renewable power generations, IET Renew. Power Gener. 8 (3) (2014) 240–248.
[26] L.A. Serpa, S. Ponnaluri, P.M. Barbosa, J.W. Kolar, A modified direct power control strategy allowing the connection of three-phase inverters to the grid through LCL -filters, IEEE Trans. Ind. Appl. 43 (2) (2007) 1388–1400.
[27] J. Jongudomkarn, J. Liu, T. Ise, Comparison of current-limiting strategies of virtual synchronous generator control during fault ride-through, in: IFAC Papers On Line, 51(28), 2018, pp. 256–261.
[28] J. Jongudomkarn, J. Liu, T. Ise, Virtual synchronous generator control with reliable fault ride-through ability: a solution based on finite-set model predictive control, IEEE J. Emerg. Selected Topics Power Elect. (2020), https://doi.org/10.1109/JESTPE.2019.2942943.

[29] J. Liu, Y. Miura, H. Bevrani, T. Ise, Enhanced virtual synchronous generator control for parallel inverters in microgrids, IEEE Trans. Smart Grid 8 (5) (September 2017) 2268−2277.
[30] S. Golestan, J.M. Guerrero, Conventional synchronous reference frame phase-locked loop is an adaptive complex filter, IEEE Trans. Ind. Electron. 62 (3) (March 2015) 1679−1682.
[31] J. Liu, Y. Miura, T. Ise, Fixed-parameter damping methods of virtual synchronous generator control using state feedback, IEEE Access 7 (2019) 99177−99190.
[32] H. Bevrani, J. Liu, T. Kato, Robust optimal damping control design for grid-forming converters, in: 9th International Renewable and Sustainable Energy Conference (IRSEC'21), Nov. 23-27, 2021.
[33] J. Liu, Y. Miura, H. Bevrani, T. Ise, A unified modeling method of virtual synchronous generator for multi-operation-mode analyses, IEEE J. Emerg. Sel. Topics Power Electron. (2020), https://doi.org/10.1109/JESTPE.2020.2970025.
[34] P. Gahinet, P. Apkarian, A linear matrix inequality approach to H∞ control, Int. J. Robust Nonlinear Control 4 (1994) 421−448.
[35] T. Iwasaki, R.E. Skelton, All controllers for the general H∞ control problem: LMI existence conditions and state space formulas, Automatica 30 (1994) 1307−1317.
[36] M. Fathi, H. Bevrani, Optimization in Electrical Engineering, Springer, 2019, pp. 113−136.
[37] Y.Y. Cao, Y.X. Sun, W.J. Mao, Static output feedback stabilization: an ILMI approach, Automatica 34 (12) (1998) 1641−1645.
[38] P. Gahinet, A. Nemirovski, A.J. Laub, M. Chilali, LMI Control Toolbox, The MathWorks, Natick, MA, 1995.
[39] H. Bevrani, Robust Power System Frequency Control, second ed., Springer, Gewerbestrasse, Switzerland, 2014.

Grid connected converters for grid dynamics shaping

9

9.1 Flexible grid connected converters for dynamics emulations

Modern power grids face new technical challenges arising from the increasing penetration of grid connected converter (GCC)-based loads and distributed generators (DGs). Among these, the most important one is the inertia and damping requirement challenge, as the GCC-based renewable sources are increasingly replacing synchronous generators (SGs). Unlike conventional SGs, GCC-based DGs do not provide physical inertia and damping support for the grid, and thus, the system is faced with the issue of lack of inertia and damping characteristic, which intrinsically leads to a large rate of change of frequency (RoCoF) and low frequency oscillations in the grid.

To address this issue, the concept of the virtual synchronous generator (VSG), virtual synchronous machine, or synchronverter has been proposed [1]. It is shown in Chapters 2, 3, 6, and 7 that by adding short-term energy storage to emulate the kinetic energy of a rotating mass and mimicking the swing equation of an SG in the control scheme, GCCs can also provide inertia support for the grid to restrain its frequency fluctuation, in the same way as an SG.

GCC-based dynamics emulation is not only limited to inertia and damping characteristics, and thus improving the inertia-based power grid dynamics such as frequency and active power control. It also applicable to other power grid dynamic issues such as improving voltage regulation, reactive power control, power sharing, and power quality [2,3]. Therefore, one can use a GCC as a *dynamic emulator* and it is true to say that flexible controlled GCCs may provide a significant capability to shape the various desirable dynamics in a power grid in different directions.

Additional flexibility may be required from various control levels so that the system operator can continue to balance supply and demand in the modern power grids. The contribution of DGs and renewable energy sources (RESs) in regulation task refers to the ability of renewable integrated power grids to regulate their power output, by appropriate control action. This can be effectively realized by injecting active and reactive powers as well as the virtual building of required properties and dynamics such as inertia and damping characteristics. To follow each characteristic, the GCC must be controlled in a specific way by the flexible shaping of its output active and reactive powers. For example as discussed before, a virtual inertia emulation system requires the GCC to be able to store or release an amount of energy depending on the grid frequency's deviation from its nominal value, analogous to the inertia of a conventional SG, which this system is known as VSG [4].

The GCC-based dynamic emulation, which is conceptually shown in Fig. 9.1, can be used as a promising solution toward the grid stability and performance issues

Grid Connected Converters. https://doi.org/10.1016/B978-0-323-99902-1.00010-0

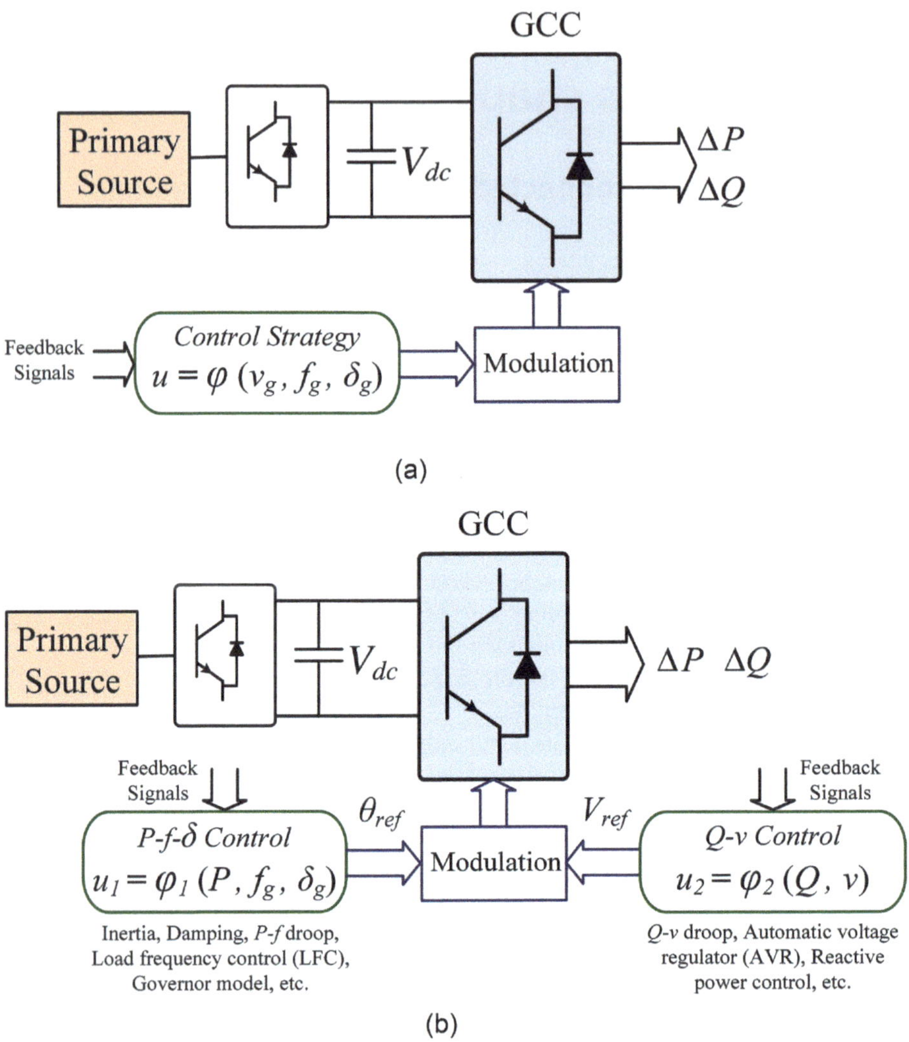

Figure 9.1 GCC-based dynamic emulator; (A) conceptual scheme, and (B) overall control framework.

caused by high penetration RESs in the electrical grid. As can be seen depending on the applied control strategy to emulate a function of voltage, frequency, angle, or a combination of them, $u = \phi\left(v_g, f_g, \ \delta_g\right)$ and feedback signals, the output power of the GCC can be properly controlled through one or a set of required dynamics. Fig. 9.1B with more detail shows that depending on the objective target, to perform the required magnitude and phase references for the converter pulse generator (modulation block), the proposed control strategy may emulate a combination of various control loops for power grid active power/frequency $\phi_1\left(P, f_g, \delta_g\right)$ and reactive power/voltage regulation $\phi_2(Q, v_g)$ or other desirable dynamic systems.

These control loops may cover active power-frequency (P–f) droop, SG governor model, inertia and damping property using swing equation formula, load-frequency control (LFC), reactive power-voltage (Q–v) droop, automatic voltage regulator (AVR), automatic reactive power regulator (AQR), and other desired dynamic functions. From the control theory point of view, the GCC-based dynamic emulator can be considered as a reference model-based control system to shape the dynamic behavior of a given power grid based on a desirable dynamic model.

For example, the virtual impedance loop and the VSG can be considered as special cases that emulates the physical impedance and conventional SG dynamics, respectively. The VSG itself may emulate several dynamic functions such as the P–f droop, swing equation (inertia and damping characteristics), and Q–v droop. All mentioned characteristics can be realized in the control system of a GCC as shown in Fig. 9.2 [1,4]. This block diagram was briefly explained in Section 6.4 and will be discussed with more detail in the next section.

9.2 Virtual dynamic shaping

In order to show how a GCC affects the tie-line and grid behaviors using virtual dynamics in this section, two examples of virtual impedance and virtual SG (VSG) are discussed using the shown GCC control system in Fig. 9.2.

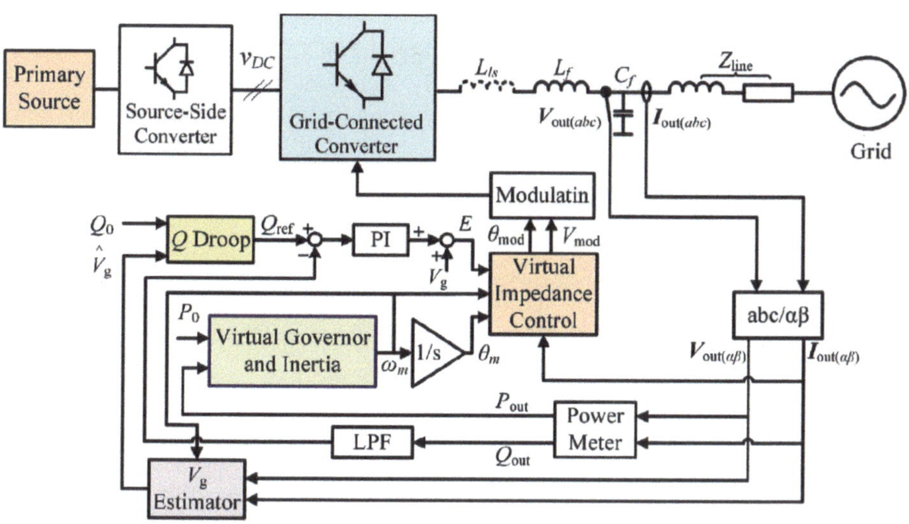

Figure 9.2 A closed-loop GCC system (VSG) with impedance and SG dynamics emulator.

9.2.1 Virtual impedance loop

As mentioned in Chapter 7, to enhance oscillations damping, and to decrease the negative impact of the output filter resonance on the GCC performance, the virtual impedance can be used. This issue can be realized by providing virtual resistor(s) R_i in series with the LCL filter components as shown in Fig. 7.3, which unlike an actual resistor does not cause losses. In general, in addition to a pure resistor, an inductance and/or a capacitance term in the form of an impedance element Z_v can be virtually produced by the GCC control system. An example for implementation of the virtual impedance concept in a GCC (working with $P\text{–}Q$ droop control method) is shown in Fig. 9.3 [1].

This concept is also successfully implemented for regulating the power sharing among parallelized GCCs and limiting over currents under grid disturbances. The virtual impedance updates the power converter output voltage reference as given in Eq. (9.1), where new voltage reference $\left(v_{ref}^*\right)$ is obtained by subtracting the virtual voltage drop across the virtual impedance from the reference value originally provided by the droop equations $\left(v_{ref}\right)$. The transfer function of virtual impedance block $Z_v(s)$ must be selected according to the rated power of the GCC.

$$v_{ref}^* = v_{ref} - v_g^* = v_{ref} - Z_v i_g \tag{9.1}$$

In Fig. 9.2, the "Virtual Impedance Control" block plays the same role. A realization scheme for this block is shown in Fig. 9.4. In addition to above applications of virtual impedance, it can also be used to limit the output overcurrent [5]. The idea is to increase virtual impedance proportional to the size of the current magnitude when the current magnitude exceeds a predefined value as a threshold. The increased impedance reduces the voltage command and thus preventing overcurrent. If the current magnitude is lower than the threshold, then the virtual impedance will be reverted back to the original value.

In order to achieve the current-limiting ability, the GCC's output impedance will be increased during an overcurrent condition. By increasing the output impedance, the voltage reference is reduced and thereby preventing the voltage controller from commanding an excessively large current reference [6]. Unlike multiloop control, where

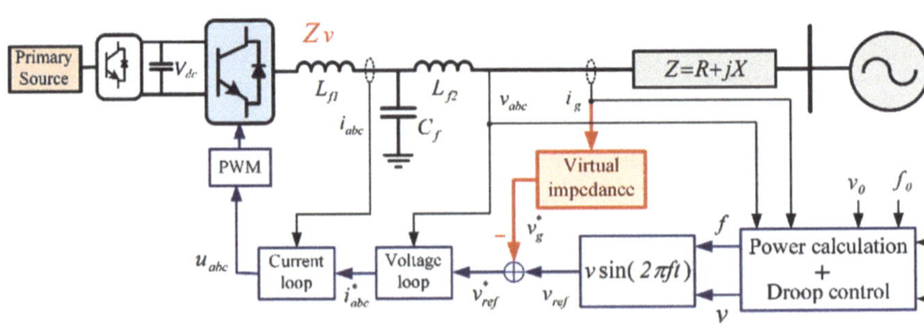

Figure 9.3 A GCC control system with virtual impedance loop.

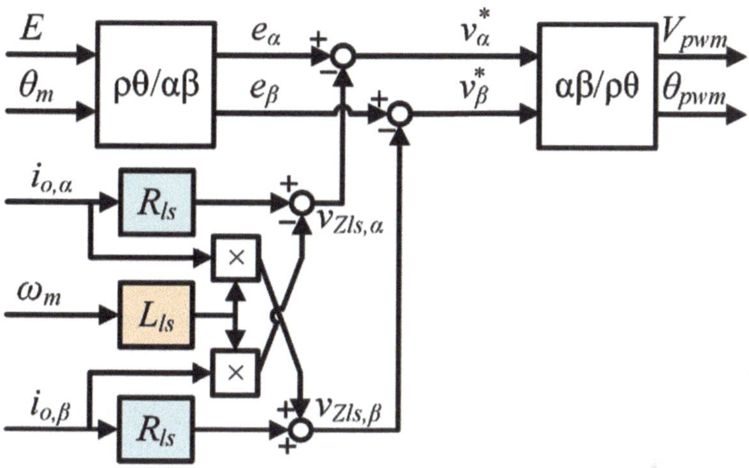

Figure 9.4 A realized scheme for the "Virtual Impedance Control" block.

the current limiter is embedded in the current control loop, the virtual impedance control can be integrated into the voltage reference controller, and hence, the current-limiting ability can be achieved without actually controlling current.

In the GCC control scheme shown in Fig. 9.2, a constant virtual inductance (L_{ls0}) is generated by the "Virtual Impedance Control" block. This inductance exists to assist in transient active power-sharing and system damping increasing [2]. In the GCC control with virtual impedance current limiting, the transient virtual impedance term ($\Delta Z_{ls} = \Delta R_{ls} + \omega \Delta L_{ls}$) is also generated in addition to L_{ls0} for the purpose of DG over-current limiting during a fault [5].

Hence, the overall virtual inductance (L_{ls}) becomes $L_{ls} = L_{ls0} + \Delta L_{ls}$. The effect of increased impedance to reduce peak current can be easily understood with the following equation.

$$V_o = I_{\max} \sqrt{\Delta R_{ls}^2 + (\omega L_{ls})^2} \tag{9.2}$$

where V_o is the magnitude of GCC output voltage during fault and $I_{\max}$ is the desired maximum allowed current magnitude. If the voltage drop is constant, $I_{\max}$ becomes smaller as virtual impedance increases. The increase in impedance was achieved by adding ΔZ_{ls} to the original virtual impedance. The value of ΔZ_{ls} is defined according to Eq. (9.3) and Fig. 9.5.

$$\Delta R_{ls} = \max\{K_z(i_{o,\,rms} - I_{th}),\ 0\}\ ;\ \Delta L_{ls} = \Delta R_{ls}\left(\frac{X}{R}\right) \tag{9.3}$$

where K_z is limiting gain, and X/R is the current limiting virtual impedance ratio. According to Eq. (9.3), if the output current $(i_{o,rms})$ exceeds the predefined threshold

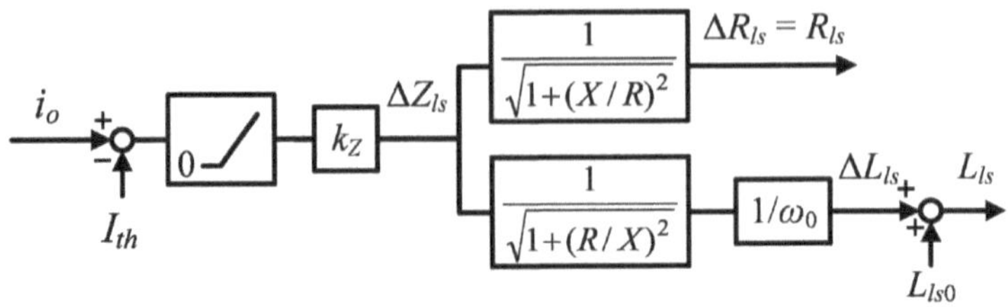

Figure 9.5 Virtual impedance calculator.

value (I_{th}), ΔZ_{ls} proportional to the amount of overcurrent will be applied to the GCC. In this strategy, by using the amount of overcurrent to generate extra output impedance, the control enters the current limiting mode automatically once overcurrent occurs, and it exits the mode as soon as the overcurrent condition ceases to exist.

Considering the impact of constant and transient parts of the virtual impedance, as it is later shown in the experimental results, the constant part (L_{ls0}) improves the damping and transient power sharing of the system, while the transient part (ΔR_{ls} and ΔL_{ls}) limits the overcurrent.

Suitable values of K_z and X/R should be chosen corresponding to the level of happening faults. The voltage drop V_o should equal the voltage magnitude command during the event of faults [5]. This can be ensured by determining K_z and X/R that generates the desired V_o according to Eq. (9.2). It should be noted that the parameter K_z must be determined according to the amount of voltage drop. However, in a real system, different kinds of disturbance exist, and it is hard to foresee precisely how significant the voltage drop will be. This makes the tuning of this limit gain complicated in practice. If the gain is too small, current limiting fails. On the other hand, if the gain is too large, the amount of drop in voltage command generated by virtual impedance could surpass the magnitude of the original voltage command. This can lead to instability in the control system, therefore using an online adaptive tuning can be considered as a suitable solution to fix this problem.

9.2.2 Experimental verification

Experiments are executed in an islanding microgrid (MG), which its circuit is shown in Fig. 9.6A. The MG includes two GCC-based DGs, which its configuration and parameters are same as the simulated MG used in Ref. [2], except that instead of dc sources, AC supply rectified by diode bridges to imitate the dc sources of GCCs. The setup of experiment system is shown in Fig. 9.6B.

As mentioned, the virtual impedance control block in the control system is used to adjust the output reactance of the GCCs. This function is realized by multiplying output current to the virtual inductor in a stationary frame. To keep more accuracy, the inductor current through L_f is used for this purpose. However, this increases the number of current sensors, which is not necessary. As the current flowing into C_f at fundamental frequency is too small in comparison with the inductor current, using

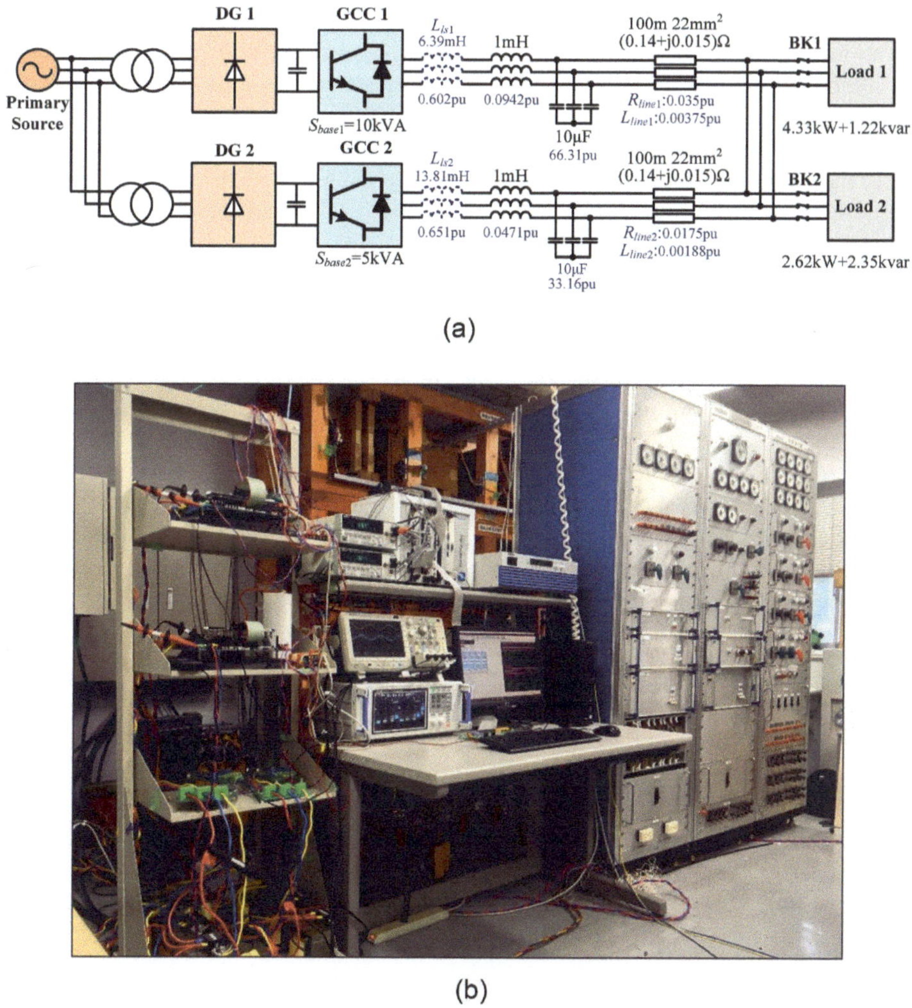

Figure 9.6 Case study system; (A) MG circuit, and (B) experimental setup.

output current instead of inductor current does not affect the performance of the control scheme.

According to Eq. (9.4), for each GCC, tuning of virtual inductor L_{ls} is suggested to set total output reactance X for both GCCs in same large per unit value. This approach increases active power damping ratio and shares transient load without oscillation. The target value of X is set at 0.7 pu because it is a typical value for the total direct-axis transient reactance of a real SG.

$$X = S_{base}\, \omega_m \left(L_{ls} + L_f + L_{line}\right) / E_0^2 \tag{9.4}$$

The L_f and Z_{line} $(R_{line} + jL_{line})$ are considered as known parameters here. With the mentioned design of reactance adjustment, oscillation in the GCC-based MG should be almost eliminated during a loading transition. It can be also effective to damp the oscillation due to other disturbances, for example, change of active power reference of GCCs, connection/disconnection of a GCC, etc.

In the performed experimental study, the applied test scenarios and disturbance sequences are as follows: for $t < 0.5$ s only Load 1 is connected and the active power reference $(P_{0,i})$ of each GCC is fixed at 1 pu. At $t = 0.5$ s, Load 2 is also connected, and finally at $t = 3$ s, the $P_{0,2}$ is changed to 0.6 pu. The experimental results are shown in Fig. 9.7. The system responses for without and with constant virtual reactance as the main term of virtual impedance are shown in Fig. 9.7A and B, respectively. The root loci for dominant eigenvalues of GCC-based DG1 respect to change of its output impedance also verifies the damping improvement (Fig. 9.7C).

Experimental results verify the effectiveness of the virtual impedance control loop in the system performance. First, the oscillation due to loading transition at 0.5 s is eliminated, and the oscillation due to change of the active power reference at 3.0 s is also damped. It implies that the proposed control scheme is able to track the loading transition rapidly without oscillation.

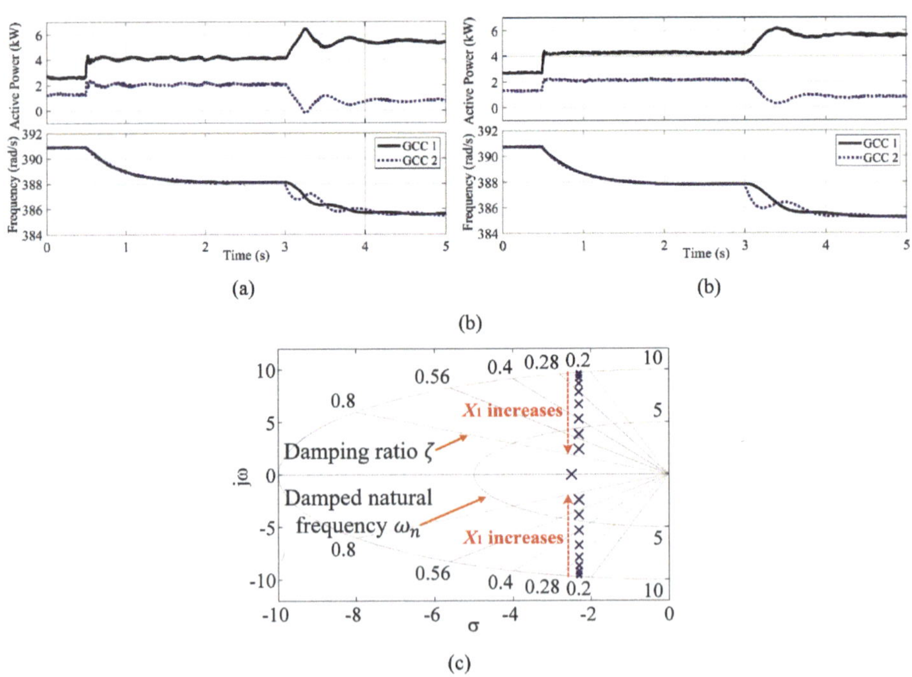

Figure 9.7 System responses; (A) without constant virtual reactance, (B) with constant virtual reactance, and (C) root loci for dominant eigenvalues due to change of output impedance of GCC-based DG1.

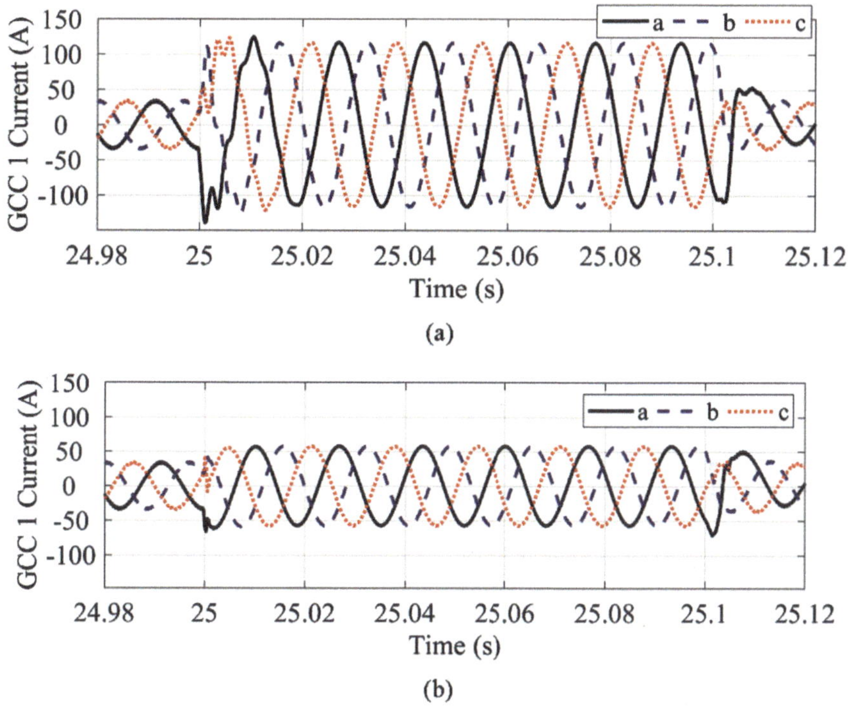

Figure 9.8 GCC 1 response following a three-phase to ground fault; (A) without transient virtual impedance, and (B) with transient virtual impedance.

As mentioned above, the transient part of virtual impedance may limit the current amplitude in the case of a fault. For the given case study, the impact of a three-phase to ground fault on the output current of GCC 1 without and with transient part of virtual impedance is shown in Fig. 9.8A and B (simulation results), respectively. The fault is removed after a while. The result shows a significant reduction in the GCC overcurrent.

In addition of the "Virtual Impedance Control" block, the other control blocks in Fig. 9.2 provide some useful virtual dynamics that have positive effects on the grid performance and stability. Using the experimental results, some examples are briefly discussed in the next subsections.

9.2.3 Virtual Q-droop and voltage estimator

The "Q Droop" and "V_g Estimator" blocks in Fig. 9.2, as discussed in Refs. [1,2], enhance voltage behavior and properly share the reactive power according to the assigned power rating ratio and is immune to active power sharing change and line impedance mismatch in per unit values. The dynamic impact of Q droop is discussed in previous chapters, and it is well known in the literate. Therefore, only the impact of voltage estimator is emphasized here. For this purpose, some experimental results for the same test scenarios discussed for Fig. 9.7 are obtained for the cases of with and without mentioned

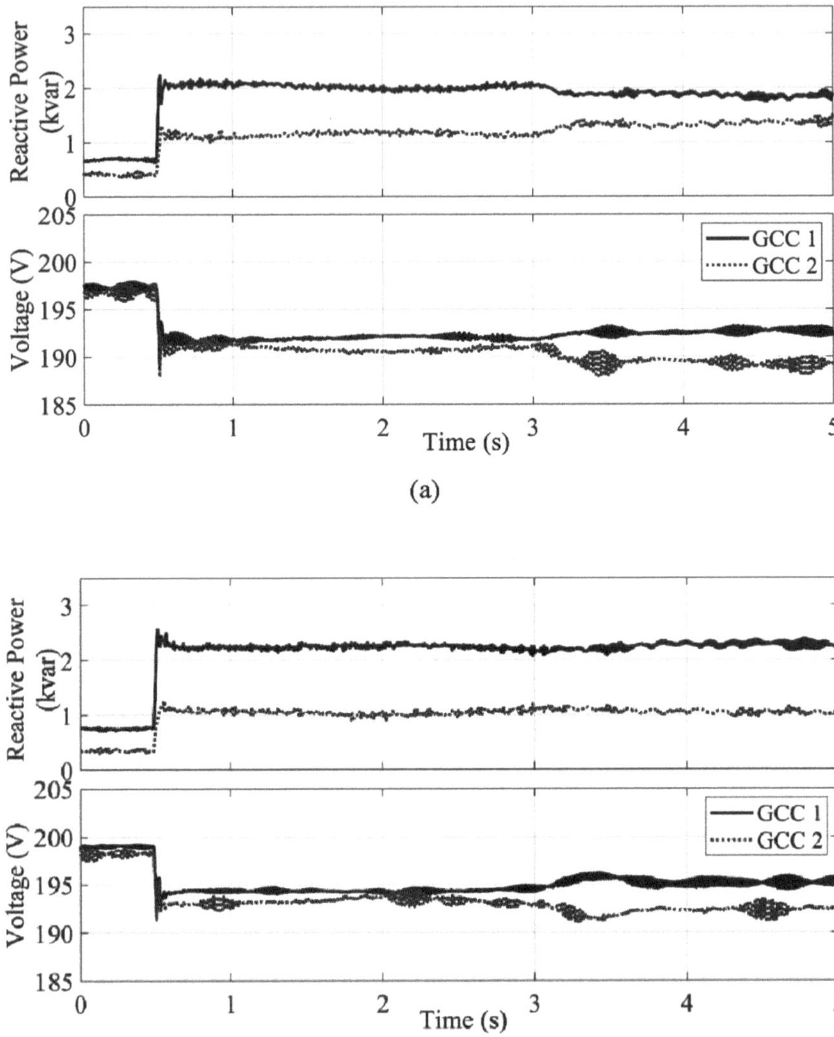

Figure 9.9 Experimental reactive power and voltage of the GCCs; (A) without voltage estimator, and (B) with voltage estimator.

functions that are shown in Fig. 9.9. This figure confirms the significant constructing effect of voltage estimator on the reactive power sharing between two GCCs [2].

9.2.4 Virtual governor and inertia

The "Virtual Governor and Inertia" block in Fig. 9.2 provides virtual $P-w$ droop, damping and inertia characteristics that can properly shape the active power and frequency response of the grid. Fig. 3.3 shows a block diagram for realization of this block. As discussed in Chapter 3, the $P - \omega$ droop makes a mechanism for frequency regulation and proper active power sharing. Virtual inertia reduces the RoCoF and system sensitivity to disturbances, and thus improves the grid stability. Virtual damping characteristic attenuates the undesirable oscillations on the active power and frequency of the grid. The positive dynamic impact of adding virtual damping term to the "Virtual Governor and Inertia" of the GCC control system (Fig. 9.2) following a sequence of test scenarios is shown in Fig. 9.10 [2]. At the beginning, the MG starts with only GCC 1 and Load 1, GCC 2 is connected at 2 s. Then, the active power reference of GCC 2 is changed at 8 s, and finally at 8 s, Load 2 is connected to the grid.

9.2.5 Virtual dynamics for power quality enhancement

Power quality of GCC-based power grids is a challenging issue due to nonlinearities of converters, multiple resonance modes of network impedance and unbalanced and nonlinear loading condition. In this direction, an important problem of using power electronic converters in a renewable integrated power grid is producing voltage harmonics.

The harmonic distortion has created new problems in power networks, including equipment malfunction, reduced life and efficiency of devices, loss increase, power capacity reduction, neutral line overload, and distortion of voltage-current waveforms. Therefore, several standards are performed for different grids by the world-wide relevant committees. For example, according to IEEE standard 519–2014 [7], the total harmonic distortion (THD) and individual harmonic levels of voltage for power grids with voltage levels less than 1 kV must be maintained less than 8% and 5%, respectively.

A straightforward solution for harmonic distortion problem is to use shunt active power filter [8]; however, dedicated inverters of active power filters increase the cost of grids. A preferable approach is to use GCCs control systems for unbalance and harmonics compensation, and meanwhile, to share the compensation current between GCCs without communication. Therefore, coordinated control of parallel GCCs to have an appropriate level of power quality plays a major role in the grid dynamic performance.

A well adopted decentralized approach to deal with unbalance and harmonics is to add virtual impedance to the GCCs [9,10]. However, this method may not be so effective if in addition to the load/grid dynamics, nonlinearities of GCCs to be considered. In fact, grid voltage quality and mitigation of GCC nonlinearities seem to be an intrinsic trade-off, if virtual-impedance-based method is applied.

Another communication-less approach is to share the harmonic power in a similar manner to that of fundamental power, such as harmonic droop controller proposed in Ref. [11]. However, grid voltage quality is not intentionally controlled in this method. A generalized droop proposed in Ref. [12] mixes the concept of harmonic droop and virtual impedance. This droop is equivalent to a virtual resistance in parallel with the GCC impedance, in order to reduce the overall GCC impedance. However, the resulted reduction of GCC impedance is quite limited [13].

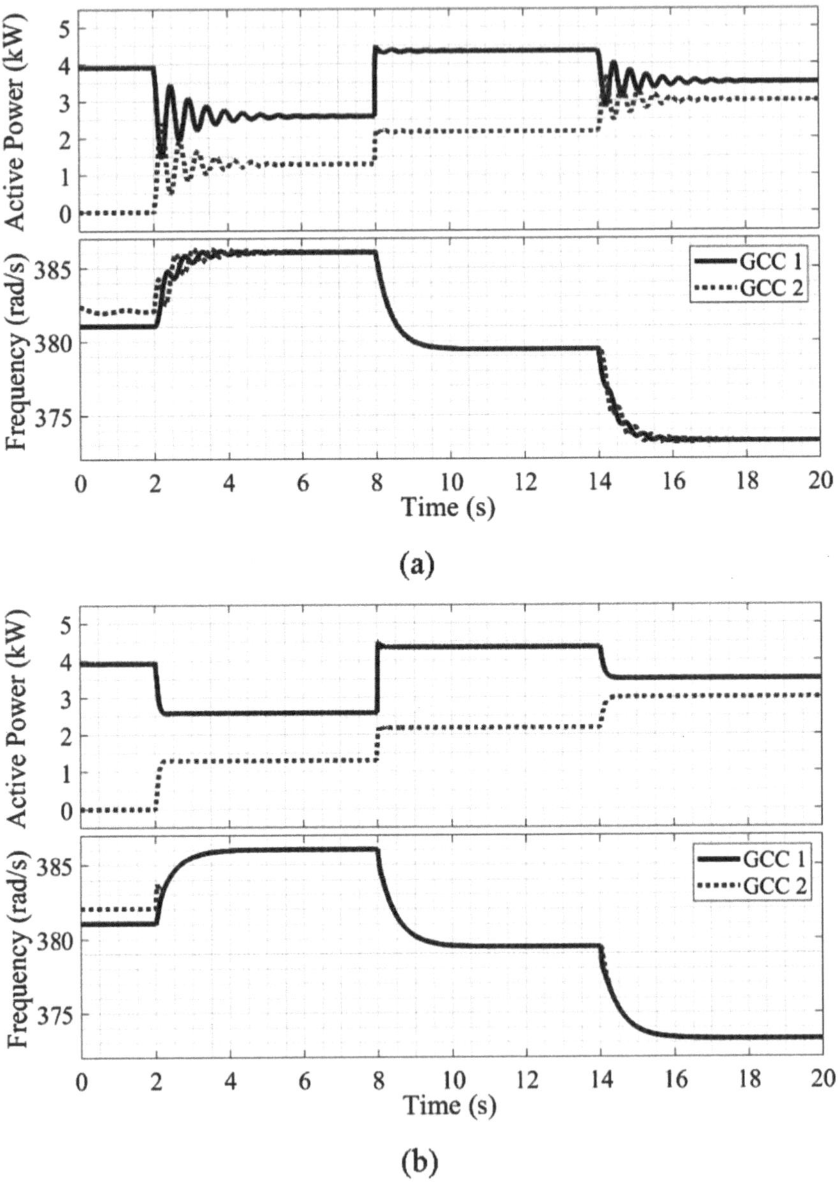

Figure 9.10 Experimental active power and frequency of the GCCs; (A) without virtual damping, and (B) with virtual damping.

In this section, a control framework for reducing the voltage harmonics is briefly explained. Full details are available in Ref. [14]. Many methodologies have been performed to reduce the grid voltage harmonics in MGs and GCC-based DGs. Most of them use the virtual harmonic impedance and/or droop methods. The addressed method in this section uses the VSG concept in the GCC control system to remove the specified voltage harmonics.

9.2.5.1 Control methodology

To improve the power quality in a grid, this section addresses a control method to reduce main voltage harmonics selectively. This method uses a VSG-based control that for each voltage harmonic, considers an appropriate harmonic voltage as the GCC voltage reference (V_{rh}) to compensate the harmonic voltage generated at the output due to the corresponded harmonic current. The VSG-based harmonic control loops are finally added to the inner control level of the GCC control system and significantly reduce the voltage harmonics and THD.

The control method can be explained by considering the harmonic equivalent circuit of the GCC, which are shown in Fig. 9.11A and B for h^{th} harmonic and all harmonics, respectively (for fundamental frequency, $h = 1$). From h^{th} harmonic perspective, a GCC can be modeled as an AC source, with voltage reference of V_{rh}, and the grid can be modeled as a combination of voltage and current sources

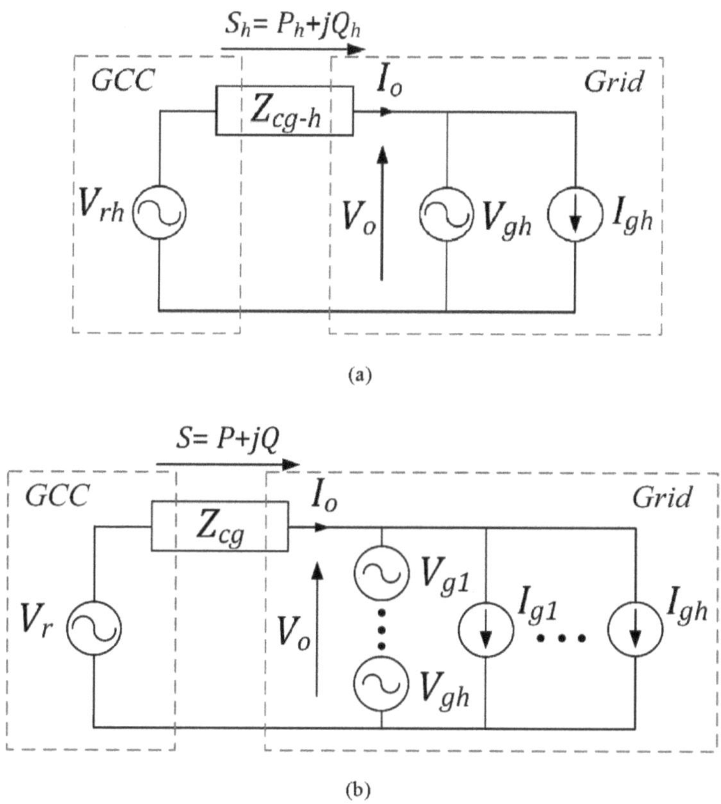

Figure 9.11 Harmonic-based equivalent model of a GCC; (A) for h^{th}-harmonic, and (B) for all voltage and current harmonics.

(V_{gh}, I_{gh}) at harmonic frequencies with an equivalent impedance Z_{cg-h}. The active and reactive harmonic powers delivered to the grid are calculated based on the power delivered to the harmonic voltage source (V_{gh}) and the power delivered to the harmonic current source (I_{gh}) [11].

Having an inductive network impedance, for the fundamental frequency, the grid is considered as a voltage source. Thus, a desirable GCC's $P-\omega$ characteristic can be realized via a virtual swing equation as Eq. (9.5), and $Q - V_r$ characteristic can be formed as Eq. (9.6) [14].

$$P_1 - P^* + D_1(\omega_1 - \omega^*) = J_1\omega^* \frac{d(\omega_m - \omega^*)}{dt} \tag{9.5}$$

$$V_{r1} = V_r^* + n_1(Q_1 - Q^*) \tag{9.6}$$

where ω^* and V_r^* show the nominal values of frequency and voltage, P^* and Q^* are the reference active and reactive powers, P_1 and Q_1 are the active and reactive output powers, n_1 is $Q - V$ droop factor, and J_1 and D_1 are the inertia and damping coefficients of the VSG, respectively.

However, for the harmonic frequencies (e.g., h^{th} harmonic), the grid is considered as a current source, and to minimize harmonic voltages at the output, the characteristics given by Eqs. (9.7) and (9.8) are used, that are different from Eqs. (9.5) and (9.6), which were depended to the type of impedance [14].

$$V_{rh} = -k_{ph}(P_h - P_h^*) \tag{9.7}$$

$$Q_h - Q_h^* - D_h(\omega_h - h\omega^*) = J_h\omega^* \frac{d(\omega_h - h\omega^*)}{dt} \tag{9.8}$$

where P_h and Q_h are the active and reactive h^{th} harmonic powers, J_h and D_h are the inertia and damping coefficients of the hth harmonic VSG, which have a major impact on the performance and stability of the controller. The reference voltage V_{rh} at the hth harmonic frequency can be formed by the effective value V_{rh} and the phase angle produced by ω_h.

Therefore, the used control strategy for harmonic voltages attenuations is different from the strategy of the power delivery to the grid by the control of the VSG in fundamental frequency, that is, in Eqs. (9.5) and (9.6). The $P_h - V_{rh}$ and $Q_h - \omega_h$ control loops have negative feedback in Eqs. (9.7) and (9.8), so the VSG controller can adjust the voltage and frequency, respectively. As mentioned, the main advantage of this controller is independency to the type of impedance. This facilitates the controller design without the need to consider the type of impedance in harmonic voltages attenuation point of view.

As discussed, the control system must operate such that for the hth harmonic, the resulted voltage at the GCC output (V_{gh}) must be closed to zero (Fig. 9.11A). In other words, the active and reactive powers delivered to the harmonic current source (I_{gh}) must be zero. For this purpose, the frequency set-point should be set as the hth harmonic frequency and the voltage set-point (V_{rh}^{*}) in the VSG control Eq. (9.7) should be set at zero for h^{th} harmonic. It leads to control of hth harmonic voltage by the VSG, where its control laws are given as Eqs. (9.7) and (9.8). The parameters J_h, D_h, and k_{ph} must be determined by designer.

The block diagram of the VSG-based harmonic control strategy is shown in Fig. 9.12. In order to reduce several harmonics in the output voltage, several harmonic controllers related to the harmonic commands can be included in the control scheme to produce a summation of the required harmonic voltages, that is, $\sum_h V_{rh}$. The voltage reference V_r is obtained by adding $\sum_h V_{rh}$ to V_{r1}, which is generated by the controller of the VSG at the fundamental frequency.

It should be noted that the fundamental VSG control block depends on the type of output impedance (the fundamental VSG control block in Fig. 9.12 is designed for a pure inductive output impedances at the fundamental frequency). If the output impedance of the GCC at the fundamental frequency is not predominantly inductive, then the

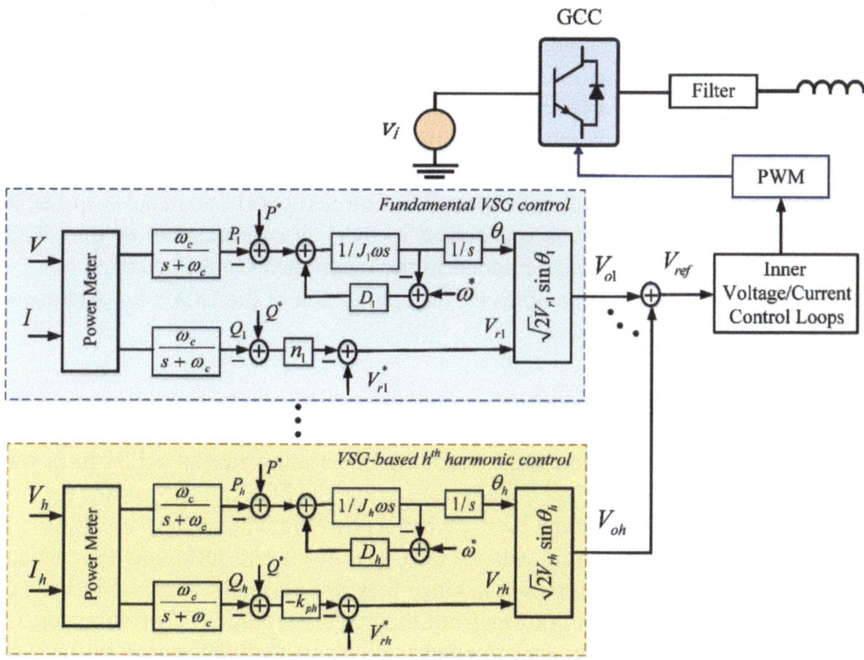

Figure 9.12 Control block diagram of the VSG-based voltage harmonic compensator.

fundamental VSG control scheme must be updated accordingly, but the VSG-based h^{th} harmonic control block is independent of the output impedance.

9.2.5.2 Simulation results

Some simulations for an MG with two GCC-based DGs are performed. The GCC's control systems are equipped with the above discussed method for removing fifth and seventh harmonics. The closed-loop system efficiency of the proposed VSG-based control method is compared with the droop-based harmonic rejection method [11]. Fig. 9.13 shows the system responses following adding a load unit at $t = 1$ s. The frequency in both methods is settled on the nominal value, 60 Hz. The active and reactive powers for VSG and droop controls are properly shared. However, the proposed VSG control performs better transient response and harmonic voltage reduction.

9.3 Grid ancillary service support

Rotational inertia and damping reduction in the grid may negatively affect the existing system frequency characteristic, stability, and control performance. Accordingly, the increasing penetration of GCC-based DGs motivates the need to develop additional ancillary control services to improve the power grid dynamics and stability. Owing to the grid codes imposed to the generating units and to the connected HVDC link to the power grid, all these devices are asked to provide ancillary services to the grid. The ancillary services are mainly used for supporting the voltage/reactive-power and frequency/active power functionalities.

Both grid-forming and grid-following GCCs can be used for the grid ancillary service support [15]. Since a grid-following GCC operates as a current source, the voltage or reactive control support command is transferred to a reactive current reference and then in the low control level of the GCC, it is translated to an appropriated switching signal. In the same way, the frequency or active control support command is linked to an active current reference and is transferred to the low control level of the GCC. Furthermore, an inertial term can be added to the mentioned control system.

In the present section, two examples on the application of the GCCs for grid ancillary service support are discussed.

9.3.1 Frequency regulation support from a motor drive system

Nowadays, to address environmental issues, the penetration rate of RESs in power grids is rapidly growing. In conventional power grids, the SGs can slow down the frequency deviation by their rotational inertia when a power imbalance occurs. However, most of RESs connected to power grids via GCCs are from grid-following type which do not provide any frequency regulation, since they are normally operated as current sources to extract the maximum power from the primary sources. Therefore, further penetration of RESs may make grid frequency prone to fluctuation due to the lack of system inertia.

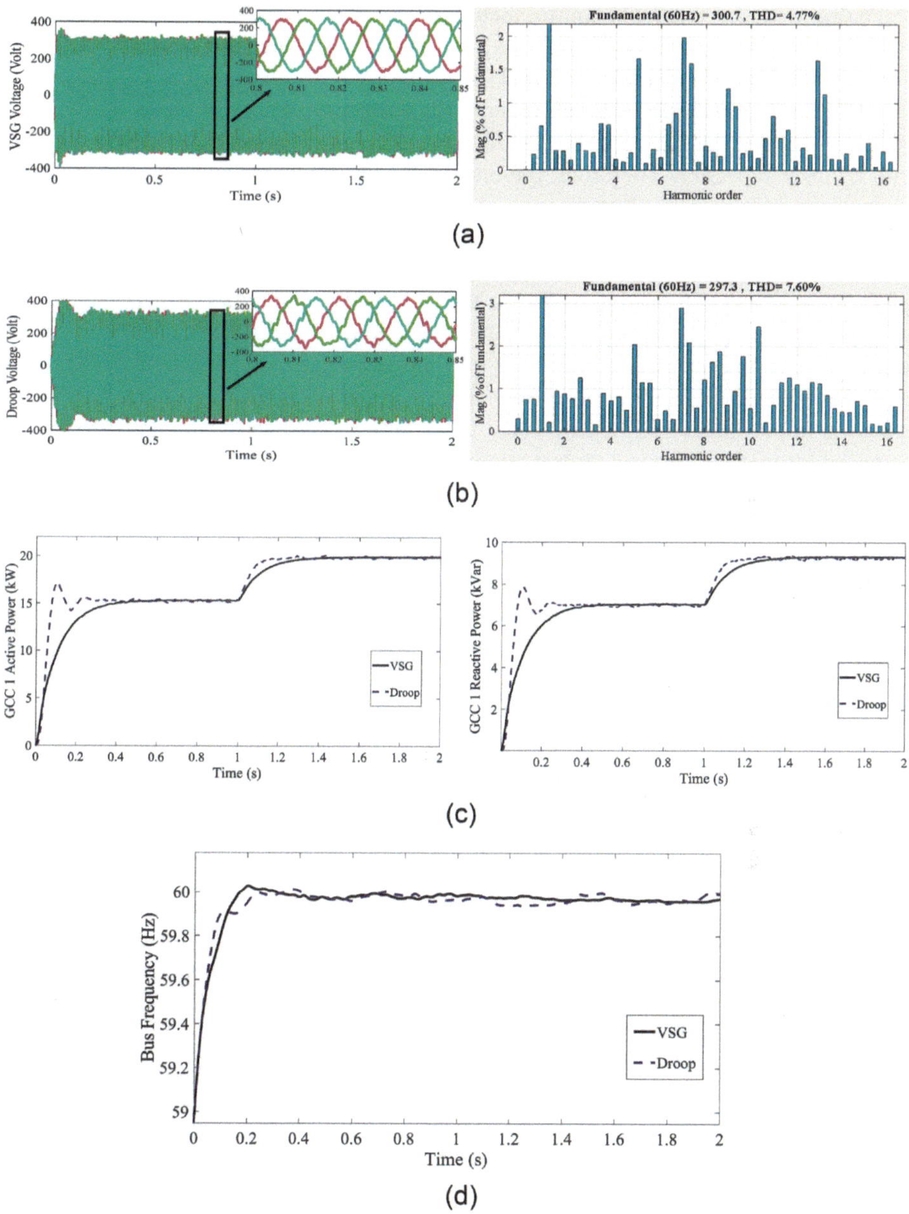

Figure 9.13 Comparison of droop and VSG-based controllers; (A) Output voltage and THD with VSG control, (B) Output voltage and THD with droop control, (C) GCC 1's active and reactive power, and (D) common bus frequency.

To deal with this issue, the control methods of producing virtual inertia as discussed in previous section have been proposed. In these control methodologies, the swing equation of SGs is calculated so that GCC-based DGs emulate inertia feature of SGs to provide frequency regulation supports for power grids. In these studies, the DC-side of converter is simplified by an ideal DC-voltage source or a diode rectifier with a bulky capacitor. Therefore, the energy for virtual inertia and speed governors comes from idealized dc sources.

However, considering practical applications of the VSG control, if no energy storage is involved in power converter systems, the energy for inertia emulation and frequency regulation is restricted to that in their DC-link capacitors [16]. However, this energy is generally quite limited, and a trend to minimize the dc-link capacitor can be found in the literature due to concerns of volume, weight, and life cycle. Thus, the VSG control schemes including energy storage involved in GCC systems have been proposed in the literature [3]. However, the installation of energy storage systems requires additional cost, thus a large-scale inertia support using energy storage systems leads to a considerable investment. The VSG control has also been studied for the application in converters of transmission systems such as low-frequency AC (LFAC) [17]; however, it only facilitates the inertial response interaction between interconnected systems and does not provide additional energy reservoir.

To avoid expensive additional backup in the generation side, demand-side management attracts attention as a solution for grid frequency stabilization [18,19]. Therefore, this section addresses a frequency regulation support scheme by application of the VSG control to demand-side rotating loads, which is discussed in Ref. [20] in detail. Compressors, fans, blowers, and pumps are the candidates for dynamic demand-side management since considerable kinetic energy is stored in their rotating mass. If a short-time motor speed variation is permitted by the owners, their kinetic energy can contribute to the grid frequency regulation through proper control of their motor drives.

Regarding the potential benefit of a motor drive with an active switching front end, which commonly uses a back-to-back (BTB) configuration, in addition to the regeneration capability, more benefits can be obtained if the VSG control concept is applied to the active switching front end, which is also known as the grid-side converter, and is called as GCC here. First, the VSG control enables a phase locked loop (PLL)-free inertia emulation and frequency regulation, thus the aforementioned concern on the use of PLL is avoided. Second, a VSG-controlled converter behaves as a voltage source, thus it provides synchronizing power like an SG. This is beneficial for the initial active power sharing during a transient state. By further considering the fact that the diode rectifier front end requires a bulky grid filter inductor, or a power factor correction circuit to meet the grid code requirements for harmonics injection, a properly controlled BTB motor drive may become a favorable solution.

9.3.1.1 Overall control scheme

The circuit configuration of the motor drive system employing the proposed control system is shown in Fig. 9.14. The induction motor (IM) coupled with the load is

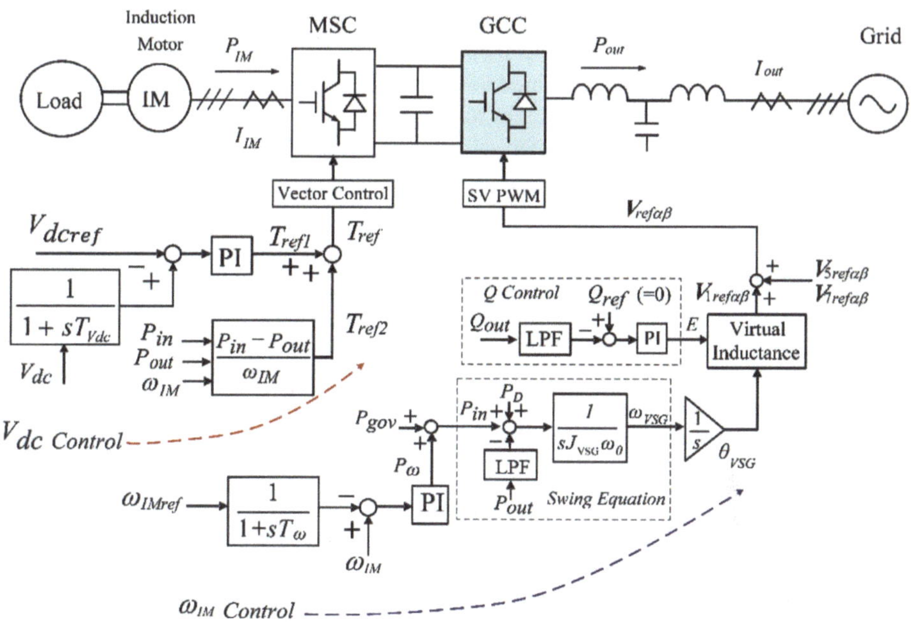

Figure 9.14 Overall control scheme.

connected to the grid via the motor side converter (MSC), the GCC, and an LCL filter. In this work, squirrel-cage IMs are studied since they are widely used for fans, blowers, and pumps. The VSG control is applied to the GCC, and the motor speed control, which is a slip frequency type vector control, is applied to the MSC. The VSG control regulates the power flow between DC-link and the grid to provide frequency regulation support during grid frequency fluctuation. The MSC accelerates or decelerates the IM to keep the DC-link voltage V_{dc} constant. Therefore, the kinetic energy of the motor and the load is utilized for the grid frequency control support. In addition, motor speed ω_{IM} is controlled to follow its reference ω_{IMref} at a steady state. A coordination system between GCC and MSC (or VSG and Vector Control) is required which is fully explained in Ref. [20] and is not discussed here.

9.3.1.2 VSG-based GCC control

Fig. 9.14 also shows the block diagram of the VSG control applied to the GCC. The used VSG control scheme is almost same as introduced one in the previous section. In the VSG control, the swing equation of an SG represented by Eq. (9.9), is calculated to estimate the VSG angular frequency ω_{VSG} and the phase θ_{VSG} so that the GCC emulates the inertia feature of an SG.

$$J\omega_0 \frac{d\omega_{\mathrm{VSG}}}{dt} = P_{in} + P_D - P_{out} \tag{9.9}$$

where J, ω_0, P_{out}, and P_{in} are the VSG moment of inertia, nominal frequency ($= 2\pi 50$) output active power of the GCC, and input power reference of the VSG, respectively. Here, the positive power flow is defined as the direction from the DC-link to the grid as shown in Fig. 9.14. The time constant of low-pass filter (LPF) is properly tuned, so that sufficient ripple attenuation is obtained without affecting the desired inertial response.

Since the swing equation has oscillatory features in P_{out} and ω_{VSG}, a damping power P_D is added to P_{in} for oscillation damping. It should be noted that at the steady state, P_D becomes zero owing to the integral term. P_{in} is the sum of P_{gov} and P_ω, where P_ω is generated by the control scheme shown in Fig. 9.14 and the detail of coordination mechanism is discussed in Ref. [20]. P_{gov} is generated from a virtual governor model which realizes $P{-}\omega$ droop characteristic as discussed before. Considering steady-state error between the grid frequency and ω_0, the P_{gov} is set to be zero within the dead-band $\omega_{th\_l} \leq \omega_{\text{VSG}} \leq \omega_{th\_h}$ since the proposed system cannot continue releasing or absorbing active power at a steady state. The threshold ω_{th_l} and ω_{th_h} are determined by the users, based on their willingness to how frequently they want to help the grid frequency regulation. In the proposed system, the thresholds are set at $2\pi \times (50 \pm 0.1)$ rad/s, which imply the proposed system is used for grid frequency regulation only when a 0.1 Hz deviation is detected. It is worthwhile to mention that the percentage of operation time during which the frequency deviation is kept within 0.1 Hz is an important index for a power system.

The GCC reactive power output Q_{out} is controlled by regulating the virtual induced electromotive force of the VSG as shown in Fig. 9.14. Q_{ref} denotes the reference value of Q_{out} and is set to be zero for a unity power factor operation. The virtual inductance block can be realized using the $\alpha\beta$ components of the GCC output current I_{out} (Fig. 9.2).

Since the VSG control employed in this study has no current control loop, -fifth and +seventh-order harmonic currents tend to appear due to the nonlinearity of converter such as the dead time [13]. These harmonic currents will further produce ripples in the DC-link voltage V_{dc}. Consequently, the torque of the IM will oscillate due to the proposed control as shown in Fig. 9.14. Therefore, an elimination method for -fifth and +seventh-order harmonic currents is considered, where $V_{5ref\alpha\beta}$ and $V_{7ref\alpha\beta}$ in Fig. 9.14 are the relevant output voltage references [20].

With the presented VSG control, the GCC can provide frequency regulation support for the connected power grid by the virtual inertia concept and governor characteristics.

9.3.1.3 Experimental results

To verify the addressed control system, some experiments are conducted using the configuration shown in Fig. 9.15A. The setup of the experiment system is shown in Fig. 9.15B. As shown in Fig. 9.15, a 6 kVA scale-down weak grid is emulated by an inverter, to which a 2 kVA system and a Y-connected 47 Ω pure resistive load (851 W at 200 V) is connected. The IM load of the proposed drive system is a commercial blower, which uses a squirrel cage two-pole IM. As the power rating of the proposed system is comparable with that of the weak grid, the impact of its frequency regulation support effect can be observed clearly.

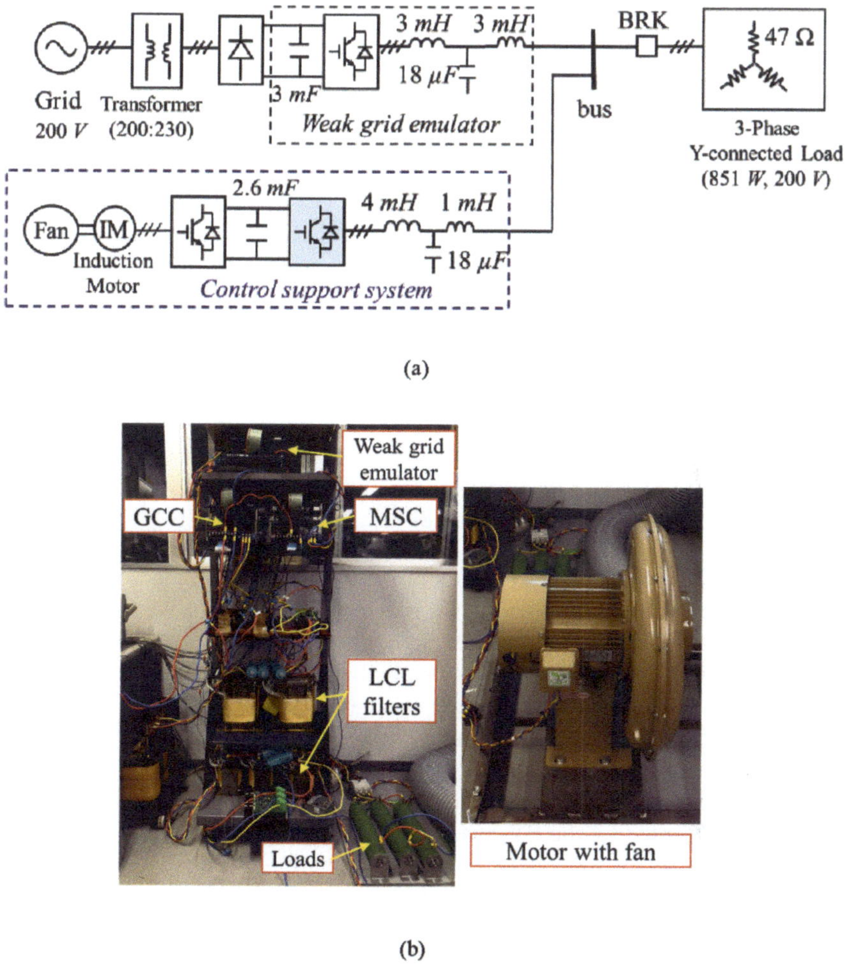

Figure 9.15 Experimental system; (A) circuit, and (B) laboratory setup.

The inverter which mimics a weak grid is controlled to emulate a thermal power plant using a reheat turbine dynamics [21]. The output reactance of this inverter including the virtual inductance is set to 0.8 pu to emulate the synchronous reactance of a real SG. The parameters of the experiment system are given in Ref. [20].

Here, the suppression of the grid frequency fluctuation by the proposed system is evaluated. By closing the breaker BRK, a step load increase occurs in the grid. The response of the grid frequency in this case is shown in Fig. 9.16. It is clear that for the proposed control system the frequency nadir is about 49.6 Hz and frequency drop is 0.4 Hz smaller than the conventional system. Moreover, the proposed system can slow down grid frequency deviation right after the disturbance, thus results in a lower RoCoF.

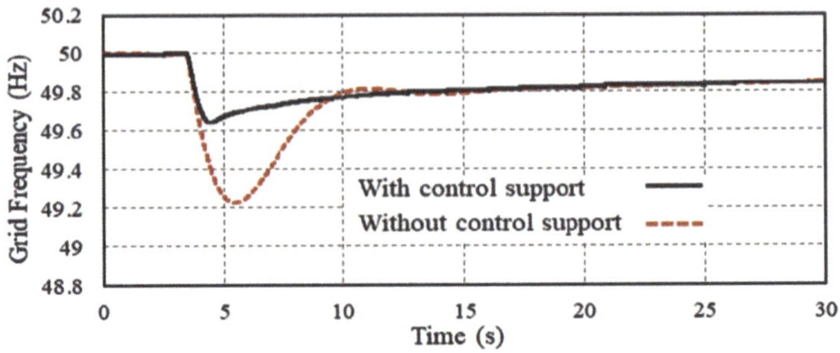

Figure 9.16 Grid frequency response.

9.3.2 PV-based $P-\omega$ control support

The increased penetration of RESs leads to various challenges for the stable operation of an electric power system. For example, utilizing PV power to maintain the demand-supply balancing in the daytime with a high PV power supply is a challenging issue. Despite the increased fluctuation, high penetration of RESs leads to less contribution of the conventional thermal power plants in the overall power generation, and hence their ability to regulate power will also be reduced. To maintain the balance between supply and demand, when a high power supply of PV is predicted, the power output *curtailment* approach is a well-known applied method. Although curtailment has proved its ability in supporting frequency through different methods, more flexibility in active power control strategies is required by PV systems in supporting the grid frequency regulation in all time frames [22].

The active power-frequency ($P-\omega$) droop control providing fast frequency response is represented by the PV-based GCCs. For instance Ref. [23], has focused on providing frequency reserve using autonomous $P-\omega$ functions that modifies active power according to the measured grid frequency. It also discussed the design of the $P-\omega$ curve as for the curtailment, dead band, and droop setting and hence the performance of different penetration of PV is tested along the aforementioned parameters during contingencies. Another strategy of power adjustment and frequency regulation is stated in Ref. [24] where two control designs are created to include the frequency droop control mode for PV to provide primary frequency support to power systems and emergency control mode to prevent system frequency collapse. Also, a predictive PV inverter control method for very fast and accurate control of active power is represented in Ref. [25]. Moreover, the use of grid support functions to improve grid frequency response using a frequency-watt function for an island grid is investigated in Ref. [26]. The proposed approach dampens frequency disturbances associated with variable irradiance conditions as well as contingency events without utilizing expensive energy storage systems or supplemental generation. These studies confirm the importance of $P-\omega$ droop control of PV to regulate frequency during contingencies, yet they lack $P-\omega$ droop control design that mitigates the normal frequency deviations caused by frequent disturbances in load/generation.

Using a flexible GCC control system, a power system operator can fully control the PV power output of different capacities. This control system must autonomously adjust the active power in response to the frequency deviations in renewable integrated power systems. To realize such an active power control scheme in GCC-based PV systems and contributing to both fast and slow frequency responses in a power grid, [22] addresses an autonomous active power control based on a dual $P-\omega$ droop control scheme. The mentioned control scheme, which is shown in Fig. 9.17, is capable to change the droop characteristics depending on the magnitude of frequency change. As shown in Fig. 9.17, a combination of two different droop characteristics for slow and fast dynamic responses is used. Two different modes in terms of frequency deviation separated between small and large frequency deviation ranges are determined by the threshold value of $\pm\,|df_{th}|$. Based on the measured frequency deviation used in slow df_{slow} (or fast df_{fast}) frequency control, the corresponding active power change can be given as follows [22]:

$$
\begin{aligned}
dP_{slow} &= -\frac{1}{D}\times\frac{P_m}{f_0}\times df_{slow} \\
dP_{fast} &= -\frac{1}{D}\times\frac{P_m}{f_0}\times df_{fast}
\end{aligned}
\tag{9.10}
$$

where D is the droop setting of 4%, P_m is power due to maximum power point tracking, and f_0 is the nominal frequency of the grid. Although the formula of dP_{slow}

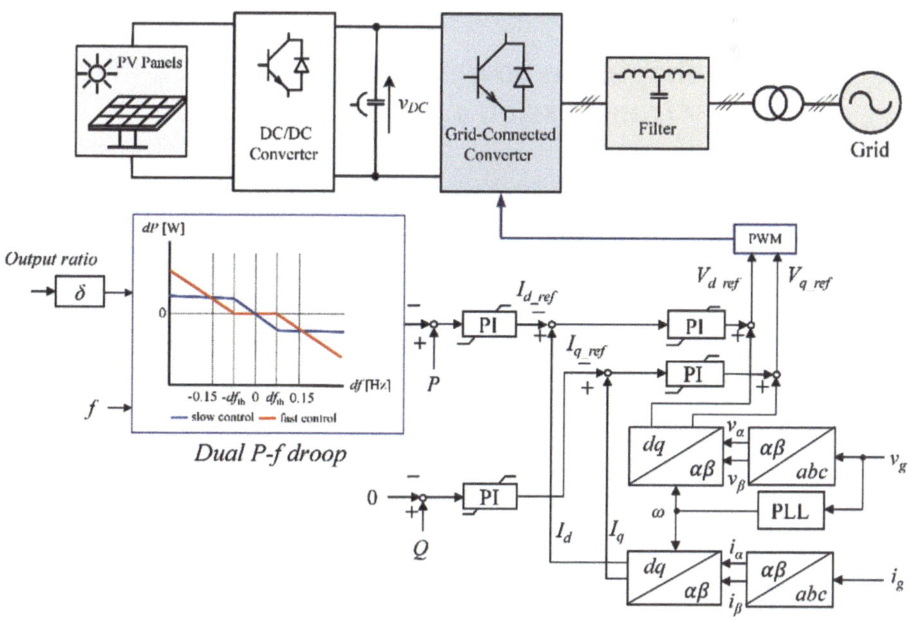

Figure 9.17 Dual P–f droop control for grid control support.

and dP_{fast} is the same, the time-series change is different because of the difference in measurement method of df_{slow} and df_{fast}.

The "*Dual P–f droop*" block in Fig. 9.17 shows that only the slow response is activated within the threshold value $\pm|df_{th}|$ and both fast and slow responses operate beyond this threshold. In particular, when df_{fast} exceeds the threshold, dP_{fast} is expressed as linear equation as stated in (9.10) while dP_{slow} is expressed in the same equation but df_{slow} will be equal to df_{th} and hence dP_{slow} will be a constant value all through beyond the threshold. Finally, the regulation power will be represented in terms of both dP_{fast} and dP_{slow}.

In transition state from one to other operation mode, the sudden change in power output should be avoided. To reduce the sudden change in regulation power and to have a smooth recovery phase, one solution is providing a different regulation power value as a transient power (dP_{trans}) during the transition from fast to slow for t seconds as follows:

$$dP_{trans} = \left(1 - \frac{1}{3}t\right) \times dP_{fast} + \frac{1}{3}t \times dP_{slow} \tag{9.11}$$

However, if only df_{slow} is activated within the threshold $\pm\, df_{th}$, and both df_{fast} and df_{slow} are enabled when df_{fast} is more than df_{th}, a transient signal like Eq. (9.11) is not necessary since there is not discrete change in the fast and slow droop characteristics. The interested readers can find more detail, as well as several validation results in Ref. [22].

9.4 Dynamic shaping in power grids with HVDC and low-frequency transmission systems

The large penetration of RESs has added an increased power density and efficiency to the modern power systems, which have brought many challenges to be emerged in power quality, power delivery, and transmission fields. Numerous topologies and schemes have been suggested to deal with power control and transmission issues of the renewable integrated power grids. This section is focused on the stability and performance improvement of the weak power grids with HVDC and low-frequency transmission systems. In this direction, dynamic shaping via creating virtual desirable dynamics by the existing GCCs is considered as a key solution.

9.4.1 Dynamic shaping in power grids with HVDC

Consider a power grid, as shown in Fig. 9.18A, in which the wind and PV farms are located far from the load centers and thus require a large distance of transmission. For farther wind/PV farms installations, current trends in research and practice point toward the use of HVDC transmission with GCC-based HVDC transmission being a

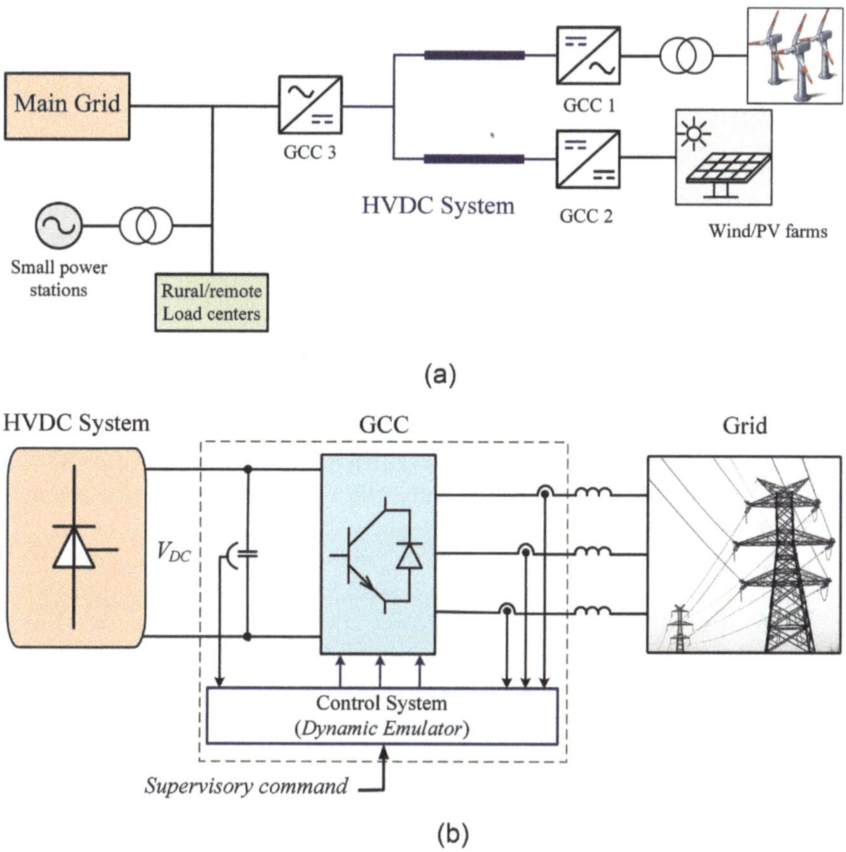

Figure 9.18 Power grid with HVDC system; (A) a network example, and (B) overall scheme of a GCC-based HVDC system with dynamic emulation.

promising approach as it displays different control advantages over traditional technologies [27]. In such installations, as shown in Fig. 9.18A, the wind/PV farm collector network is then converted to HVDC system using wind/PV farm converter stations for transmission to a load center converter station.

The GCC-based HVDC is currently considered as the market leader for wind/PV integration at distances greater than 60–80 km largely due to its established use in point to point bulk power transfer [28]. The HVDC system is broadly applied for point-to-point interconnection of two power systems. It is also well used for interconnecting the offshore wind farm with the onshore stations using a submarine cable [29].

An HVDC link may improve the system stability of a wide area power system by decoupling of some areas. For instance, consider two interconnected ac grids by an HVDC link. This system exhibits a natural decoupling between the AC grids in terms of both voltage and frequency (the AC grids are interconnected in an asynchronous manner). These kinds of interconnections are primarily aimed at preventing the excursions of oscillations between ac systems, for instance, between a stronger and a weaker electric network [30]. Furthermore, the HVDC transmission technology is currently deemed a desirable solution for long-distance power transmission and offshore RESs.

In contrast, the GCC-based HVDC systems have some weak points. The HVDC system is sensitive to faults on the DC lines. When a fault occurs on the DC side of a GCC-based HVDC system, the insulated-gate bipolar transistor cannot control and freewheeling diodes work as a bridge rectifier. It is not able to withstand large surge currents and may be damaged before the fault is cleared. Some solutions are proposed; however, additional control and switching devices are needed [29].

An HVDC system uses the GCC stations that from physical inertia point of view, their dynamic behavior is quite different from that of the SGs. Thus, it makes the frequency stability problem in the connected AC grid more serious than the past. For this reason, the HVDC systems are expected to participate in the grid frequency support. This issue to be more significant, when it is known that the integration of RESs instead of SGs reduces system inertia. Consequently, the system frequency stability and dynamic performance are affected [29].

This issue can be considered in a different perspective. One typical application of HVDC transmission systems is to supply a weak or passive/islanded grid far away from the main grid. With the constant power control, the output power from the (HVDC) GCC station is controlled as a constant, and no frequency support is provided to the weak grid. When a significant part of power in a local gird is supplied by this station, the system inertia becomes lower, and changes in generation or load may cause large frequency deviations, and even leads to system instability [31]. In response to this challenge, it could be a promising solution to use the GCC station to provide frequency support to the local grid.

There are several works investigating system frequency support through the HVDC systems [32,33]. The main idea is to use the virtual dynamic concept in the control system of the grid-tied HVDC GCCs to emulate desirable dynamics, like inertia emulating by VSG for the system frequency support, as shown in Fig. 9.18B.

9.4.2 Dynamic shaping in power grids with low-frequency transmission systems

The connection to a remote ac grid via high voltage AC (HVAC) system requires a long transmission line, which is expensive, with poor efficiency and low transient performance. If a remote AC grid operates as an isolated grid, it will be suffered from a large frequency fluctuation during transition due to relatively small power rating of generators and low inertia. This problem becomes more important if SGs in the remote AC grid are replaced by power-electronics-interfaced generators. GCC-based HVDC systems are proposed to connect the remote AC grid with the main grid in order to overcome the conventional HVAC system limitations.

Despite all merits offered by an HVDC system, still, its protection reliability, as well as its control complexity level are questionable. For example, due to the absence of the zero-crossing points, it is difficult to quickly disconnect a faulty line. Moreover, the HVDC circuit breakers are based on power electronic technology, in which they are vulnerable to thermal failure. Therefore, the protection of an HVDC system is still an immature technology that requires further development and research [34].

To respond to the limitations of both HVAC and HVDC systems, an LFAC transmission system, also referred to as fractional frequency transmission system (FFTS) [35,36], is being introduced as an additional transmission solution [37]. This system transmits the power at a lower frequency ranging about 1−60/3 Hz and offers significant advantages. Compared with the HVDC system, the LFAC system has zero-crossing points; thus, the protection is more advantageous. The LFAC system reactance is lower than that of the conventional HVAC transmission, which can be beneficial in two ways. First, the LFAC system allows more maximum transmission power, and second, it improves transient stability compared to the conventional HVAC system, owing to comparatively larger power angle curve [38]. Due to the mentioned technical advantages, the LFAC system is ongoing to be studied in many works. These works employ the LFAC system using various AC/AC GCCs, circuit configurations, and topologies.

This subsection addresses a control scheme for frequency regulation support of a remote ac grid via LFAC transmission system [39]. Here it is assumed that the LFAC transmission system uses GCCs from type of modular multilevel matrix converter (M3C) for a direct AC-AC conversion which is an alternative solution for a long-distance transmission system [39]. One of its potential applications is the system connection between a large ac grid and a remote ac grid, whereas the latter usually suffers from a fluctuant frequency due to lack of inertia. A dynamic emulating-based control scheme can enhance the frequency regulation in the remote AC grid, which cannot be achieved by the conventional control schemes.

9.4.2.1 LFAC system

An example for a point-to-point LFAC configuration is presented in Fig. 9.19. It is a combination of two GCC stations, and each station is driven by a GCC that acts as a frequency converter to convert the low-frequency ac (10 Hz) to 60 Hz in order to interface with the grid. It is explained in Ref. [36] that the selection of the operating low frequency can affect the transmission distance, the transformer design on the LFAC

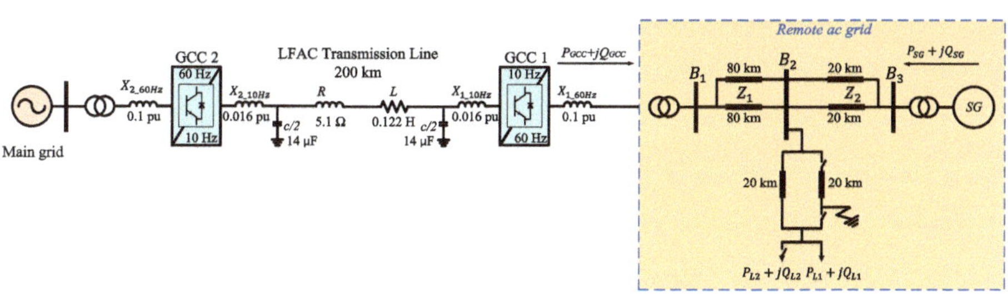

Figure 9.19 Single-line diagram of a point-to-point LFAC system.

side, and the capacitive charging current. Here, the 10 Hz frequency is selected merely to operate the LFAC system.

It is noteworthy that this configuration can be easily extended to a multi-terminal LFAC system as proposed in Ref. [37]. The focal point here is the remote ac grid (60 Hz) side of GCC 1 as an interface to the LFAC system to show the capability of the frequency control support issue. Therefore, the control scheme of the main grid side GCC (GCC 2) is a conventional control scheme [37], which is not discussed here.

9.4.2.2 Addressed control scheme

Fig. 9.20 shows the circuit configuration and the entire control scheme of GCC 1. The control scheme can be divided into two parts; the grid side (60 Hz side) control part, and the LFAC side (10 Hz side) control part. From Fig. 9.20, the grid side is controlled by using either a virtual dynamic control or a conventional (proportional-integral (PI)-based) control scheme.

As shown in Fig. 9.20, virtual dynamic control system in form of VSG is used for both the LFAC and the remote AC grid sides. Each VSG system comprises $P-\omega$ control part, and a reactive power control part. The GCC 1 employs a $Q-V$ droop controller on its remote ac grid side, whereas an AVR on its LFAC side. The $P-\omega$ control part used for both LFAC and the remote AC grid sides of the GCC 1 is almost identical except that we chose to include the droop characteristic of the average DC capacitor voltage and active power ($V_{ave}-P$) to be an additional part of the LFAC side control system.

It is noteworthy that applying a dual VSG for controlling both GCC sides is not straightforward. Due to the intentional slow response of VSG control in both sides, it is difficult to apply conventional average DC capacitor voltage control, which requires a fast inner current control loop. Therefore, here an average DC capacitor voltage versus active power ($V_{ave}-P$) droop control method integrated into the LFAC side VSG control is used, to achieve the balance between input and output

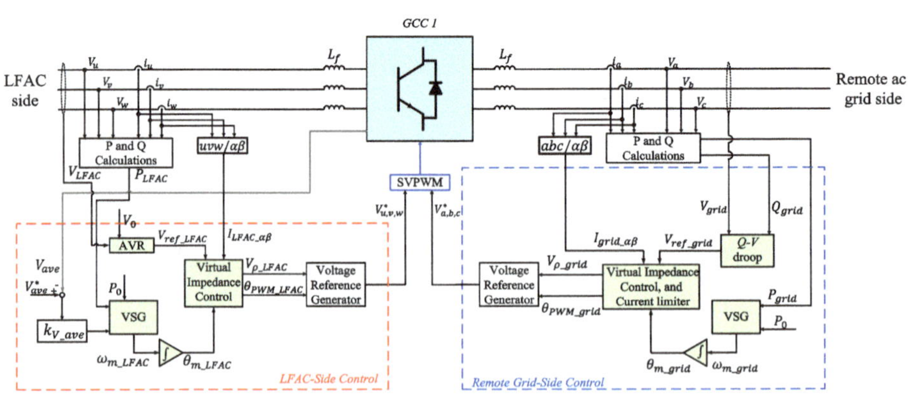

Figure 9.20 The control system of GCC 1.

GCC active powers and thus to maintain the capacitor voltage while reacting fast enough to support the dynamics of a remote ac grid. The used VSG control scheme is shown in Fig. 9.21A, which the input signals are included with the subscript i, referring to the LFAC or the remote AC grid side measured signals. The VSG control function comprises the swing equation, governor, and damping units. The swing equation of the VSG is

$$P_{in\_i} - P_{out_i} + P_{d_i} = J\omega_{0_i}\frac{d\omega_{m_i}}{dt} \ ; \ J = \frac{MS_{base}}{\omega_0^2} \tag{9.12}$$

where ω_m is the virtual rotating angular frequency, ω_0 is the nominal frequency, and P_{in}, P_{out}, P_d are the virtual shaft power produced by the governor control, the output

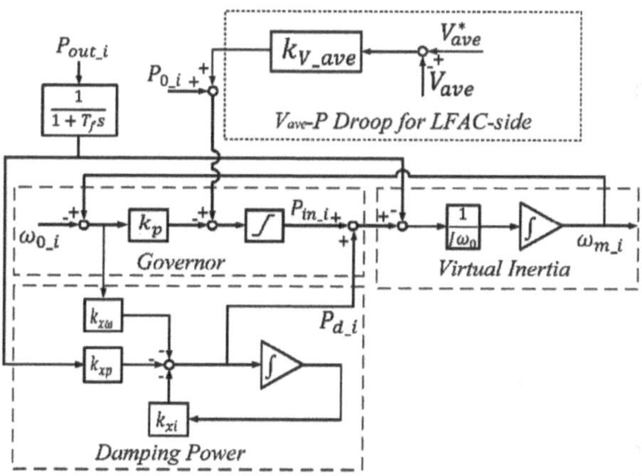

(a)

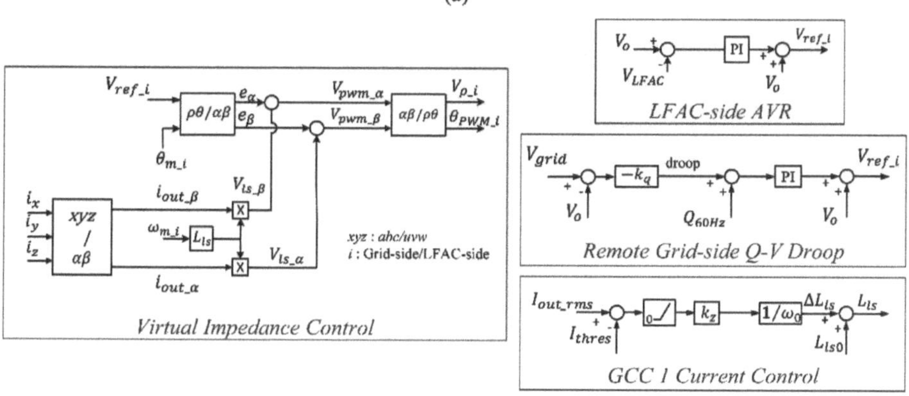

(b)

Figure 9.21 GCC control system blocks; (A) VSG, (B) virtual impedance control, LFAC side AVR, remote grid side $Q-V$ droop, and GCC 1 current control for remote ac grid side.

power, and the damping power, respectively. J is the moment of inertia of rotating mass, M is the inertia time constant, and S_{base} is the rated MVA of the GCC.

The damping power P_d is generated by a state feedback control loop as follows [40].

$$P_{d\_i} = -k_{x\omega}(\omega_m - \omega_0) - k_{xP}\frac{1}{1+T_f s}P_{out} - k_{xi}\frac{1}{s}P_{d\_i} \tag{9.13}$$

where $k_{x\omega}$, k_{xp}, and k_{xi} are the feedback gains of virtual rotor frequency, output active power, integral term of P_d; and T_f is the time constant of the LPF, VSG control parameters are given in Ref. [39]. The used damping method allows the designer to avoid using the PLL for detecting the frequency, as applied in some previous methods [1,2]. By rearranging Eq. (9.12), the virtual mechanical frequency is thus

$$\omega_{m\_i} = \frac{1}{J\omega_0}\int \left(P_{in\_i} - P_{out\_i} + P_{d\_i}\right)dt \tag{9.14}$$

where the obtained virtual angular frequency ω_m is then integrated to generate the phase angle θ_{m_i} to form a three-phase reference voltage for the LFAC and the grid sides. The design of VSG control parameters is explained in Refs. [1,2].

The governor control part of the VSG, shown in Fig. 9.21A, is working based on the $P-\omega$ droop characteristic. It is preferable to use the $P-\omega$ droop control especially when the system is highly inductive like traditional SG. Moreover, conventional droop control is used for the sending terminal (GCC 2), whereas a reversed droop control is used for the receiving terminal (GCC 1) as explained in Ref. [37]. The conventional (with $-k_p$) and reversed (with $+k_p$) governor control produce the shaft power P_{in} as follows

$$P_{in} = P_0 \pm k_p(\omega_m - \omega_0) \tag{9.15}$$

where P_0 is the commanded value of active power, k_p is the droop coefficient, ω_0 is the nominal frequency [1]. Fig. 9.21B represents the $Q-V$ droop control on the remote ac grid side, as well as the AVR of the LFAC side. The $Q-V$ droop controller is preferable when two parallel DGs are connected at the point of common coupling in order to share the reactive power autonomously. The term "droop" is provided by the voltage difference block with a droop coefficient k_q that is selected to allow 10% variation when producing 1 pu reactive power block; while the droop term in $P-\omega$ droop control is chosen to allow 5% change of frequency as a response to a full active power change. The droop terms are set in per unit value as follows

$$k_{p\_pu} = \frac{k_p\omega_0}{S_{base}}\ ;\ k_{q\_pu} = \frac{k_q S_{base}}{V_{base}} \tag{9.16}$$

The AVR is used in the LFAC in order to maintain the voltage level by varying the reactive power. Whereas, the other terminal(s) in the LFAC network use an AQR in order to minimize the reactive power sharing errors, which is not discussed here. The result of all controllers is V_{ref_i} (the subscript *i* refers to either remote AC grid or LFAC sides parameters of each GCC), which indicates the generated reference voltage by these controllers to be used later as an input to each corresponding virtual impedance control block as shown in Fig. 9.21B.

The virtual impedance control block is incorporated along with the dual VSG control in the GCC-based LFAC system. Each one has a different function; for example, for the LFAC side, due to the low frequency, the X/R ratio will be 6 times smaller than that of the 60 Hz for a given rated voltage level, in which it is 1.5 in this system. Thus, the system will become less inductive, and therefore, the virtual impedance control is introduced in order to increase the total reactance of the GCC station in order to avoid the coupling between P and Q. Based on [2], a virtual impedance control loop is applied to adjust the total output reactance of each terminal, which is calculated as

$$X^* = \frac{XS_{base}}{V_{base}^2} = \frac{\omega_0\left(L_{ls} + L_f + L_{line}\right)S_{base}}{V_{base}^2} \tag{9.17}$$

where L_{ls} is the virtual inductance, L_f and L_{line} are the inductance of the L filter and the transmission line, respectively.

On the other hand, another function of the virtual impedance control is used to increase the GCC's output impedance temporarily during an overcurrent condition (e.g., occurrence of fault), as shown in Fig. 9.21B. This function provides the fault ride-through ability for a VSG control scheme [39]. Finally, this controller results on generating the voltage references for the GCC space vector pulse width modulation as shown in Fig. 9.20.

9.4.2.3 Simulation results

To verify the proposed system, some simulation scenarios were carried out for the test system shown in Fig. 9.19. GCC 2 regulates the active power received from the main grid and sends this power through a 200 km LFAC link to GCC 1 by integrating the virtual mechanical speed from the VSG to provide the phase angle. Then the LFAC side VSG control of GCC 1 regulates the active power and uses its AVR to maintain the voltage by varying the reactive power. The remote AC grid includes a conventional SG and associated loads, which represents a relatively weak system. The fixed load P_{L1}+Q_{L1} is 400 MW + 100 MVar, whereas a disturbance load $P_{L2} + Q_{L2}$ represents 5% of the fixed load and it is set to be turned on and off at specific times. These loads are being supplied by both SG and the GCC-based LFAC system. The system parameters and detailed system characteristics are given in Ref. [39]. Generally, from the frequency regulation point of view, the most common disturbance in power system is load change [21]. If the power system lacks the inertia, this will lead to a large frequency deviation. This deviation may lead to system instability, particularly if the load varies significantly.

Fig. 9.22 shows the simulation results for a step load increase. In order to perform the load change, a switchable load $P_{L2}+Q_{L2}$ as can be seen from Fig. 9.19 is turned on or off, at $t = 20$ s, to perform load increase or decrease. Fig. 9.22A illustrates the remote grid frequency. It can be observed that when the system is under the VSG

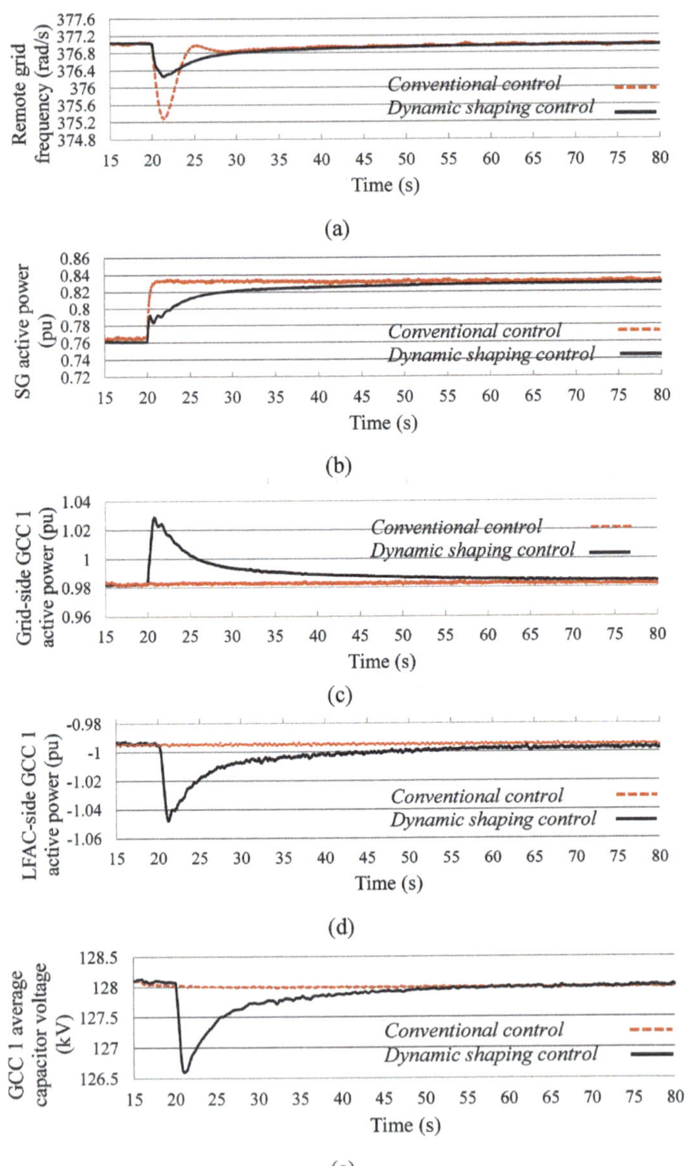

Figure 9.22 Simulation waveforms of the system with the dynamic shaping and conventional control schemes under 5% load increase: (A) remote grid frequency, (B) SG active power, (C) grid-side GCC 1 active power, (D) LFAC-side GCC 1 active power, and (E) GCC 1 average DC capacitor voltage.

control, the frequency nadir is reduced, and the frequency fluctuation becomes slower and smoother than that under the conventional control. After a short period of transient (which is decided by the LFC controller of the SG), the frequency of the grid can be restored to the nominal value, and the system stability is improved.

Fig. 9.22B–D illustrate the active powers of the SG and the GCC 1 (in both grid and LFAC sides). It can be noticed that the VSGs of both grid side and LFAC side are reacting in order to inject power into the remote AC grid to support the load increase demand. This confirms how the GCC-based LFAC link can provide fast power injection response. From Fig. 9.22C, it is clear that the GCC 1 injects the required power until the SG increases its active power as shown in Fig. 9.22B to respond to the power mismatch. Similarly, the VSG on the LFAC side will request this power to be provided from GCC 2 through the LFAC link. As a result, both VSGs respond accordingly to inject the required power. Once the steady-state operation is restored, the GCC-based LFAC system restores its active power to the pretransient value. It is shown that the VSG-based control system has enhanced the GCCs performance with a fast response to support the frequency regulation of the remote ac grid.

In load change scenario, the VSG control has succeeded in mitigating the frequency deviation and the average DC capacitor voltage deviation to keep within 2% during transient states. Moreover, from Fig. 9.22E, the proposed V_{ave}–P droop on the LFAC side managed to maintain the power balance between the input and the output and thus succeeded in keeping the average DC capacitor voltage constant. Interested readers can find more detailed explanations and test scenarios in Ref. [39].

9.5 Summary

The advanced control schemes in modern power grids rely on virtual manipulation of desirable dynamics; for example, inertia and damping characteristic using flexible GCCs in the grid to enhance the system performance and stability. The successful development of an advanced control scheme highly depends on the adequacy of flexibility and capability of the GCCs in the emulation of desired dynamics.

In the present chapter, the general concept of flexible GCC-based grid dynamic shaping is explained and some applications of this concept in improving the performance and dynamics of renewable integrated power grids are emphasized. First, an overall scheme for a flexible GCC-based dynamic emulator is briefly introduced, and then, some new perspectives to enhance the power grid stability and performance are discussed. The role of GCC-based dynamic shaping in the grid ancillary service support is addressed and supplemented by two examples for the grid frequency regulation support. Finally, the application of GCC-based dynamic shaping for HVDC and low-frequency transmission systems is discussed.

References

[1] H. Bevrani, B. Francois, T. Ise, Microgrid Dynamics and Control, Wiley, NJ, USA, July 2017.
[2] J. Liu, Y. Miura, H. Bevrani, T. Ise, Enhanced virtual synchronous generator control for parallel inverters in microgrids, IEEE Trans. Smart Grid 8 (5) (September 2017) 2268–2277.
[3] H. Bevrani, T. Ise, Y. Miura, Virtual synchronous generators: a survey and new perspectives, Int. J. Electr. Power Energy Syst. 54 (2014) 244–254.
[4] H. Bevrani, M. Imanaka, T. Kato, Grid dynamics shaping using flexible controlled power converters, Energy Rep. 6 (Suppl. 9) (2020) 1490–1495.
[5] A.D. Paquette, D.M. Divan, Virtual impedance current limiting for inverters in microgrids with synchronous generators, IEEE Trans. Ind. Appl. 51 (2) (March–April 2015) 1630–1638.
[6] J. He, Y.W. Li, Analysis, design, implementation of virtual impedance for power electronics interfaced distributed generation, IEEE Trans. Ind. Appl. 47 (6) (November–December 2011) 2525–2538.
[7] D.M.F.G. Roger, IEEE recommended practice and requirements for harmonic control in electric power systems, in: IEEE Std. 519-2014, vol. 2014, 2014.
[8] M.M. Hashempour, M. Savaghebi, J.C. Vasquez, J.M. Guerrero, A control architecture to coordinate distributed generators and active power filters coexisting in a microgrid, IEEE Trans. Smart Grid 7 (5) (September 2016) 2325–2336.
[9] X. Wang, F. Blaabjerg, Z. Chen, Autonomous control of inverter-interfaced distributed generation units for harmonic current filtering and resonance damping in an islanded microgrid, IEEE Trans. Ind. Appl. 50 (1) (January–February 2014) 452–461.
[10] P. Sreekumar, V. Khadkikar, A new virtual harmonic impedance scheme for harmonic power sharing in an islanded microgrid, IEEE Trans. Power Deliv. 31 (3) (June 2016) 936–945.
[11] Q.C. Zhong, Harmonic droop controller to reduce the voltage harmonics of inverters, IEEE Trans. Ind. Electron. 60 (3) (March 2013) 936–945.
[12] T.-L. Lee, P.-T. Cheng, Design of a new cooperative harmonic filtering strategy for distributed generation interface converters in an islanding network, IEEE Trans. Power Electron. 2 (5) (July 2007) 1919–1927.
[13] J. Liu, Y. Miura, T. Ise, Cost-function-based microgrid decentralized control of unbalance and harmonics for simultaneous bus voltage compensation and current sharing, IEEE Trans. Power Electron. 34 (8) (August 2019) 7397–7410.
[14] D. Hosseini, M. Naderi, Q. Shafiee, M. Savaghebi, H. Bevrani, Virtual synchronous generator controller for power quality enhancement in microgrids, in: 8th International Conference on Control, Instrumentation and Automation (ICCIA), Tehran, Iran, 2022, pp. 1–6.
[15] T. Qoria, X. Guillaud, Control of Power Electronics-Driven Power Sources, Chapter 8 in Converter-Based Dynamics and Control of Modern Power Systems, Elsevier, 2021.
[16] J. Fang, H. Li, Y. Tang, F. Blaabjerg, Distributed power system virtual inertia implemented by grid connected power converters, IEEE Trans. Power Electron. 33 (10) (October 2018) 8488–8499.
[17] M. Al-Tameemi, Studies on Low Frequency AC Transmission System Using Modular Multilevel Matrix Converter Based Virtual Synchronous Generator Control (Ph.D. Dissertation), Osaka University, 2020.

[18] P. Babahajiani, Q. Shafiee, H. Bevrani, Intelligent demand response contribution in frequency control of multi-area power systems, IEEE Trans. Smart Grid 9 (2) (March 2018) 1282–1291.
[19] A. M-Garcia, F. Bouffard, D.S. Kirschen, Decentralized demand-side contribution to primary frequency control, IEEE Trans. Power Syst. 26 (1) (February 2011) 411–419.
[20] D. Terazono, J. Liu, Y. Miura, S. Sakabe, H. Bevrani, T. Ise, Grid frequency regulation support from back-to-back motor drive system with virtual-synchronous-generator-based coordinated control, IEEE Trans. Power Electron. 36 (3) (March 2021) 2901–2913.
[21] H. Bevrani, Robust Power System Frequency Control, second ed., Springer, Gewerbestrasse, Switzerland, 2014.
[22] N. Harag, M. Imanaka, M. Kurimoto, S. Shigeyuki, H. Bevrani, T. Kato, Autonomous dual active power-frequency control in power system with small-scale photovoltaic power generation, J. Moder. Power Syst. & Clean Energ. (2021), https://doi.org/10.35833/MPCE.2020.000700.
[23] H. Bevrani, T. Hyama, Intelligent Automatic Generation Control, CRC Press, NY, USA, 2011.
[24] H. Xin, Y. Liu, Z. Wang, D. Gan, T. Yang, A new frequency regulation strategy for photovoltaic systems without energy storage, IEEE Trans. Sustain. Energy 4 (4) (October 2013) 985–993.
[25] A.F. Hoke, M. Shirazi, S. Chakraborty, E. Muljadi, D. Maksimovic, Rapid active power control of photovoltaic systems for grid frequency support, IEEE J. Emerg. & Select. Topic. Power Electr. 5 (3) (September 2017) 1154–1163.
[26] J. Neely, J. Johnson, J. Delhotal, S. Gonzalez, M. Lave, Evaluation of PV frequency-watt function for fast frequency reserves, in: IEEE Applied Power Electronics Conference and Exposition (APEC), Long Beach, CA, 2016, pp. 1926–1933.
[27] A. Madariaga, J. Martín, I. Zamora, I.M. De Alegría, S. Ceballos, Technological trends in electric topologies for offshore wind power plants, Renew. Sustain. Energy Rev. 24 (2013) 32–44.
[28] S. Rodrigues, C. Restrepo, E. Kontos, R.T. Pinto, P. Bauer, Trends of offshore wind projects, Renew. Sustain. Energy Rev. 49 (2015) 1114–1135.
[29] H. Golpira, A.R. Messina, H. Bevrani, Renewable Integrated Power System: Stability and Control, IEEE-Wiley, 2021.
[30] L.M. Castro, E. Acha, On the provision of frequency regulation in low inertia AC grids using HVDC systems, IEEE Trans. Smart Grid 7 (6) (2015) 2680–2690.
[31] M. Guan, W. Pan, J. Zhang, Q. Hao, J. Cheng, X. Zheng, Synchronous generator emulation control strategy for voltage source converter (VSC) stations, IEEE Trans. Power Syst. 30 (6) (2015) 3093–3101.
[32] A.E. Leon, Short-term frequency regulation and inertia emulation using an MMC-based MTDC system, IEEE Trans. Power Syst. 33 (3) (2017) 2854–2863.
[33] E. Rakhshani, D. Remon, A.M. Cantarellas, J.M. Garcia, P. Rodriguez, Virtual synchronous power strategy for multiple HVDC interconnections of multi-area AGC power systems, IEEE Trans. Power Syst. 32 (3) (2016) 1665–1677.
[34] C.M. Franck, HVDC circuit breakers: a review identifying future research needs, IEEE Trans. Power Deliv. 26 (2) (April 2011) 998–1007.
[35] X. Wang, X. Wang, Feasibility study of fractional frequency transmission system, IEEE Trans. Power Syst. 11 (2) (May 1996) 962–967.
[36] X. Wang, C. Cao, Z. Zhou, Experiment on fractional frequency transmission system, IEEE Trans. Power Syst. 21 (1) (February 2006) 372–377.

[37] M. Al-Tameemi, Y. Miura, J. Liu, H. Bevrani, T. Ise, A novel control scheme for multi-terminal low-frequency ac electrical energy transmission systems using modular multi-level matrix converters and virtual synchronous generator concept, Energies 13 (3) (February 2020) 747.

[38] Q. Nguyen, G. Todeschini, S. Santoso, Power flow in a multi-frequency HVac and HVdc system: formulation, solution, and validation, IEEE Trans. Power Syst. 34 (4) (July 2019) 2487−2497.

[39] M. Al-Tameemi, J. Liu, H. Bevrani, T. Ise, A dual VSG-based M3C control scheme for frequency regulation support of a remote AC grid via low-frequency AC transmission system, IEEE Access 8 (2020) 66085−66094.

[40] J. Liu, Y. Miura, T. Ise, Fixed-parameter damping methods of virtual synchronous generator control using state feedback, IEEE Access 7 (2019) 99177−99190.

Index

Note: 'Page number followed by *"f"* indicate figure and *"t"* indicate table.'

T

FSC
www.fsc.org
MIX
Papier | Fördert
gute Waldnutzung
FSC® C083411

Kolibri 360 GmbH
Ferdinand-Jühlke-Straße 7
99095 Erfurt, Deutschland
produktsicherheit@kolibri360.de